Advances in Industrial Control

Advances in Industrial Control is a series of monographs and contributed titles focusing on the applications of advanced and novel control methods within applied settings. This series has worldwide distribution to engineers, researchers and libraries.

The series promotes the exchange of information between academia and industry, to which end the books all demonstrate some theoretical aspect of an advanced or new control method and show how it can be applied either in a pilot plant or in some real industrial situation. The books are distinguished by the combination of the type of theory used and the type of application exemplified. Note that "industrial" here has a very broad interpretation; it applies not merely to the processes employed in industrial plants but to systems such as avionics and automotive brakes and drivetrain. This series complements the theoretical and more mathematical approach of Communications and Control Engineering.

Indexed by SCOPUS and Engineering Index.

Proposals for this series, composed of a proposal form downloaded from this page, a draft Contents, at least two sample chapters and an author cv (with a synopsis of the whole project, if possible) can be submitted to either of the:

Series Editors

Professor **Michael J. Grimble**
Department of Electronic and Electrical Engineering, Royal College Building, 204 George Street, Glasgow G1 1XW, United Kingdom
e-mail: m.j.grimble@strath.ac.uk

Professor **Antonella Ferrara**
Department of Electrical, Computer and Biomedical Engineering, University of Pavia, Via Ferrata 1, 27100 Pavia, Italy
e-mail: antonella.ferrara@unipv.it

or the

In-house Editor

Mr. **Oliver Jackson**
Springer London, 4 Crinan Street, London, N1 9XW, United Kingdom
e-mail: oliver.jackson@springer.com

Proposals are peer-reviewed.

Publishing Ethics

Researchers should conduct their research from research proposal to publication in line with best practices and codes of conduct of relevant professional bodies and/or national and international regulatory bodies. For more details on individual ethics matters please see:
https://www.springer.com/gp/authors-editors/journal-author/journal-author-helpdesk/publishing-ethics/14214

More information about this series at http://www.springer.com/series/1412

Chuei-Tin Chang · Hao-Yeh Lee ·
Vincentius Surya Kurnia Adi

Process Plant Operating Procedures

Synthesis, Simulation and Abnormal Situation Management

Springer

Chuei-Tin Chang
Chemical Engineering
National Cheng Kung University
Tainan, Taiwan

Vincentius Surya Kurnia Adi
Chemical Engineering
National Chung Hsing University
Taichung, Taiwan

Hao-Yeh Lee
Chemical Engineering
National Taiwan University of Science
and Technology
Taipei, Taiwan

ISSN 1430-9491 ISSN 2193-1577 (electronic)
Advances in Industrial Control
ISBN 978-3-030-70977-8 ISBN 978-3-030-70978-5 (eBook)
https://doi.org/10.1007/978-3-030-70978-5

Mathematics Subject Classification: 90B06, 90B35, 93C65, 93C83, 93C85, 93C95

This Springer imprint is published by the registered company Springer Nature Switzerland AG
The registered company address is: Gewerbestrasse 11, 6330 Cham, Switzerland

Chuei-Tin Chang · Hao-Yeh Lee ·
Vincentius Surya Kurnia Adi

Process Plant Operating Procedures

Synthesis, Simulation and Abnormal Situation Management

Springer

Chuei-Tin Chang
Chemical Engineering
National Cheng Kung University
Tainan, Taiwan

Vincentius Surya Kurnia Adi
Chemical Engineering
National Chung Hsing University
Taichung, Taiwan

Hao-Yeh Lee
Chemical Engineering
National Taiwan University of Science and Technology
Taipei, Taiwan

ISSN 1430-9491 ISSN 2193-1577 (electronic)
Advances in Industrial Control
ISBN 978-3-030-70977-8 ISBN 978-3-030-70978-5 (eBook)
https://doi.org/10.1007/978-3-030-70978-5

Mathematics Subject Classification: 90B06, 90B35, 93C65, 93C83, 93C85, 93C95

This Springer imprint is published by the registered company Springer Nature Switzerland AG
The registered company address is: Gewerbestrasse 11, 6330 Cham, Switzerland

Series Editor's Foreword

Control engineering is viewed rather differently by researchers and those that must implement and maintain control systems. Researchers develop general algorithms with a strong underlying mathematical basis, whilst practitioners have more immediate concerns over the limits of equipment, quality of control, safety and security, and plant downtime. The series *Advances in Industrial Control* attempts to bridge this divide and hopes to encourage the adoption of advanced control techniques when they are likely to be beneficial.

The rapid development of new control theory and technology has an impact on all areas of engineering and applications. This monograph series has a focus on applications that are now so important as the rate of new technological developments poses new challenges. There is also a gradual morphing of the traditional view of control engineering design into the more computer-focused subject of systems engineering. These developments require new solutions and stimulate the development of new control paradigms.

A focus on applications is also desirable if the different aspects of the "control design" problem are to be explored with the same dedication that "control synthesis" problems have received in the past. *Advances in Industrial Control* provides an opportunity for researchers to present new work on industrial control and applications problems and solutions. It raises awareness of the substantial benefits that advanced control can provide whilst not ignoring the difficulties that can sometimes arise.

This monograph is concerned with process plant operating procedures, and the first chapter introduces basic tools such as process flow, piping, and instrumentation diagrams. The second chapter involves a heuristic approach to procedure synthesis. The examples included provide a reminder that a real process is much more than a set of equations and the control engineer needs practical skills to understand vendors' software and hardware. The examples are very relevant to many process plants. The approach taken is one based on rather intuitive ideas and this is what was often used. In fact, in the past, control systems were often specified by chemical engineers that had built a pilot plant, and final full-scale designs were simply scaled up. When the costs of energy and pollution were not monitored so closely, advanced control was not so essential but this is no longer the case. Modern environmental standards have had a huge impact on industries like petrochemicals.

Chapter 3 turns to a model-based approach to procedure synthesis and the use of discrete-event models. It also covers an essential approach for large systems, namely the use of hierarchical structures and supervisory control. Chapter 4 addresses the use of simulation and startup operations. The subject of safe startup and shutdown is often neglected, and this is a useful topic to be included.

Chapter 5 turns to Petri net-based operating procedures. This is a mathematical topic that may not be so familiar, and the authors have included useful and rather practical examples to illustrate their ideas. Chapter 6, on normal operating procedures obtained with untimed automata, is an attempt to introduce a systematic approach to achieving production goals. It relates to the essential components in a production process but with a very different perspective on the framework for analysis. Chapter 7 is concerned with a systematic approach for generating normal operating procedures with timed automata, and two major examples are included.

Chapter 8 involves the generation of test plans for fault diagnosis using untimed automata. The importance of safe operation in chemical plants is of course self-evident, and this chapter illustrates how such an approach provides a standardized methodology to diagnose problems.

The synthesis of diagnostic tests based on timed automata is covered in Chap. 9. If these methods are successful, the need for additional sensors can be reduced with obvious advantages. Apparently, the procedure can differentiate between various time delays caused by faults of the same type and different intensities. Finally, Chap. 10 involves the synthesis of diagnostic test plans based on hybrid automata. Hybrid models are used and feasibility is demonstrated in two examples involving startup operations.

This text is unusual in dealing with the many practical problems that arise in operating chemical processes but using a scientific basis for the strategies. It should provide a valuable perspective on what may be possible to engineering managers in process plants and of course researchers in process operations.

January 2021 Michael J Grimble

Preface

This book presents an introduction to both theoretical and application aspects of *procedure synthesis*, which is primarily concerned with the task of conjecturing the sequence of controller (or operator) actions needed to achieve the designated operational goal(s) in a given system. The formal problem statement, two alternative synthesis approaches, their assessment methods, and a series of realistic examples are provided in sufficient detail to facilitate practical implementation. It is thus appropriate for use as an industrial reference, a senior-level design course, or a graduate course in chemical process design, operation, and/or fault diagnosis.

Standard Operating Procedures (SOPs) are widely adopted for startup and shutdown of continuous processes, normal operations in batch processes, and other indispensable activities, such as sampling, maintenance, and emergency response, in various process plants. At the present time, the related issues have not been discussed thoroughly in typical undergraduate process design and/or control courses and textbooks. On the other hand, procedure synthesis is an indispensable task often encountered by process engineers in various design projects and, therefore, a research topic that received considerable attention in recent years. This book attempts to bridge the gap between academia and practice by exposing the practitioners (i.e., researchers and engineers) to an important area of plant operations, teaching them effective approaches for procedure synthesis, enabling them to construct and solve scheduling models, and providing them with software tools for simulation, validation, and assessment of procedures and schedules.

The contents in each chapter are briefly summarized in the following sequence.

Chapter 1 provides a general introduction to the problem at hand. All available data for the procedure synthesis problem are assumed to be embedded in the Process Flow Diagram (PFD) and the Piping and Instrumentation Diagram (P&ID) of a given plant, while the operating procedure is regarded as the "solution" and can usually be expressed with a Sequential Function Chart (SFC). The above graphical representations of the given system and also the resulting procedure are described in detail, and a formal problem statement is also presented accordingly in Chap. 1.

Two alternative solution strategies are provided next in this book. A heuristic approach for procedure generation is outlined in Chap. 2, whereas a model-based approach in Chap. 3. In the former case, the operation steps are supposed to be conjectured according to past experiences obtained by running a pilot setup or an actual plant, and the resulting procedure can be tested with dynamic simulation studies and/or experimental studies. The software and hardware tools used for validation are also described to facilitate practical applications. On the other hand, the suggested model-based synthesis strategies are presented in Chap. 3. Two different types of Discrete-Event System (DES) models, i.e., Petri nets and automata, are adopted to identify the best path(s) leading to the specified goal of the operation. The model-building methods for characterizing all components in the given system and the required control specifications are explained with simple examples. The sequential control actions and the corresponding time schedule can then be identified accordingly with readily accessible free software.

The aforementioned solution strategies are applied to various realistic problems in Chaps. 4 through 7 to generate normal operating procedures. Aided with the commercial package Aspen Plus Dynamics®, the heuristic approach is adopted in Chap. 4 to produce procedures for the startup operations of simple and extractive distillation columns. By making use of Petri nets, the feasibility and effectiveness of the model-based procedure synthesis approach is demonstrated with a wide variety of industrial case studies in Chap. 5. The specific examples used in this chapter are concerned with the cleaning operation of a pipeline network, the material transportation in a batch plant, a beer filtration process, and a semiconductor manufacturing process. Two variant DES models, i.e., the untimed and timed automata, are then utilized in Chap. 6 and Chap. 7, respectively, to generate SFCs for a semi-batch reaction process, and the startup operations of a flash drum and a simple distillation column.

The solution strategies given in Chaps. 2 and 3 are also applicable for Abnormal Situation Management (ASM). More specifically, in the presence of fault(s) or failure(s), there may be urgent needs for inventing and implementing temporary procedures to perform diagnostic tests and/or emergency response actions. These applications are described in Chaps. 8 through 10.

Finally, this book is written primarily for those who have a basic understanding of process design and control activities, and its presentation is oriented toward a multidisciplinary audience and thus should appeal to engineers in diversified fields with an interest in synthesizing operating procedures in process plants.

Tainan, Taiwan — Chuei-Tin Chang
Taipei, Taiwan — Hao-Yeh Lee
Taichung, Taiwan — Vincentius Surya Kurnia Adi

Contents

1 Introduction ... 1
1.1 Graphical Representations of Chemical Processes ... 1
1.1.1 Process Flow Diagram ... 2
1.1.2 Piping and Instrumentation Diagram ... 2
1.2 Graphical Representations of Operating Procedures ... 5
1.3 Problem Statement ... 10
1.4 Framework and Organization ... 10
References ... 10

2 Heuristic Approach to Procedure Synthesis ... 11
2.1 Heuristic Rules ... 11
2.2 Dynamic Simulation Software ... 13
2.3 Pilot Setups ... 16
2.4 Illustrative Examples ... 17
References ... 27

3 Model-Based Approach to Procedure Synthesis ... 29
3.1 Hierarchical Structure of Chemical Processes ... 30
3.2 Discrete-Event System Models ... 32
3.2.1 Languages ... 33
3.2.2 Automata ... 34
3.2.3 Petri Nets ... 46
3.3 Supervisory Controllers ... 54
3.4 Control Specifications ... 56
3.5 Systematic Synthesis of Operating Procedures ... 57
3.5.1 Process Description ... 57
3.5.2 Components ... 58
3.5.3 Control Specifications ... 60
3.5.4 Supervisor Generation ... 62
3.5.5 Procedure Synthesis ... 70
References ... 71

4 Normal Operating Procedures Based on Dynamic Simulations 73
4.1 Startup Operations of Simple DCs 73
4.2 Startup Operations of Extractive DCs 78
References 92

5 Petri Net-Based Operating Procedures 95
5.1 Cleaning Operation of a Pipeline Network 96
5.1.1 Representation of Material-Transfer Paths 97
5.1.2 Enumeration of Possible Routes 99
5.1.3 Route Selection Procedures 102
5.1.4 Equipment Models 107
5.1.5 Generation of Operation Steps 108
5.1.6 Execution of Multiple Tasks 110
5.2 Material Transportation in a Batch Plant 113
5.2.1 Component Models 115
5.2.2 Generation of Operation Steps 119
5.2.3 Identification of Feasible Routes 121
5.2.4 Execution of Multiple Tasks 124
5.3 Automatic Synthesis of Batch Operating Procedures in Beer Filtration Plant 130
5.3.1 Stage-Based Operating Procedures 130
5.3.2 Time-Based Operating Procedures 139
5.4 Semiconductor Manufacturing Process Scheduling Strategy 154
5.4.1 Petri Net Models 156
5.4.2 Optimal Scheduling Strategy 157
5.4.3 Token Movements 159
5.4.4 Shared Resources 160
5.4.5 Linearization of Minimum-Horizon Model 161
5.4.6 Final Test Process 161
5.5 Concluding Remarks 163
References 164

6 Normal Operating Procedures Obtained with Untimed Automata 167
6.1 Extended Finite Automata 168
6.2 An Illustrative Example 169
6.3 Component Models 170
6.3.1 Actuators 171
6.3.2 PID Controller 172
6.4 Intrinsic Stages and Their Control Specifications 172
6.5 Procedure Synthesis 174
6.6 Dynamic Simulation 175
6.7 Additional Examples 177
6.7.1 Semi-batch Reaction 177
6.7.2 Distillation Startup 181
6.8 Concluding Remarks 189

Appendix 6.1 Component Models Used in Flash Startup Example 189
Appendix 6.2 Control Specifications Used in Flash Startup Example 194
References 200

7 Normal Operating Procedures Obtained with Timed Automata 203
7.1 Timed Automata 204
7.2 Process Structure 204
7.3 Construction of Component Models 205
7.3.1 Actuators and PID Controllers 206
7.3.2 Flash Drum 207
7.4 Intrinsic Stages and Their Control Specifications 209
7.4.1 Control Specifications for Stage 1 210
7.4.2 Control Specifications for Stage 2 210
7.4.3 Control Specifications for Stage 3 213
7.5 Procedure Synthesis 213
7.6 Dynamic Simulation Studies 215
7.7 Another Example 217
7.7.1 Operation Stages 221
7.7.2 Feasible Operating Procedures 221
7.7.3 Unsafe Operating Procedures 225
7.7.4 Comparison with Aspen Built-in Procedure 227
7.8 Concluding Remarks 228
Appendix 7.1 Exploratory and Test Runs on Aspen Plus Dynamics 234
Appendix 7.2 Component Models and Layer Model Used in the Flash Startup Example 237
Appendix 7.3 A Detailed Listing of the Guards and Updates of Every Edge in the Component Model of Flash Drum 245
Appendix 7.4 Task File Used for SFC-7.3 Simulation 251
References 252

8 Generation of Test Plans for Fault Diagnosis with Untimed Automata 253
8.1 Model-Building Methods 254
8.2 Observable Event Traces in Diagnoser 258
8.3 Control Specifications for Diagnostic Tests 261
8.3.1 Test Target 262
8.3.2 Auxiliary Constraint 262
8.4 Test-Plan Synthesis 263
8.5 Case Studies 269
8.5.1 A Three-Tank Buffer System 269
8.5.2 A Beer Filtration Plant 272

8.6 Concluding Remarks 280
References 280

9 Synthesis of Diagnostic Tests Based on Timed Automata 283
9.1 Model-Building Approach 284
9.2 Exhaustive Enumeration of Fault Propagation Scenarios 288
9.3 Construction of Test Plans 290
9.4 Case Studies 295
9.5 Concluding Remarks 308
References 309

10 Synthesis of Diagnostic Test Plans Based on Hybrid Automata 311
10.1 Illustrative Example 312
10.2 Automata Built with Engineering Knowledge 314
10.2.1 PID Controller and Actuator 314
10.2.2 Process Configuration 314
10.2.3 Processing Units 315
10.2.4 Sensors 316
10.2.5 PLC or Operator 317
10.2.6 System Hierarchy 318
10.2.7 Path Explosion 319
10.3 Automata Built with Simulation and/or Historical Data 320
10.4 Hybrid Models 323
10.5 Observable Event Traces 325
10.6 Test Plan Synthesis 327
10.7 Validation of Test Plans 331
10.7.1 Trace $Tr_{02.1}$ 331
10.7.2 Trace $Tr_{02.2}$ 333
10.7.3 Trace $Tr_{02.3}$ 334
10.8 Additional Case Studies 335
10.8.1 Process Description 335
10.8.2 Aggregated Hybrid Model 338
10.8.3 Diagnoser 343
10.8.4 Test Plans 344
10.9 Concluding Remarks 346
References 346

Index 349

Chapter 1
Introduction

Standard operating procedures (SOPs) are indispensable for running chemical plants. They are needed in performing a wide variety of essential tasks for continuous processes, such as the startup and shutdown operations of all processing units, the emergency response actions under abnormal circumstances, and equipment maintenance routines, and for almost all production activities in batch processes. Therefore, other than the process flow diagram (PFD) and piping and instrumentation diagram (P&ID), the sequential function chart (SFC) of every SOP should also be documented accurately in the process design stage. However, although the modern plants are becoming more complex than they used to be, their operating procedures are still created manually on the basis of the designer's experience in most cases. Manual synthesis of an operating procedure can be a very difficult undertaking since it is both time-consuming and error-prone. It is thus desirable to develop systematic approaches to automatically conjecture viable steps so as to achieve a specific operation goal.

As mentioned above, any batch or sequential operation can be fully characterized with a PFD or a P&ID and an SFC. All related hardware items and their interconnections can be found in the PFD or P&ID, while the operation steps and their activation conditions are incorporated in an SFC. In this chapter, a detailed description of these needed documents is given in the sequel.

1.1 Graphical Representations of Chemical Processes

A typical chemical process may consist of various different hardware items such as the reactors, distillation columns, and flash tanks. For communication convenience, the graphical representations of a chemical plant are usually adopted to make the essential information easily available to all users. Several versions of flow diagrams are utilized by chemical engineers, such as the block flow diagram (BFD), the PFD, the P&ID, the layout drawing, the mechanical drawing, and the construction drawing.

C.-T. Chang et al., *Process Plant Operating Procedures*, Advances in Industrial Control,
https://doi.org/10.1007/978-3-030-70978-5_1

Since, among them, the PFD and P&ID are the most commonly used, these two diagrams are described below.

1.1.1 Process Flow Diagram

The information embedded in a PFD is much more than that in a BFD. Also, the PFD is the most used in practical applications. The unique features of a typical PFD can be summarized below.

1. All main units of the chemical process are shown in the PFD, along with their respective tag names or numbers and the corresponding brief descriptions.
2. Each stream is identified with a directed line segment and a unique ID number in PFD. The corresponding temperature and pressure are also specified next to this stream line. Additional information, such as its flowrate and composition, could also be provided.
3. The utility streams and their operating conditions should be shown in a PFD.
4. Every stream originated from or directed to another PFD should be clearly marked.
5. The symbols and tag names should be consistent. Usually, every company has its own system.
6. On each PFD, its author, reviewer, approver, and the document's serial number, date, edition, etc., should be clearly stipulated.

Let us next consider the simple PFD in Fig. 1.1 as an example for illustration convenience. In this diagram, each unit is labeled with a systematic code, which consists of one or two characters and a serial number. More specifically, the initial character C represents compressor, TK tank, E heat exchanger, H utility heater, P pump, and R reactor. Notice that pressure and temperature are also given on the streams. Notice also that the codes of both pump and compressor are attached with the notation "A/B". It means there are two identical units. Only one is utilized during normal operation, while the other is on standby. Finally, the information about the author, reviewer, approver, edition number, etc., should be given at the lower right corner.

1.1.2 Piping and Instrumentation Diagram

P&ID is also known as the mechanical flow diagram (MFD) or engineering line diagram (ELD), and it can be considered as an extended version of the PFD. This diagram shows all the items, e.g., pipes, equipment, instruments and actuators, and their interconnection required for the nominal operation and all other auxiliary operations, e.g., startup, shutdown, and maintenance, of a chemical plant. Each PFD is often expanded into several P&IDs due to the need to incorporate a vast amount of

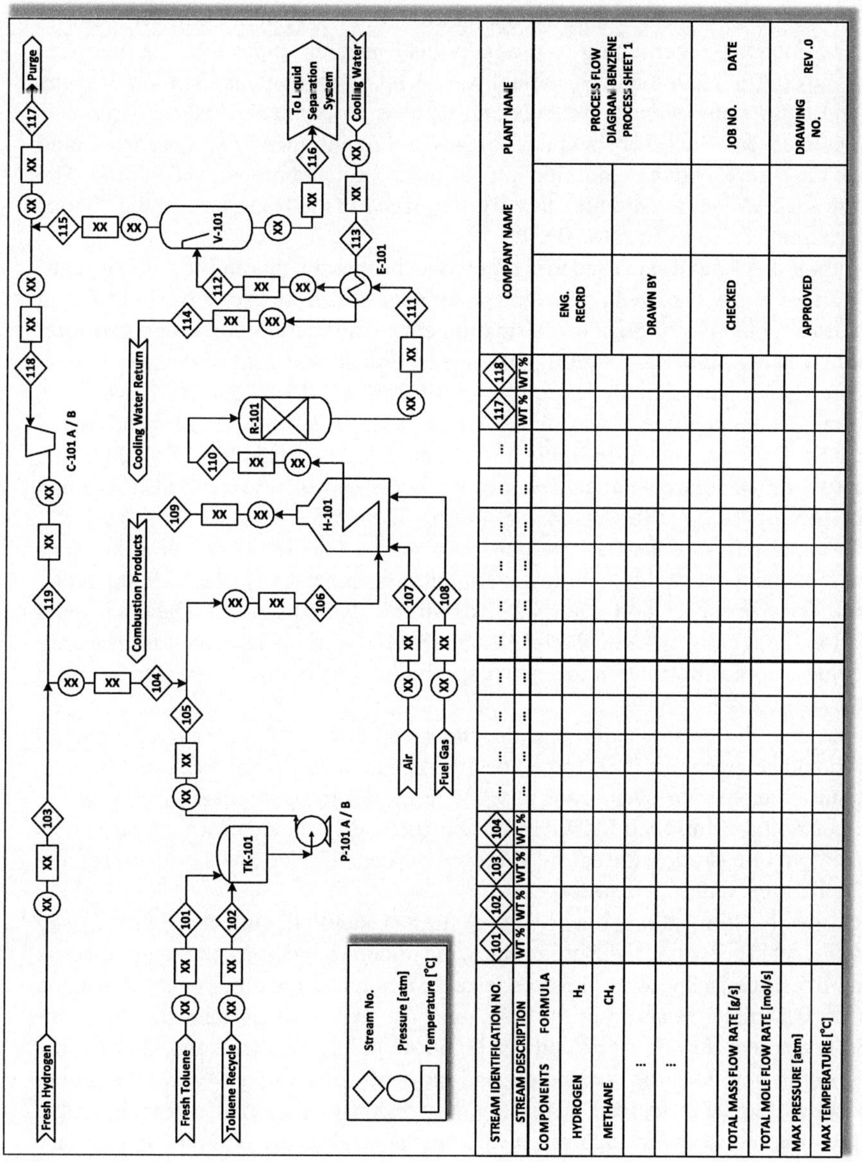

Fig. 1.1 An example of a process flow diagram

extra information. For the same reason, a P&ID is usually drawn by several engineers of different backgrounds from separate departments in a company.

Notice that the P&ID is the most important reference document for both construction and operation of a chemical plant. In addition to the fact that P&ID is the most efficient means of understanding an operation, its index system allows the user(s) to locate documents pertaining to the individual unit, the pipework, the measurement, and control equipment. Specifically, a P&ID might contain a border, drawing lists and equipment indices, symbols and legends, equipmental data, instruments and pipelines, and the other special messages and requirements. The border should include the company name, plot title, plot number, project number, and edition. The drawing list and equipment index usually are placed in the front page of the "book" of approximately 50 to 100 P&IDs.

Symbols and legends are used to characterize the instruments and control systems. Many different standards can be adopted, for example, BS 1646, DIN 28,004, and ANSI/ISA-5.1-2009 (2009). For the instrument identification table, one can refer to Table 4.1 in ANSI/ISA-5.1-2009. The measurement and control device or functions are also indicated in Tables 5.1.1 and 5.1.2 in ANSI/ISA-5.1-2009, while the measurement elements and transmitters are in Tables 5.2.1–5.2.5. The standards of instrument lines are shown in Tables 5.3.1 and 5.3.2 in ANSI/ISA-5.1-2009. The standards of final control element and signal processing function could also be found in ANSI/ISA-5.1-2009 (Tables 5.4.1–5.4.4 and Table 5.6).

The upper part of a P&ID is in general reserved for the towers, vessels, tanks and heat exchangers, and the pumps or compressors are usually placed in the lower section. The size of each unit is not scaled down exactly in the P&ID. The tracing and insulation for equipment shell should also be shown clearly. Although it is necessary to provide specific information for each type of unit, every company tends to establish its own standards.

Other than the symbols and legends, additional information about instrumentations should be given in a P&ID: specifically, the measuring point on a unit, the size, type, and other data of every control valve (e.g., fail to open position, FO; fail to close position, FC; and fail locked in last position, FL), the control loop structures (e.g., simple or cascade), the safety valve specs (i.e., I/O size and set pressure), the flow meter type, and miscellaneous info.

The specifications of pipelines and pipe fittings should include those of the main (or auxiliary) process lines, the utility lines, the underground lines, the steam/electric tracing lines, the jacket lines, the type, size, location and action of every valve, the class breaks (material and rating), the slopes (or free draining and gravity flow), the drainage and vent of every unit (or high/low point), drainage and vent collection systems, safe locations (e.g., vent and blow-off), the strainer types, the piping connections (e.g., the flanged, cap, plug, expansion joint, and the flexible hose), the line size reducing (or enlarging), the spare nozzles, the equipment auxiliary piping, special items (e.g., strainer, hose, spray nozzle, and sight flow indicator), and special requirements like air cooler piping manifold (symmetrical piping arrangement), anchor securely (two-phase flow), and slurry or dirty services (minimum distance). For the distribution system, plot plant orientation shall be used for the schematic presentation of the system piping information should also include those concerning the special service lines, such as steam-out, purge, flush, wash, snuffing, decoking, pump-out, startup, shutdown and circulation, and the bypass lines (usually next to the heat exchanger, compressor, and pump). The other information and special requirements may also include the scope boundary, notes (e.g., the detail design, safety measures, and operation goal), the thermosiphon required static head, the seal height, and distance of condensate pot to heat exchange, etc.

Figure 1.2 is the typical P&ID of the upper part of a distillation system. This diagram includes the upper half of a distillation column (T-101), a partial condenser (E-101), a reflux drum (D-101), and two pumps (P-101 A/B). The instruments are labeled according to a systematic coding system. The first letter always represents the type of signal. For example, F denotes flow, P denotes pressure, L denotes level, and T denotes temperature. Then, the next letter represents the corresponding function, e.g., A is an alarm, I is an indicator, C is a controller, E is an element, S is a switch, and T is a transmitter. In addition, CV is a function which represents the control valve. Therefore, TT, FIC, and PCV mean the temperature transmitter, flow controller with the function of indication, and pressure control valve in this P&ID, respectively. It can also be observed that there are 4 control loops. At the vapor output section of the reflux drum (D-101), there is a pressure control loop which consists of PE-10314 (sensor), PT-10314 (transmitter), PICA-10314 (alarm), and PCV-10314 (control valve). After measuring the drum pressure and transmitting the measurement signal to PICA-10314 (located in the central control room), the PCV-10314 should adjust the opening of the control valve according to the controller output.

Another example is presented in Fig. 1.3. For confidentiality reasons, the details of this P&ID have been hidden. However, since the same principles mentioned above have been adopted to draw this diagram, the readers should be able to read and understand the process without difficulties.

1.2 Graphical Representations of Operating Procedures

Since the time order of steps in an operating procedure is of critical importance, the sequential logic is usually adopted to ensure that a batch process or an unsteady operation (e.g., startup, shutdown or maintenance) in a continuous process attains the proper sequence of states. This logic can be implemented with a microprocessor or a computer. For each process step, each discrete-acting device in the given system is expected to be in a specified device state, which is usually represented with a binary value (0 or 1). The sequential logic is directly applied to these device states. For example, state 0 may be a valve closed, an agitator off, and so on, while state 1 represents the opposite conditions. Basically, the sequential logic determines when the process should proceed from one set of states to the next. There are various different ways to depict the sequential logic, e.g., the information flow diagram, the SFC, the ladder logic diagram, and the binary logic diagram (Seborg et al. 2004).

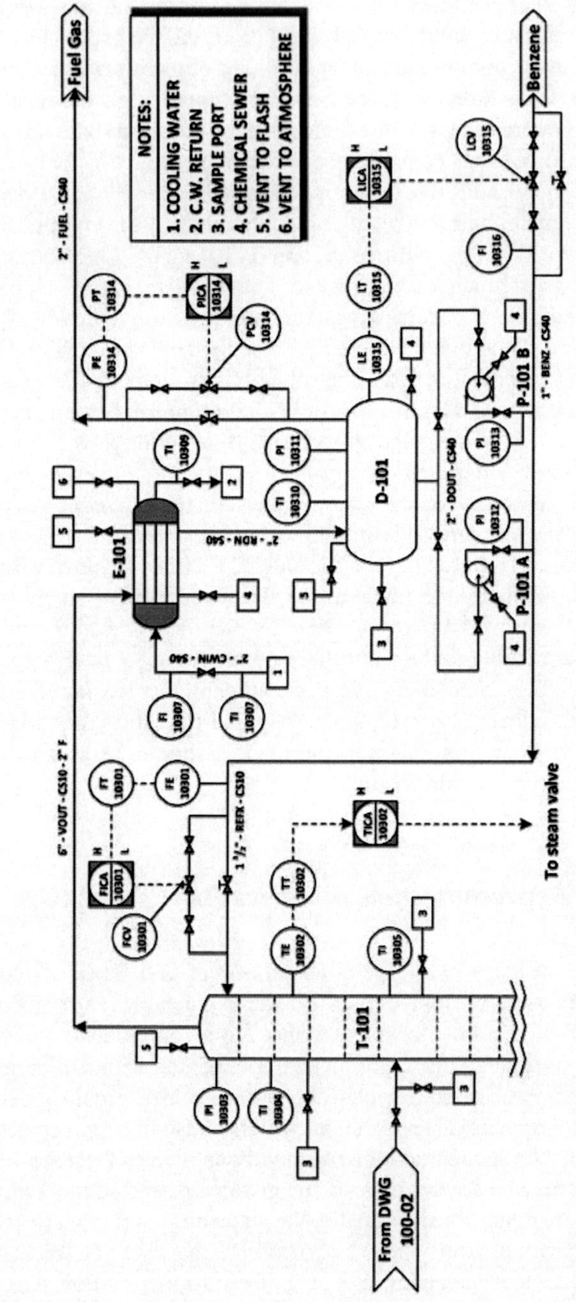

Fig. 1.2 Example piping and instrumentation diagram

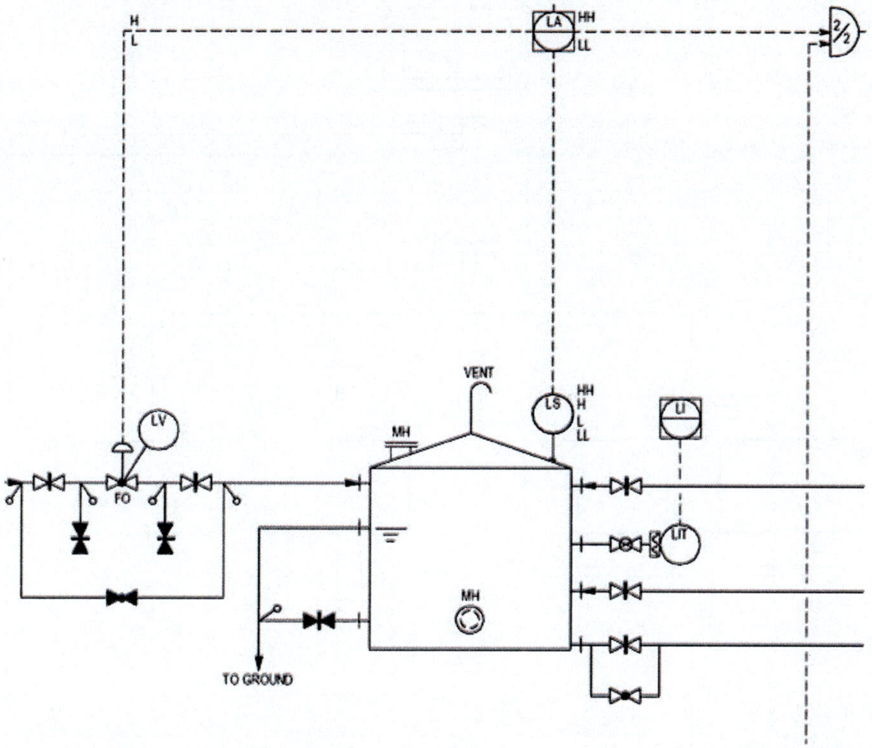

Fig. 1.3 P&ID of a storage tank with level control loops

Since the SFC is most commonly utilized in practice, this chart is explained in detail in the sequel.

Generally speaking, the SFCs are used to "program" the operation steps in a defined order sequentially (Fleming and Pillai 1999). An SFC consists of such steps and transitions, while in some cases each step may in turn point to another SFC. Transitions cause the control execution to move from one step to the next. Figure 1.4 shows a conceptual example. The boxes correspond to steps, and the horizontal bars correspond to transitions. The SFC is always processed from the top down and left to right. Multiple single horizontal bars at the same level correspond to the conditional path. As mentioned above, these conditions are processed from left to right. The first satisfied condition causes the actions after that transition to be carried out, and all transitions and paths to the right of this first true transition should be ignored. Double horizontal lines correspond to parallel processing, and all paths underneath should be performed simultaneously. Due to their sequential nature, SFCs are very suitable for use in procedural control. The following is a more concrete example.

Let us consider the liquid transfer system shown in Fig. 1.5. The system is made of a storage tank, a supply system, two 3-way valves (V-1 and V-3), and two gate

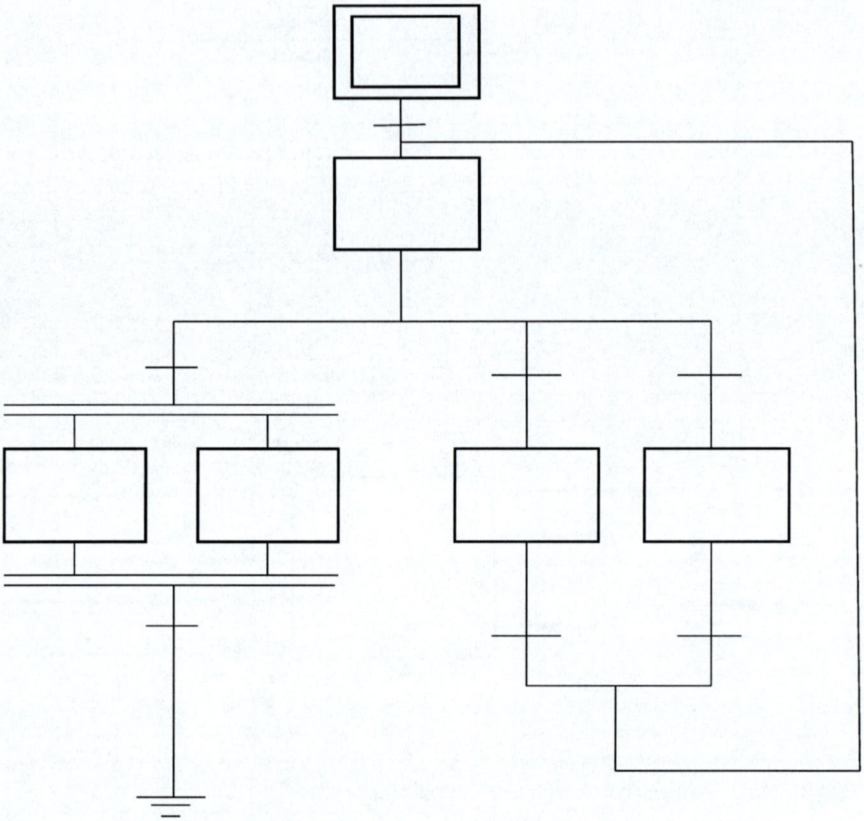

Fig. 1.4 Conceptual example of sequential function chart

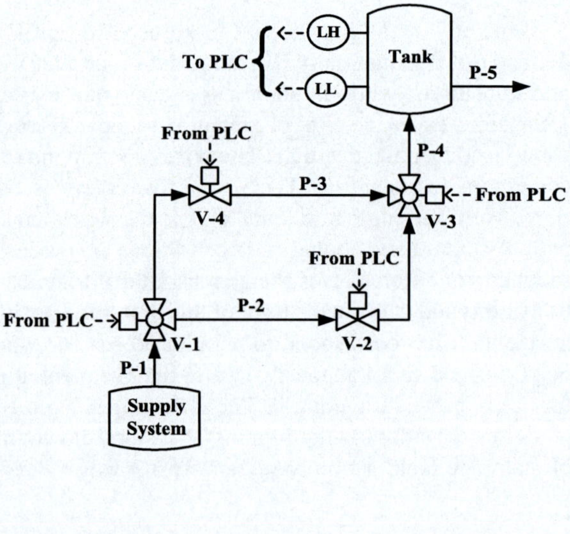

Fig. 1.5 A liquid transfer system (Reprinted with permission from Yeh and Chang 2012. Copyright 2012 Elsevier)

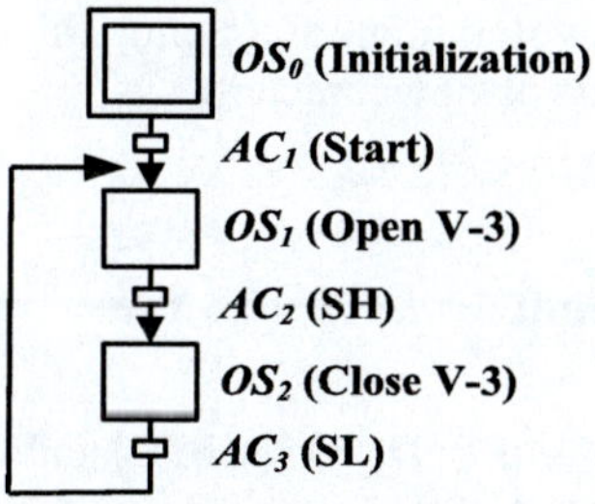

Fig. 1.6 Sequential function chart of the normal operating procedure in the liquid transfer example system (Reprinted with permission from Yeh and Chang 2012. Copyright 2012 Elsevier)

valves (V-2 and V-4). Notice that each 3-way valve can be switched to one of two alternative positions, i.e., open or close, to manipulate the flow directions. The fluid in vertical pipeline P-1 is allowed to flow into the horizontal line P-2 via the opened V-1, while the horizontal flow in P-3 can join the vertical flow in P-4 via the opened V-3. On the other hand, if V-1 or V-3 is switched to the close position, the valve connection to/from the horizontal pipeline must be blocked completely. While the inlet flow to the storage tank can be controlled with these valves, this tank is drained continuously via pipeline P-5 as long as it is not empty. The height of the liquid level is monitored online with a sensor, and two distinct signals, i.e., (1) SH (level signal high) and (2) SL (level signal low), are transmitted to a programmable logic controller (PLC) to actuate the aforementioned four valves in this system. Under the assumptions that the initial liquid level in the storage tank is low and valve V-4 is at the open position initially while the others are all closed, the SFC in Fig. 1.6 can be stipulated to represent the normal periodic operating procedure. Notice that OS_i ($i = 0, 1, 2$) and AC_j ($j = 1, 2, 3$) denote the operation steps and the activation conditions of these steps, respectively. The control actions taken in each step and the sensor signals used in each condition are also specified in this chart.

The SFC instructions can then be translated into the ladder diagram (LD), which is a popular form of the programming language for the PLCs. Simplicity and easy understanding are the key reasons for the wide acceptance of LD. LD (see Fig. 1.7) is also known as ladder logic which mimics the relay logic used for batch processes.

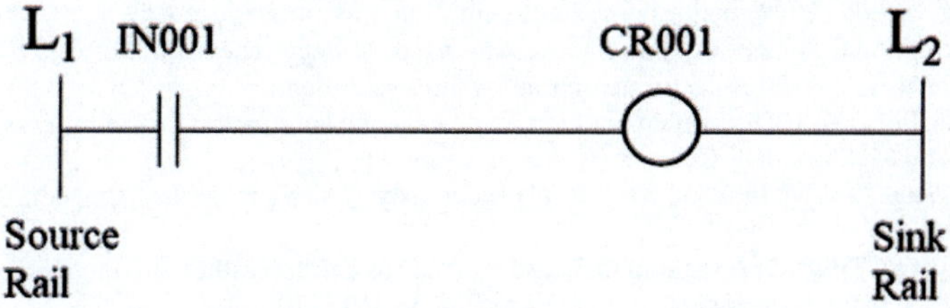

Fig. 1.7 Typical ladder diagram instruction for motor on–off control

Since it can be understood with minimum training, this programming language has become quite popular for the plant technicians.

1.3 Problem Statement

Having provided illustrations of PFD, P&ID, SFC, and LD in sufficient details, it is quite straightforward to formulate the procedure synthesis problem as follows:

> Given the PDF or P&ID of a process and the corresponding initial and target operating conditions, generate the SFCs (or LDs) that bring the given system from its initial state to the target state(s).

1.4 Framework and Organization

The remainder of this book is organized as follows. To facilitate an explanation of the current approach to synthesize operating procedures in practice, the popular heuristic rules and a few examples are presented in Chap. 2. On the other hand, the model-based approach to procedure synthesis is outlined in Chap. 3. In particular, the modeling methods based on automata and the Petri nets are described in sufficient detail, and the connections between these models and languages are also explained. Chapter 4 presents extensive industrial applications of the heuristic rules and dynamic simulation tools for generating normal operating procedures. Chapter 5 is an in-depth illustration of Petri net-based procedures, while its counterparts concerning untimed and time automata are given in Chaps. 6 and 7, respectively. The synthesis methods for generating test plans for fault diagnosis using untimed and time automata are provided in Chaps. 8 and 9, respectively. Finally, Chap. 10 lays out the implementation steps for conjecturing diagnostic tests based on hybrid automata.

References

ANSI/ISA-5.1-2009 (2009) Instrumentation Symbols and Identification, American National Standard, International Society of Automation, Approved 18 September 2009, https://www.isa.org/products/ansi-isa-5-1-2009-instrumentation-symbols-and-iden

Fleming DW, Pillai V (1999) Implementation guide strategic automation for the process industries, vol S88. The McGraw-Hill Companies Inc

Seborg DE, Edgar TF, Mellichamp DA (2004) Process dynamics and control, 2nd edn. John Wiley & Sons, Inc.

Yeh ML, Chang CT (2012) An automata based method for online synthesis of emergency response procedures in batch processes. Comput Chem Eng 38:151–170

Chapter 2
Heuristic Approach to Procedure Synthesis

2.1 Heuristic Rules

Heuristic rules are the accumulation of the experiences. Commonly, heuristic rules are intuitive and easy. For example, when we take a shower in winter, most of the people would like to have a hot shower. Then, the temperature we prefer to use is the followed question. Normally, when the temperature is too high, we would reduce the hot water flowrate to let the temperature of the tap water cool down. Identically, in the chemical plant, the reboiler duty would decrease as the temperature is higher than the set point (SP).

In chemical engineering, the design of the control scheme would also consider the heuristic rules to develop the inventory control loops and quality control loops. Such as the tank level control (LC), this kind of control loop has two different control schemes. First, the feed flowrate would be selected to maintain the level. As the level is lower than the SP, the flowrate would increase to let the level increase. The other strategy is using the outlet flowrate to control the level. This control strategy would have a response that the outlet flowrate would be increased when the level is higher than SP. Figure 2.1 illustrates these two control strategies.

In the quality control loops, the online concentration measurement is quite expensive and time-consuming for a distillation column. Besides, for the polymer process, the measurement of melt index (MI) cannot be on-time measured. Thus, the other process variables (PVs) which could also reflect the quality would be used as the control variables (CVs) to fulfill the object. For example, the temperature control (TC) is often adopted as a quality control loop of distillation. It is because a distillation column is a separation unit that uses the difference of relative volatility of each component and each tray would achieve a vapor–liquid equilibrium (VLE) state as the purifying method. According to the phase diagram, the composition would be fixed as the pressure and temperature are constants. In other words, the concentration would be maintained whenever the pressure and temperature are controlled. The demo control scheme of the distillation process is illustrated in Fig. 2.2.

C.-T. Chang et al., *Process Plant Operating Procedures*, Advances in Industrial Control,
https://doi.org/10.1007/978-3-030-70978-5_2

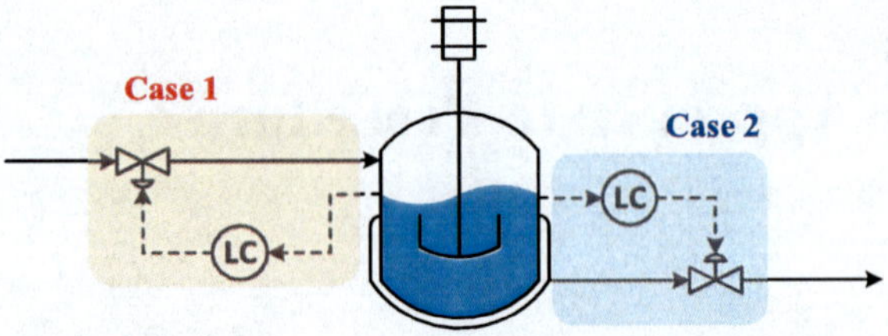

Fig. 2.1 Level control of the tank

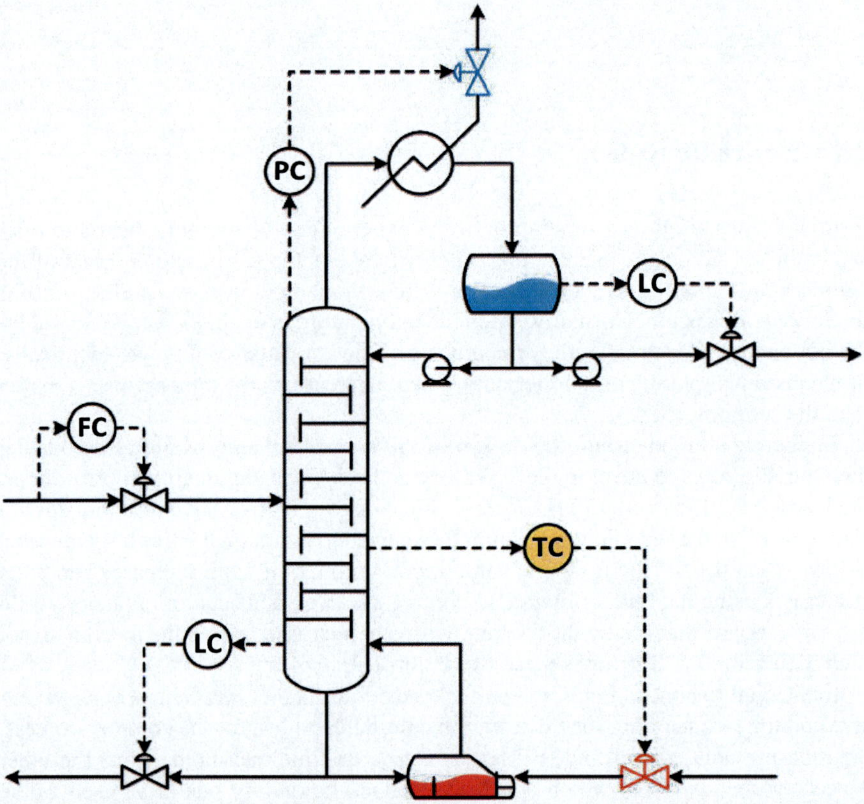

Fig. 2.2 Control scheme of the distillation process

Going a step further, the more complex operations such as startup, shutdown, and grade transition have the basic strategies to handle those situations although those strategies might not be the most effective and rapid. To shut down a distillation column, we would stop heating first and close the feed second based on heuristic rules. Until the system temperature backs to the room temperature, the holdup would be drained out. Leaving it at that, the shutdown procedure is finished.

Known from the above examples, heuristic rules are very important methods to synthesize the procedure for each object. It might not the best way to solve the problem, but it would be a start point for figuring out the real optimum procedure.

2.2 Dynamic Simulation Software

Unlike the most simulation software which can just only simulate the steady-state process, dynamic simulation software enables to imitate the transient behaviors and control responses of the process. In the dynamic simulation, natural and chemical phenomena are described by algebraic and differential equations. These kinds of software require numerical methods to solve the equation. A numerical simulation is done by stepping through a time interval and calculating the integral of the derivatives by approximating the area under the derivative curves. Some methods use a fixed-step interval, some use an adaptive step that can shrink or grow automatically to maintain the acceptable error tolerance, and the others use different time steps in different parts of the simulation model.

Several simulation software concluding open sources and commercial can be found in the chemical engineering field. The famous dynamic simulations are Aspen Plus Dynamics® and Aspen HYSYS® of AspenTech©, and DYNSIM® of SimSci©. Besides, Simulink® of MathWorks® could also execute the dynamic simulation as the mathematic model of the process is provided, or it could directly connect with the built file of Aspen Plus Dynamics via "AMSimulation" blocks. Figure 2.3 shows what the Aspen Plus Dynamics looks like, and Fig. 2.4 illustrates the outlook of Aspen HYSYS.

In Aspen Plus Dynamics, there are four integrated methods that could be selected: Implicit Euler, Explicit Euler, Runge–Kutta 4, and Gear. The Implicit Euler method is the default integrator in Aspen Plus Dynamics, and it could remain more stable than Explicit Euler whatever the value of the integration step size is. However, it would become increasingly inaccurate as the step size was increased. Usually, it is slower than the Explicit Euler method under a given simulation and step size because it must perform the iterative calculations to solve the integral equation. The Explicit Euler method would not often be selected due to the less-accurate result. But it might be an ideal selection for the operator to training simulators. Runge–Kutta 4 is a higher order method (4th) than Implicit Euler (1st). And due to that, it is more accurate than Explicit Euler. Runge–Kutta 4 is suitable for a wide range of applications. It is very effective while the problem which has a large number of disturbances needs to be solved. The Gear integrator uses the integration method based on Gear's algorithm. It could automatically vary the integration step if the step size is set as "Variable" and integration algorithm order. Hence, the estimated integration errors are within a tolerance determined by both relative and absolute variable tolerances. The Gear method is a fast and accurate integrator, and could handle stiff systems.

Figure 2.5 demonstrates an exothermic reaction process via a continuous stirred

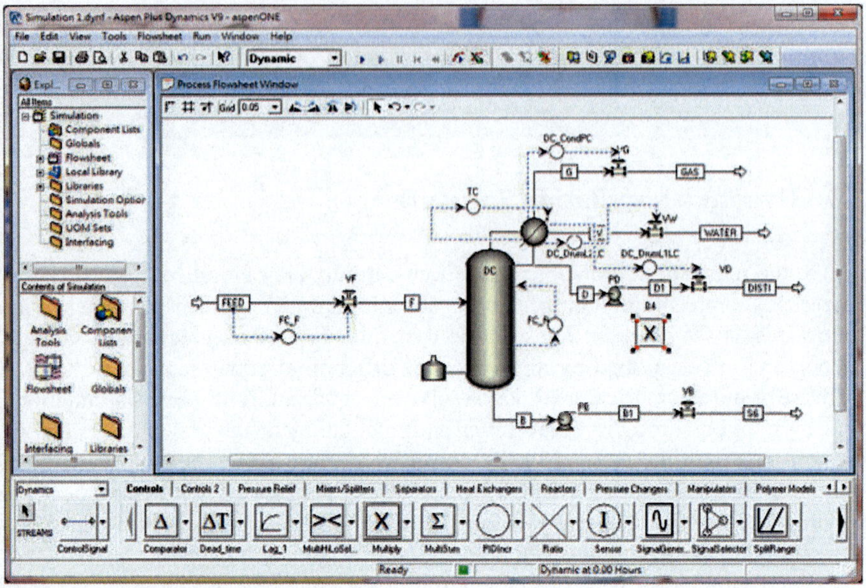

Fig. 2.3 Outlook of Aspen Plus Dynamics

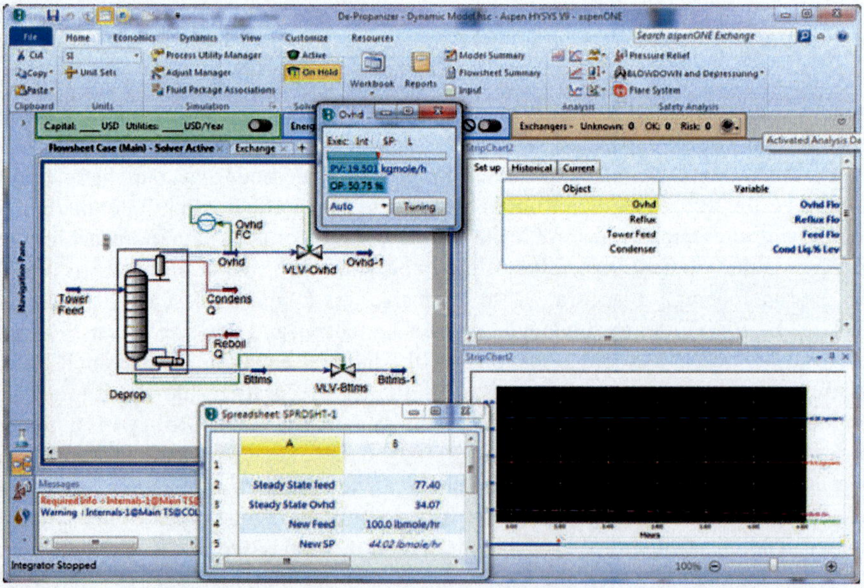

Fig. 2.4 Outlook of Aspen HYSYS

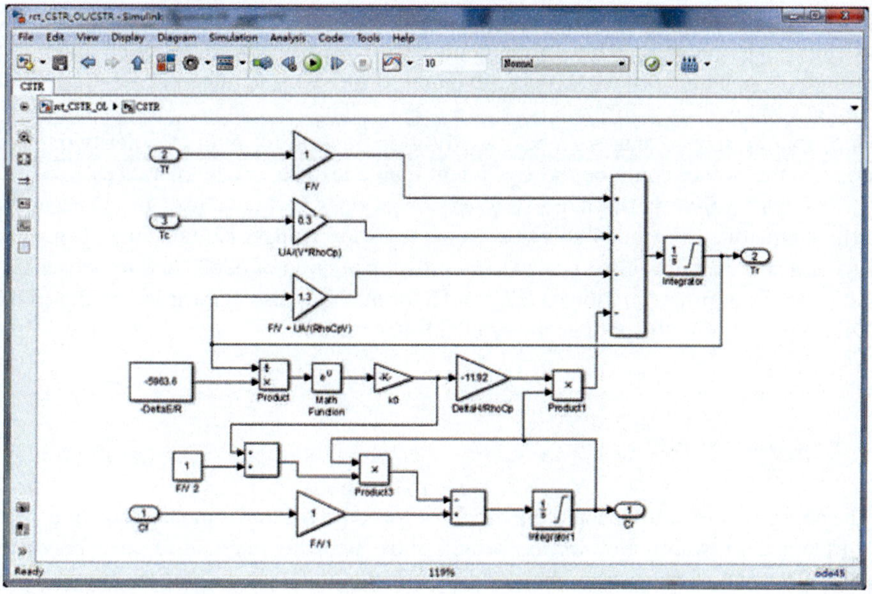

Fig. 2.5 CSTR in Simulink®

tank reactor (CSTR). It represents the plant model as a combination of several simple mathematics blocks.

In Simulink®, the integration solvers are separated into two parts, variable step and fixed step. By default, "ode45" is selected as the model that has the continuous part, and "discrete" would be chosen when there is no continuous section in the model.

ode45 (Dormand and Prince 1980) is the combination of Runge–Kutta (4,5) and Dormand–Prince; this solver only considers the former one-step result $y(t_{n-1})$. However, if the acceptable error tolerance is lower, and the problem is not complicated, ode23 (Bogacki and Shampine 1989) would be more suitable than ode45. It combines the equation of Runge–Kutta (2,3) and Bogacki–Shampine. ode113 (Shampine and Gordon 1975) is a variable-step solver based on the Adams–Bashforth–Moulton PECE method. It is more effective than the ode45 solver at crude tolerances. And when the simulation is done, it would adopt several former step results $(y(t_{n-m})\ldots y(t_{n-1}))$ to calculate the current result. ode15s (Shampine and Reichelt 1997) is a variable-order solver based on the numerical differentiation formulas (NDFs). Optionally, it uses the backward differentiation formulas (BDFs, also known as Gear's method) that are usually less efficient. When ode45 fails, is inefficient, or the problem might be stiff, this method might be useful. ode23s (Shampine and Reichelt 1997) is based on the modified Rosenbrock formula of order two. Because it is a one-step solver, it may be more efficient than ode15s at crude tolerances. It can solve some kinds of stiff problems for which ode15s cannot. ode23t (Shampine and Hosea 1996) implemented the trapezoidal rule with interpolant. This solver is suggested to use if the problem is only moderately stiff and the solution should have no numerical damping phenomenon. This solver could also deal with the differential-algebraic systems of equations (DAEs). ode23tb (Bank et al. 1985) applied the TR-BDF2, an implicit Runge–Kutta formula with two stages. The first stage is a trapezoidal rule step, and the second stage is a backward differentiation formula of order two. Like ode23s, this solver could be more efficient than ode15s at crude tolerances.

The ode5 solver is the fixed-step version of ode45. It only uses the Dormand–Prince equation. The ode4 solver is as same as the Runge–Kutta 4 in Aspen Plus Dynamics. Like ode5, ode3 is also the fixed-step version of ode23 and only uses the Bogacki–Shampine equation. ode2 is a Heun method, also known as the modified Euler method. Finally, ode1 is the typical Euler method.

2.3 Pilot Setups

The above two sections discuss the process procedure synthesis method and dynamic simulation software; this section would introduce the main difference between steady-state and dynamic simulations. In the steady-state simulation, we only give the stream condition and operated condition of each unit, etc. We do not need to specify the geometry and size of the equipment, and the steady-state simulation could be still done. However, dynamic simulation is a time-dependent affectation. The size and geometry would strongly affect the residence time of the stream. And if you consider doing a rigorous dynamic simulation, the pressure distribution of each unit would be especially important. In this section, Aspen Plus would be used to present the transiting step of steady state to dynamics.

Aspen Plus is well known for the steady-state simulation, but it cannot directly do the dynamic simulation. It should via Aspen Plus Dynamics do the dynamic simulation. Although you can directly build your process in Aspen Plus Dynamics, it would not have higher accuracy and precision than the model built-in Aspen Plus. Due to that, we would simply describe the step of pilot setups for distillation columns, reactors, tanks, pumps, and valves. Before introducing the setting of pilot setups, there is another option of the dynamic simulation named "Flow Driven" in Aspen Plus Dynamics. This kind of simulation did not need to set the pilot setups, but it might not be perfectly accurate and precise.

For executing a rigorous dynamic simulation in Aspen Plus Dynamics, the pressure difference between each unit is very important. If you did not establish the pressure changing unit in your simulation at first, then you might add these units into your flowsheet so that the pressure difference existing in each unit is reasonable. If you do not know whether this kind of connection is correct or not, you could use a "Pressure Checker" in Aspen Plus to let you gain some suggestions and hints.

After the previous section has finished, the size and geometry of the equipment would be specified next. Firstly, we are going to introduce an important unit in the chemical process, distillation columns.

There are four parts needed to be specified as you wish to transit to dynamic simulation file, column diameter, reflux drum size, sump size, and pressure distribution on each tray. If you already know the value of the above four parameters, you just only need to key them into the fixed place. However, if you do not clearly know them, there still are some methods that you could use. The first step is sizing the diameter of the column. It could be done by Aspen Plus. Then, the pressure profile in the column should be fixed. Usually, we would specify either top pressure (condenser pressure) or bottom pressure (reboiler pressure). Finally, the size of the sump and the reflux drum would be determined. The Eqs. 2.1 and 2.2 are used to calculate the diameter and height.

$$H = 2D \tag{2.1}$$

$$D = \sqrt[3]{\frac{2\dot{V}\tau}{\pi}}. \tag{2.2}$$

H means height, D means diameter, $\dot{V}$ means volumetric flowrate, and τ means residence time. Commonly, the residence time of the sump and the reflux drum is 10 min.

Next introduced unit is the reactor; it is an important unit in the chemical engineering field. Basically, the size of the reactor is determined by the kinetics, irrespective of the CSTR or the plug flow reactor (PFR).

The followed unit is the tank; it is a unit that stores the chemicals or buffers the system flow. The size is determined by its residence time. For the decanter and the flash tank, the residence time might be 15 to 30 min. And for the stored and buffered tanks, the residence time is up to each application.

The valve and the pump are up to the user to specify if they have the detailed properties such as the valve opening versus Cv. value and the pump performance curves.

2.4 Illustrative Examples

This section would use both of the examples of Aspen Plus Dynamics (Chen 2014) and a real pilot scale reactive distillation process (Lin 2020) to demonstrate the startup procedure of the distillation column. According to the example of Aspen Plus Dynamics, the startup method could be separated into four parts. The first part is the feed control method, determining when to feed by the height of the sump level. The second is the bottom flow control method used to reduce the amount of the off-specification bottom product. The third one is the reflux control method. Before start-up procedure, the V13 to V18 valves have to close and fill the heater with hot kerosene. After this step, pour 5 L of IPAc into the collection bottle to pump into the membrane device. And then, continue to pour another 5 L of IPAc in the collection bottle. Finally, the outlet pipe of the membrane device is connected back to the collection bottle. Before simulation of the dynamic operations in Aspen Plus Dynamics, the steady-state simulation should be finished first. For this, we will demonstrate a ternary system separation on separating carbon tetrachloromethane (CCl_4), trichloromethane ($CHCl_3$), and dichloromethane (CH_2Cl_2).

The V9 version of Aspen Plus is used for the steady-state simulation. At the beginning of the simulation, the physical properties should be established first and correctly because the incorrect properties would lead to a non-accurate simulation. First, type the property name in the component specifications sheet. All of the startup cases operated in Aspen Plus software should add nitrogen as one of the components because the operation in Aspen Plus Dynamics needs N_2 to executive the "Empty" script. Second, the NRTL (Non-Random and Two-Liquid) method is selected as the thermodynamic model. The binary parameters of thermodynamic are shown in Table 2.1. Please make sure the physical properties of all the components is correct. The physical properties are shown in Table 2.2.

The steady-state operations are described as follows. The feed flow rate is 10000 kg/hr, the temperature and pressure are 20 °C and 6 bar, respectively. The mass fraction of CH_2Cl_2, $CHCl_3$, and CCl_4 is 0.06, 0.54, and 0.4, respectively. The feed stage (N_F) is at the tenth stage. There are 20 trays in the DC including the reboiler. The top flow rate is 6455.01 kg/hr, the reflux ratio (RR) is 5, the operating pressure is 2 bar, and the pressure drop of the column is 0.254 bar. The specification of the top and bottom products is 83.1 wt% of $CHCl_3$ and 99 wt% of CCl_4, respectively. The steady-state flowsheet is illustrated in Fig. 2.6. The diameter of distillation is 1.5 m and the bubble cape is selected as the tray type. The diameter of the reflux drum is 1 m and the height is 2 m. Moreover, the shape of the sump is an ellipse, and the diameter and height are 1.5 m and 2 m, respectively.

Table 2.1 Binary parameters of NRTL models for simple distillation

NRTL parameters			
Comp. i	CH_2Cl_2	CH_2Cl_2	$CHCl_3$
Comp. j	$CHCl_3$	CCl_4	CCl_4
a_{ij}	0	0	−1.2316
a_{ji}	0	0	0.7632
b_{ij}	39.1235	156.697	460.338
b_{ji}	−39.8462	−14.4866	−263.328
c_{ijj}	0.3	0.3	0.3

Table 2.2 Physical Properties of components

Component	Molecular weight	Boiling point (K)
CH_2Cl_2	84.93	312.81
$CHCl_3$	119.38	334.25
CCl_4	153.82	349.88

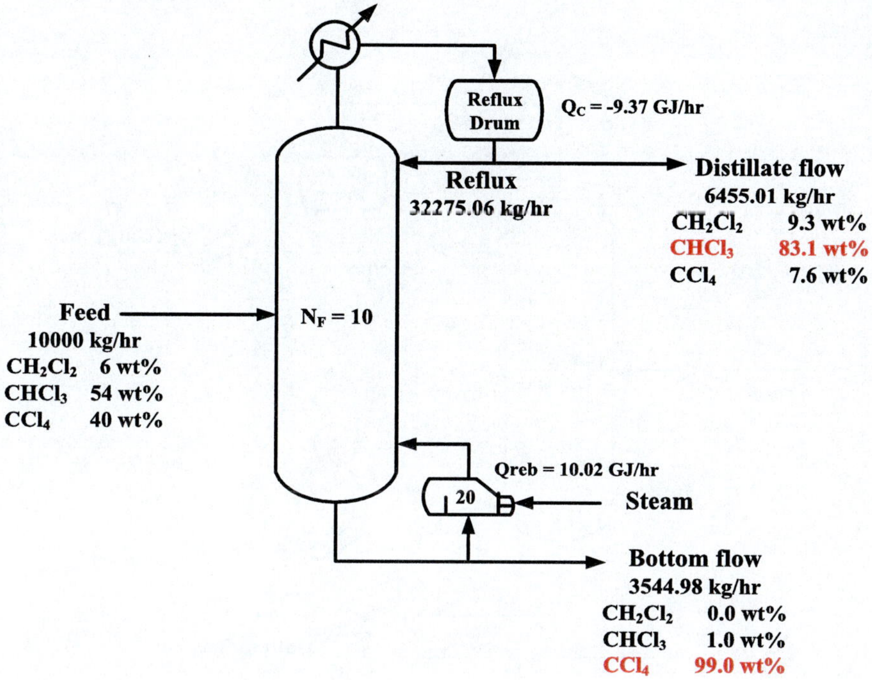

Fig. 2.6 Steady-state flowsheet of simple distillation

After finishing the simulation of the steady state, the process will be transferred to Aspen Plus Dynamics. Before running the startup operation, the column should be empty (full of N_2 at 1.013 bar and 20 °C; please refer to the help of Aspen Plus Dynamics or the example), and all of the flowrates are 0 kg/hr. The control structure is demonstrated in Fig. 2.7. The parameters of all the controllers are shown in Table 2.3. The startup operation would be simulated with the help of a task in order to reduce the negligence of humans.

The steps of the startup operation are as follows: make sure all of the controllers are set at manual mode, then feed some feed into the DC till the height of the sump level is over 2 m. Hereafter, the pressure controller is switched to auto mode, and the vent valve is opened to exhaust the N2. Next, the DC starts to be heating by raising the temperature of the heating medium and is operated under the total reflux (30000 kg/hr) situation. While the temperature of the heating medium reaches 120 °C, the vent valve would be closed.

Besides, the LC_B will be changed to auto mode; the LC_D will be switched to auto mode when the level of the reflux drum reaches 0.5 m. Finally, set all of the set points of the controller at nominal value and let the process settle steady for a while. The startup operation is finished. The startup results are illustrated in Fig. 2.8.

The next example would be demonstrated is isopropyl alcohol (IPA) esterification via reactive distillation (RD) and a membrane. The process is illustrated in Fig. 2.9.

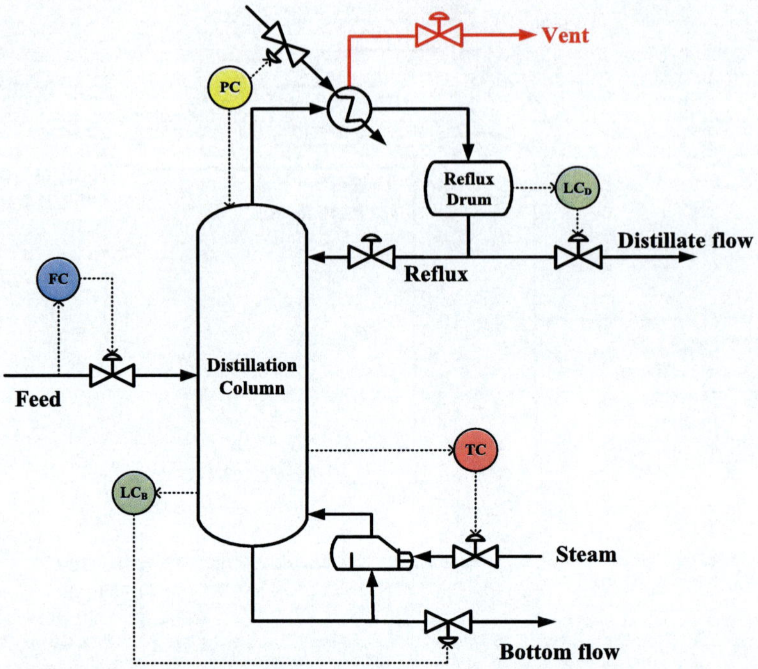

Fig. 2.7 Control scheme of simple distillation

Table 2.3 Parameters of all controllers of simple distillation

Controller	Set point	Manipulated variable	K_c	τ_I (min)
FC	10000 kg/hr	Valve position	1	0.5
PC	2 bar	Flowrate of cooling water	10	12
LC_B	1.25 m	Valve position	10	60
LC_D	1.25 m	Valve position	10	60
TC	102.75°C (the 16th stage temperature)	Temperature of heating medium	1	20

In this process, there are two main sections, RD and a membrane. This process is mainly the recycling of the wasted IPA to generate higher value products, isopropyl acetate (IPAc), via reacting with the acetic acid (HAc) and using the RD to facilitate both the reaction and separation. However, there are 4 azeotropes in this system; it would be hard to get a very pure product by just using one distillation. Hence, the membrane separation technology is adopted here to produce the high purity main product.

Unlike the previous example, this process is a real pilot plant in the campus for studying and teaching. So, this part would not only introduce the startup procedure

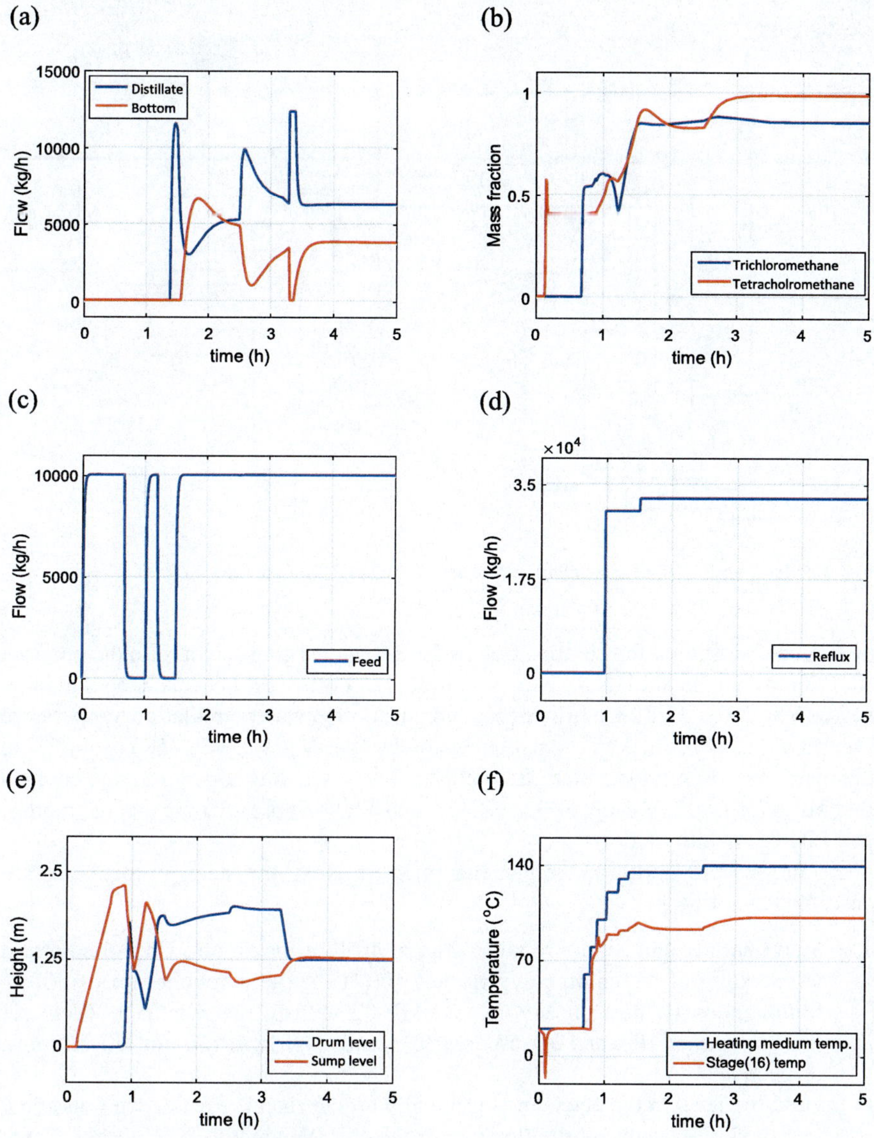

Fig. 2.8 Responses of simple distillation column startup

but also elaborate the shutdown of this system. Firstly, the RD section would be discussed. In this branch, the startup procedure could be distinguished into two parts, the pre-startup and startup procedures. Here, the pre-startup procedure means the initial charge of the distillation column, decanter, and buffer tank. According to the results of Reepmeyer et al. (2004), it would spend less time to reach the normal

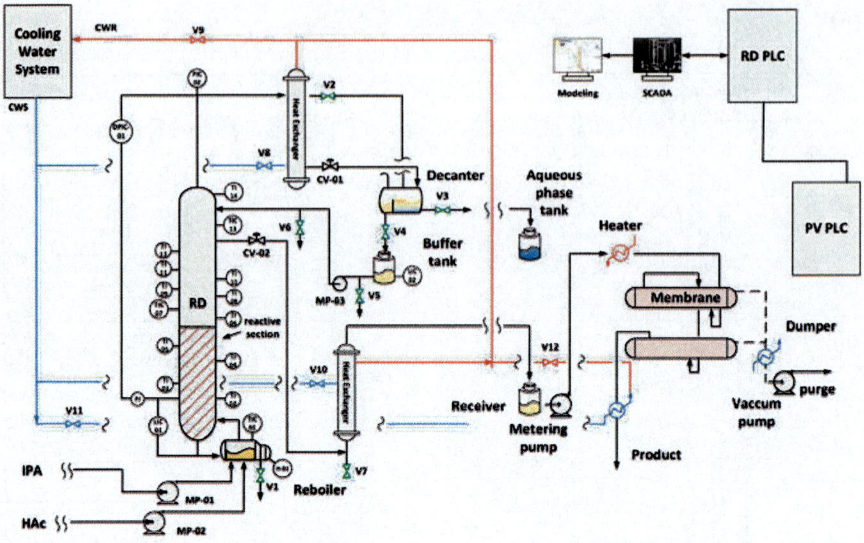

Fig. 2.9 Isopropyl alcohol esterification process via RD and a membrane

operating condition if the composition of the initial charging is closed to the nominal situation. When the initial charging is completed, confirm the heat exchanger manual valves V8, V9, and V10 are fully open, and the cooling water circulation temperature is set at 33 °C. Next, make sure that the manual valves V1, V2, V3, V5, V6, and V7 in the distillation tower system are fully closed, and V4 is fully open. Finally, confirm that MP-01, MP-02, MP-03, H-01, TIC-07, and TIC-13 are under the manual modes, and PIC-02 in auto mode.

So far, the pre-startup procedure has finished. Next, the startup procedures are described as follows:

1. Start heating, and set the reboiler duty at 40% of maximum. Until the bottom temperature of the column approaches 100 °C, set tahe reboiler duty at 50%. Simultaneously, turn on MP-01 and MP-02 to feed the reactants. Then, set the flowrate of IPA and HAc equal to 3 and 2.49 kg/h (60 and 40 ml/min), respectively.
2. Until the level in the decanter is gradually higher than the inside weir and flow into the buffer tank, set the flowrate of the MP-03 equal to 800 ml/min. Then, set the set point of LIC-02 equal to the buffer tank level at this moment and switch to auto mode. Besides, fully open V3 to let the aqueous phase exhaust out from the system.
3. As the top pressure (PIC-02) remains at 210±10 mbarg, and TIC-14 reaches 82 °C, make the set point of H-01 equal to the instant sump level, then switching to the auto mode. Meanwhile, open CV-02 at 50% to gather the sidedraw product.
4. After switching the H-01 to auto mode, measure the compositions of the sidedraw product and the liquid in the buffer tank per hour.

5. Following step 3, set the set point of TIC-07 as 104 °C while the temperature hovers at 104 ± 3 °C around 30 min. Then, switch TIC-07 and MP-02 into auto mode.
6. Till the temperature of TIC-13 reaches 86 °C, let the set point as 88 °C then switch to auto mode.
7. Finally, make sure that all the controllers in the RD system have turned to the auto mode.

Next, the startup procedure of the membrane would be explained. The detailed P&ID is illustrated in Fig. 2.10. s the pre-startup procedures are close to the V13 to V18 valves, fill the heating tank with hot kerosene in sequence, pour 5 L of IPAc into the collection bottle, then pump into the membrane device, continue to pour another 5 L of IPAc in the collection bottle, and connect the outlet pipe back to the collection bottle. Totally, there will be 10 L of IPAc circulated between the collection bottle and membrane after finishing the pre-procedures.

After the pre-procedures are completed, first open the cooling water valves V11 and V12. Second, switch to automatic heating mode to preheat the feed and set the temperature as 100 °C. After the outlet temperature of TI-02 is greater than 90 °C, the vacuum pump could be turned on to evacuate the water vapor, and the valve V17 is opened every half hour to measure the moisture content.

Finally, the startup procedure of both RD and membrane sections have finished. The experimental results are demonstrated in Figs. 2.11, 2.12, 2.13, 2.14 and 2.15.

Different from the startup procedure, shutdown operation will focus on the system safety and reduce the waste of raw materials. The shutdown procedure of RD and membrane section is demonstrated below.

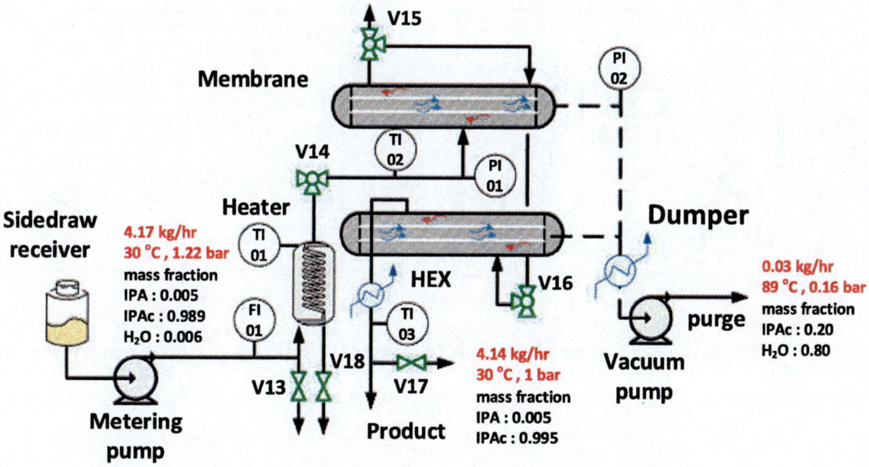

Fig. 2.10 P&ID of membrane section

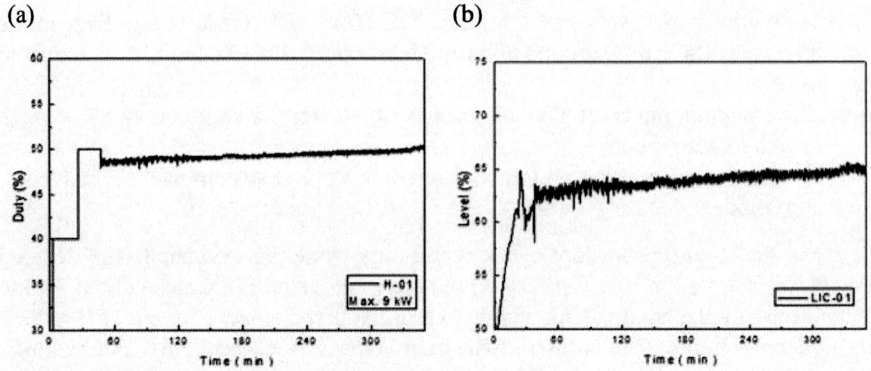

Fig. 2.11 Responses of **a** Reboiler duty **b** Sump level

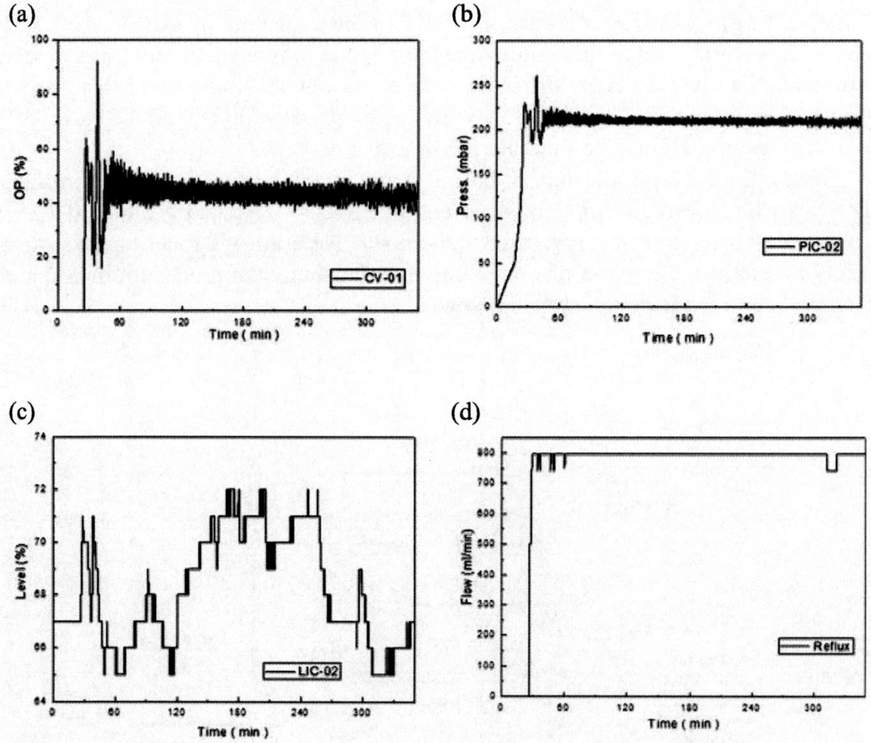

Fig. 2.12 Responses of **a** Opening of CV-01 **b** Top pressure **c** Buffer tank level **d** reflux flowrate

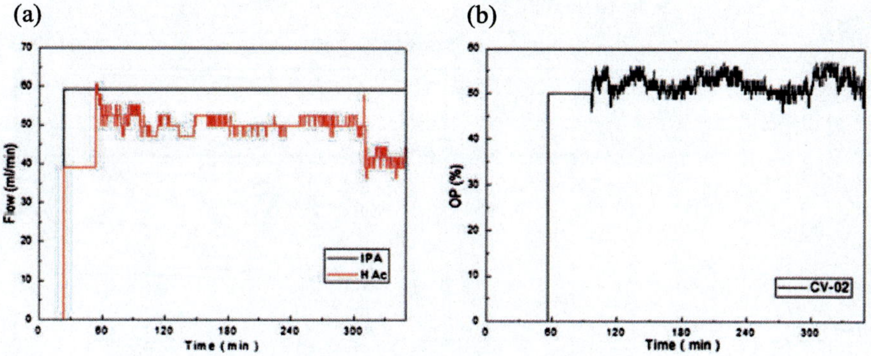

Fig. 2.13 Responses of **a** Feed flowrate **b** Opening of CV-01

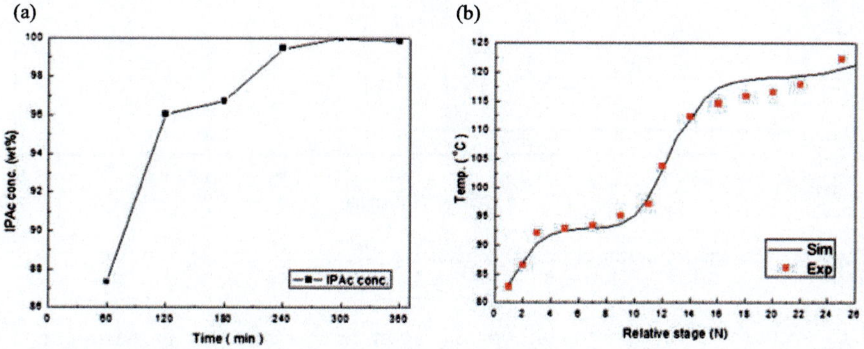

Fig. 2.14 **a** Responses of mass fraction of IPAc **b** Comparison of the temperature profile between the experiment and simulation

Reactive distillation section:

1. The MP-01/02 feed pump mode is switched to manual mode and set to 0% power and then turned off.
2. The reboiler switches to manual mode and sets the power to 0% and then turns off.
3. Switch TIC-13 to manual mode and set the valve opening to 0 (fully closed).
4. Switch TIC-07 to manual mode.
5. Switch MP-03 to manual mode and set the power to 0%, then turn it off.
6. Manually open the V2 valve to relieve pressure; the CV-01 valve opening is reduced to 0% and fully closed, and the V3 valve is fully closed.
7. Turn off H-01, MP-01, MP-02, MP-03, and the main power in the programmable logic controller (PLC) panel in order.
8. Turn off the internal water circulation system.
9. Turn off the air supply and depressurize the air regulator.

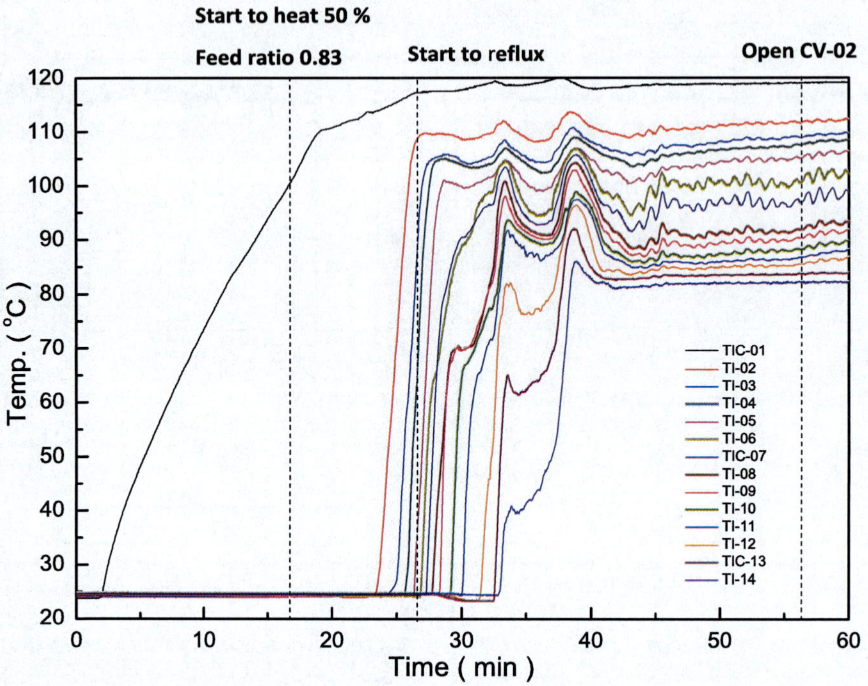

Fig. 2.15 Temperature profile of the RD column

10. Wait for the system to cool down to room temperature, empty the solution in the reboiler, decanter, and buffer tank (or store the initial packing when starting the next time).

Membrane section:

1. The feed flowrate is switched from 100 to 0 ml/min and then closed.
2. Turn off the heating tank directly.
3. Close the circulating cooling water valves.
4. Turn on V13 to V17 to drain the IPAc in the system and leave it as the initial charge for the next time.
5. Finally, turn off the main power of the PLC thin-film electric disk.

Compared with the startup procedure, it is easier and faster. However, it should be watched out whether the pressure or level in the vessel is abnormal or not during the shutdown operation to avoid the accident.

Generally, following the above step and control logics could operate the process successfully. However, there might be some issues such as power cut, trip, liquid leakage, and the reboiler duty loss of control that would make the process lose of control and make the operator dangerous. We call this situation the emergency condition. Usually, the process would also be shut down as the emergency condition occurs. This shutdown procedure is different from the normal shutdown procedure. Thus, the emergency shutdown procedure would be elaborated here.

Reactive distillation section:

1. Press the emergency button (Emergency ShutDown button, ESD) on the PLC system panel.
2. Turn off H-01, MP-01, MP-02, MP-03, and the main power in the PLC board in order.
3. Turn off the internal water circulation system.
4. The computer of SCADA (Supervisory Control and Data Acquisition) system and Modeling HMI is shut down.
5. After the system cools down to normal temperature, empty the residuals in the reboiler, decanter, and buffer tank.

Membrane section:

1. Directly turn off the main power of the PLC.
2. Press the emergency button on the PLC panel.
3. Open the V13–V17 valves to drain the residual liquid from the system.
4. After the system has cooled down to room temperature, empty the residuals in the reboiler, decanter, and buffer tank.

Please make sure all the above steps are finished before starting the troubleshooting or examination.

References

Bank RE, Coughran WC Jr, Fichtner W, Grosse E, Rose D, Smith R (1985) Transient simulation of silicon devices and circuits. IEEE Trans CAD 4:436–451

Bogacki P, Shampine LF (1989) A 3(2) pair of Runge-Kutta formulas. Appl Math Lett 2:321–325

Chen ZL (2014) Optimum startup strategies of a conventional distillation column by simulated annealing approach. Department of Chemical Engineering, National Taiwan University of Science and Technology, Master Thesis

Dormand JR, Prince PJ (1980) A family of embedded Runge-Kutta formulae. J Comp Appl Math 6:19–26

Lin CK (2020) Startup procedure of isopropyl acetate reactive distillation in pilot-scale plant and operating guidance of disturbance rejection. Department of Chemical Engineering, National Taiwan University of Science and Technology, Master Thesis

Reepmeyer F, Repke JU, Wozny G (2004) Time optimal start-up strategies for reactive distillation columns. Chem Eng Sci 59(20):4339–4347

Shampine LF, Gordon MK (1975) Computer solution of ordinary differential equations: the initial value problem. W. H. Freeman, San Francisco

Shampine LF, Hosea ME (1996) Analysis and implementation of TR-BDF2. Appl Numer Math 20

Shampine LF, Reichelt MW (1997) The MATLAB ODE suite. SIAM J Sci Comput 18:1–22

Chapter 3
Model-Based Approach to Procedure Synthesis

As mentioned before, any operating procedure in the process plant can be unambiguously described with a process flow diagram (PFD) or a piping and instrumentation diagram (P&ID) and a sequential function chart (SFC). All hardware items and their interconnections in the plant are depicted in the PFD or P&ID, while the operation steps and their activation conditions are detailed with an SFC. Although these technical drawings can be produced manually, this labor-intensive task inevitably becomes unmanageable as the system complexity increases. In order to avoid human errors and to ensure operational safety and efficiency, it is desirable to develop a collection of modeling and synthesis tools for the automatic generation of the control actions to achieve various operational goals (Patton 1997; Blanke et al. 2003; Zhang and Jiang 2003; Yamashita 2007; Hashizume et al. 2008; Tan and Yamashita 2010).

The pioneering works on the automatic synthesis of operating procedures in chemical processes were first conducted by Rivas and Rudd (1974). Subsequent studies toward the design and verification of procedural controllers under normal conditions have been carried out extensively (Kaspar and Ray 1992; Moon et al. 1992; Chen and Chen 1994; Sanchez and Macchietto 1995; Naka et al. 1997; Panjapornpon et al. 2006; Kim and Moon 2009; Hamid et al. 2010; Yang et al. 2011; Fravolini and Campa 2011). This research problem has already been tackled with numerous modeling/reasoning tools, e.g., the AI-based linear and nonlinear planning strategies (Fusillo and Powers 1987; Lakshmanan and Stephanopoulos 1988; Viswanathan et al. 1998a, b), the mathematical programming models (Crooks and Macchietto 1992; Li et al. 1997; Galán and Barton 1997), the symbolic model verifiers (Kim et al. 2009), and various different qualitative models such as the state graphs (Ivanov et al. 1980; Kinoshita et al. 1982; Hoshi et al. 2002) and Petri nets (Yamalidou and Kantor 1991; Hashizume et al. 2004; Chou and Chang 2005; Wang et al. 2005; Lai et al. 2007). As a result, there is an incentive to develop a generic and unifying approach to conjecture reliable operation steps for any given normal/abnormal system condition.

On the other hand, the model-based procedure synthesis problem for the discrete-event systems (DESs) has been long analyzed and solved rigorously on the basis of supervisory control theory. In its original framework, every discrete-event system is

C.-T. Chang et al., *Process Plant Operating Procedures*, Advances in Industrial Control,
https://doi.org/10.1007/978-3-030-70978-5_3

characterized with a set of event sequences (or the so-called “language”) which can be predicted according to an automaton model (Ramadge and Wonham 1987, 1989). An admissible “supervisor” can usually be synthesized with two distinct automata, i.e., the plant model and the specification model. The former is used to represent how a system behaves with or without hardware failures, while the latter for defining the “legal” events or actions allowed in plant operations. Although this modeling approach has already been successfully applied in many previous studies (Brandin and Wonham 1994; Koutsoukos et al. 2000; Wonham 2000; Dietrich et al. 2002; Malik and Malik 2006; Falkman et al. 2009; Ouedraogo et al. 2010; Yeh and Chang 2011), very few of them offered a specific step-by-step model-building strategy for generating realistic operating procedures and/or the emergency response procedures in the process plants. This is because the language expressiveness of untimed automata is inherently limited. There are needs to use more powerful modeling tools, e.g., the timed automata, the Petri nets, and also hybrid models.

3.1 Hierarchical Structure of Chemical Processes

To facilitate implementation of any operating procedure, the given process must be equipped with enabling components that can be classified into a hierarchy of five different levels (see Fig. 3.1), i.e., the programmable logic controller (PLC), the

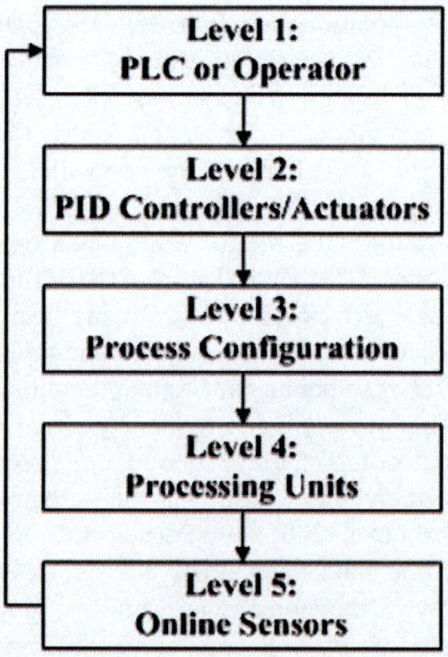

Fig. 3.1 Hierarchical structure of a chemical process (Reprinted with permission from Yeh and Chang 2011. Copyright 2011 Elsevier)

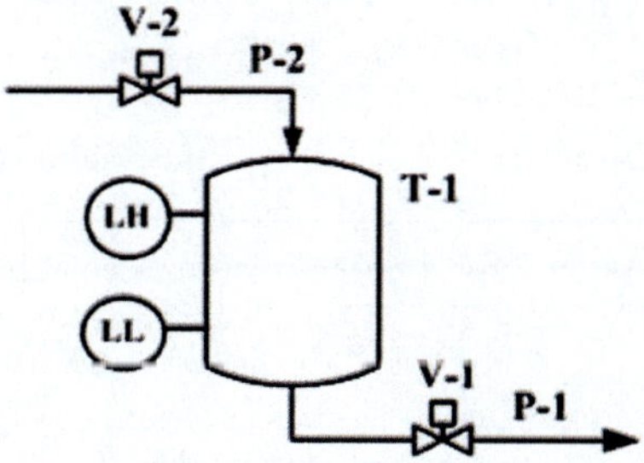

Fig. 3.2 The simple liquid storage system studied in Example 3.1 (Reprinted with permission from Yeh and Chang 2011. Copyright 2011 Elsevier)

actuators, the system configurations, the processing units, and the online sensors. To execute the operating procedure specified in an SFC, the PLC (or a human operator) performs control actions that manipulate the states of actuators which, in turn, alter the material and/or energy flows in the given system. This material and energy flow configuration further dictates the operation modes of the embedded units and subsequently causes variations in the corresponding online sensor measurements.

To further clarify the proposed model hierarchy, let us consider the simple liquid storage system presented in Fig. 3.2. This trivial problem was studied in Chen et al. (2010) and will later be referred to as Example 3.1. The height of the liquid level in the tank is monitored online, and two distinct sensor signals, i.e., (1) level high (LH) and (2) level low (LL), are sent to a PLC to actuate the control valves (V-1 and V-2) on the outlet and inlet pipelines (P-1 and P-2), respectively. Under the assumptions that the liquid level is low and both valves are at the close positions initially, an SFC can be produced to represent the needed cyclic procedure (see Fig. 3.3). Notice that OS_i $(i = 0, 1, 2)$ and AC_j $(j = 1, 2, 3)$ denote the operation steps and the activation conditions, respectively. Notice also that the control actions taken in each step and the sensor signals used in each condition are also specified in this chart. It is clear that the components in this system can be classified into the five hierarchical levels

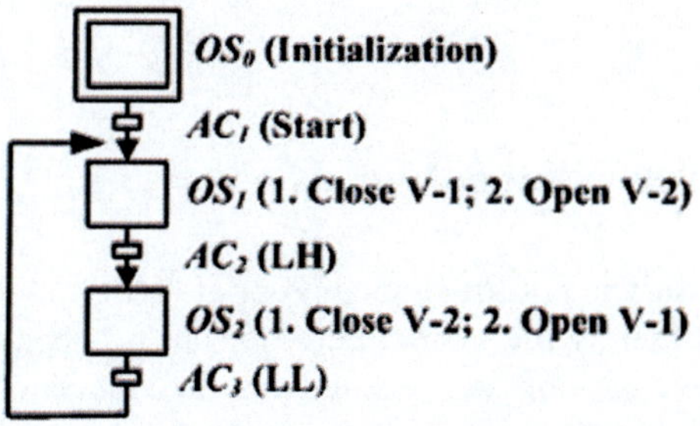

Fig. 3.3 Sequential function chart of the operating procedure in Example 3.1 (Reprinted with permission from Yeh and Chang 2011. Copyright 2011 Elsevier)

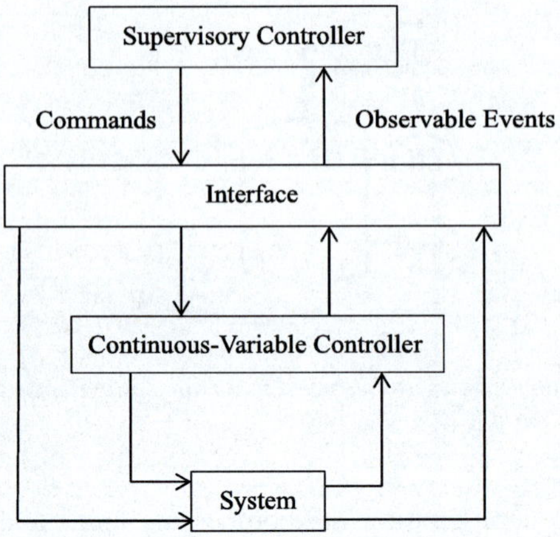

Fig. 3.4 Conceptual monitoring and control system architecture

mentioned above, i.e., the PLC, the solenoid valves (V-1 and V-2), the pipelines (P-1 and P-2), the storage tank (T-1), and the level sensor (S-1).

Conceptually, one may redraw Fig. 3.1 to represent the overall monitoring and control architecture as that presented in Fig. 3.4. This alternative sketch can be used to show the "boundary" between the traditional feedback controllers designed on the basis of the dynamic models of the given chemical processes and the operating procedures (SFCs) implemented by the supervisory controllers (PLCs). The system itself and the associated continuous-variable PID controllers at the lower levels can usually be "abstracted" as a DES for the purpose of the higher levels. This abstraction takes place at the interface in Fig. 3.4, where information from sensors of system/controller is relayed to PLC in the form of discretized states, while commands from PLC can be interpreted as events and they are converted by an interface to generate the appropriate input signals to the actuators or set points of the PID controllers. Finally, notice that the aforementioned interpretation is clearly justified by the observation that, in most cases, the SFC consists of a sequence of discrete states and events (e.g., see Fig. 3.3).

3.2 Discrete-Event System Models

It is obviously necessary to construct an appropriate DES model before generating the SFC. When considering the system behavior during operation, our concern is primarily with the sequence of states visited and the associated events causing the state transitions. To facilitate concise representation, let us first assume that it is sufficient to characterize DES behavior in terms of event strings of the form $e_1 e_2 \ldots e_n$. Such a sequence specifies the precedence order of events $e_i \, (i = 1, 2, \ldots, n)$, but it does not provide the time instances at which these events occur. In other words, the system behavior is modeled by an untimed *language*.

3.2.1 Languages

Definition. A language defined over a finite event set E is a set of finite-length strings formed from the events in E.

As examples, let us consider the event set $E = \{a, b, c\}$ and construct a few languages accordingly as follows:

- $L_1 = \{\varepsilon, aa, bb, abb, aab, aabb\}$, where ε is a string consisting of no events (i.e., an empty string). Note that there are six strings in this language.
- $L_2 = \{aac, bbc, ccc, abc, bac, cac, acc, bcc, cbc\}$, which contains all possible strings of length 3 ending with event c.
- $L_3 = \{$all possible strings of finite length which start with event $a\}$, which contains an infinite number of strings.

A language is a formal way for describing the DES behavior. It specifies all admissible event sequences that the DES is capable of "processing" or "generating". However, for some languages, such as L_3, a full enumeration of all strings is not possible. The modeling formalisms of automata and Petri nets may be adopted as the frameworks for representing and manipulating languages.

Let us next introduce another notation E^* to denote the set of all finite strings of the events in E, including the empty string ε. The symbol $*$ represents a set operation called the *Kleene closure.* Note that the strings in E^* are countably infinite since they can be arbitrarily long. For example, for the aforementioned three-element event set$E$, the Kleene closure should be $E^* = \{\varepsilon, a, b, c, aa, ab, ac, ba, bb, bc, ca, cb, cc, aaa, \cdots\}$. Therefore, a language over event set E is always a subset of E^*. Furthermore, if three strings $t, u, v \in E^*$ and, by concatenation, a new string $s = tuv$ can be built, then t is a prefix, u is a substring, and v is a suffix of this new string s. Note that both ε and s can be viewed simultaneously as the prefixes, substrings, and suffixes of string s.

On the other hand, the event set and Kleene closure of the storage system described in Fig. 3.2 may be, respectively, written as

$$E = \{o1, c1, o2, c2, lh, ll\} \tag{3.1}$$

$$E^* = \{\varepsilon, o1, c1, o2, c2, lh, ll, o1c1, o1o2, o1c2, o1lh, o1ll, c1o1, c1o2, c1c2, \cdots\} \tag{3.2}$$

where $o1$ and $c1$ denote the control actions to open and close valve V-1, respectively; $o2$ and $c2$ denote the control actions to open and close valve V-2, respectively; lh andll, respectively, denote the high- and low-level measurements in the storage tank.

Since languages are sets, the usual set operations are applicable. The following are a few additional ones:

- *Concatenation*: Let us consider two languages $L_a, L_b \in E^*$. A string is in the concatenated language $L_a L_b$ if it can be written as the concatenation of a string in L_a with a string in L_b, i.e., $L_a L_b : \{s \in E^* : (s = s_a s_b) \wedge (s_a \in L_a) \wedge (s_b \in L_b)\}$.
- *Prefix-closure*: Let us consider a language $L \in E^*$. The prefix closure of L is a language $\overline{L}$ which consists of all prefixes of all the strings in L, i.e., $\overline{L} = \{s \in E^* : (st \in L) \wedge (t \in E^*)\}$. Thus, $L \subseteq \overline{L}$ and L is said to be prefix-closed if $L = \overline{L}$.
- *Kleene-closure*: This operation is essentially the same as that defined previously for event set E, except it is now applied to a language (say L) whose elements are strings. Every element of Kleene closure L^* can be built by concatenation of two or more strings in L, i.e., $L^* := \{\varepsilon\} \cup L \cup LL \cup LLL \cup \cdots$. Note also that $(L^*)^* = L^*$.
- *Projections*: This operation takes a string formed from a larger event set (say E_l) and erases events that do not belong to another smaller event set (say E_s). Mathematically, a *projection* $\mathcal{P}$ can be written as $\mathcal{P} : E_l^* \rightarrow E_s^*$. More specifically, this mapping can be characterized as follows:

$$\mathcal{P}(\varepsilon) = \varepsilon \tag{3.3}$$

$$\mathcal{P}(e) = \begin{cases} e & \text{if} e \in E_s \\ \varepsilon & \text{if} e \in E_l \backslash E_s \end{cases} \tag{3.4}$$

$$\mathcal{P}(se) = \mathcal{P}(s)\mathcal{P}(e) \text{ for } s \in E_l^*, e \in E_l. \tag{3.5}$$

3.2.2 Automata

To facilitate a concise description of the language model, a brief review of the automaton structure is given here. Specifically, a deterministic automaton A can be viewed as a six-tuple (Cassandras and Lafortune 2008):

$$A = (X, E, f, \Sigma, x_0, X_m) \tag{3.6}$$

where X is the set of system states, E is the event set, $x_0 \in X$ is the initial system state, $X_m \subseteq X$ is the set of marked (or final) states and these states are marked if it is desired to attach a special meaning to them, $f : X \times E \rightarrow X$ represents the state-transition function, $\Sigma : X \rightarrow 2^E$ denotes the active event function, and $\Sigma(x)$ denotes the set of active events at state x. Notice that transition function $f(x, e) = x'$ characterizes a state-transition process (from $x \in X$ to $x' \in X$) triggered by event $e \in E$. Notice also that $f(x, e)$ is only a partial function on its domain, that is, it is not necessary to always provide a specific transition at each state of X in the automaton for every event in E. For convenience, the domain of f can be extended from $X \times E$ to $X \times E^*$ on the basis of the following recursive formulas:

$$f(x, \varepsilon) = x \tag{3.7}$$

$$f(x, se) = f(f(x, s), e) \tag{3.8}$$

where $x \in X$, $e \in E$, and $s \in E^*$.

Thus, an automaton is viewed in this book as a *language-generating machine*. The events in set E should be regarded as the *alphabets* of this language and an event sequence allowed in an automaton is referred to as a *string*. The language generated by an automaton (say A) can be formally written as

$$\mathfrak{L}(A) = \{s \in E^* : f(x_0, s) \text{ is defined}\}. \tag{3.9}$$

And the corresponding marked language is

$$\mathfrak{L}_m(A) = \{s \in E^* : f(x_0, s) \in X_m\}. \tag{3.10}$$

Thus, an automaton A actually represents the two languages mentioned above. The state-transition diagram of A contains all necessary information to construct them. From the definitions given in equations (3.9) and (3.10), one can deduce that

$$\mathfrak{L}_m(A) \subseteq \overline{\mathfrak{L}_m(A)} \subseteq \mathfrak{L}(A). \tag{3.11}$$

An automaton A may reach an unmarked state $x \notin X_m$ where $\Sigma(x) = \varnothing$. This is referred to as a *deadlock* since no further events are allowed to take place. On the other hand, if there is a set of unmarked states that forms a strongly connected component but with no outward transitions, then a *livelock* is formed when the system reaches any of these states. In both scenarios, the system is said to be *blocking* because it enters either a deadlock or livelock state without finishing the designated task, i.e., reaching a marked state. Mathematically, an automaton A is blocking if

$$\overline{\mathfrak{L}_m(A)} \subset \mathfrak{L}(A). \tag{3.12}$$

Thus, a nonblocking automaton can be ensured only if

$$\overline{\mathfrak{L}_m(A)} = \mathfrak{L}(A). \tag{3.13}$$

3.2.2.1 Development of Component Models

A component model is used to characterize a finite set of identifiable states of the hardware item under consideration and all possible state-transition processes. The transition from one state to another is activated by at least one event. The common features of a *normal* transition associated with an identified state are depicted in Fig. 3.5. For any component under consideration, the active events of each state should be conjectured on the basis of process knowledge and classified into two types, i.e., the state-transition events $e_i^t \in E_t$ and the state-sustaining events $e_j^s \in E_s$. The corresponding transition state functions can be written, respectively, as

$$f\left(x, e_i^t\right) = x' \tag{3.14}$$

$$f\left(x, e_j^s\right) = x \tag{3.15}$$

where $x' \neq x$. It should be noted that a state-sustaining event may be used as the state-transition event in a component model in the next level.

The automaton representations of all components in the liquid storage system in Fig. 3.2 can be briefly outlined as follows:

- Level 1: The PLC model can be constructed in a straightforward fashion according to the SFC in Fig. 3.3 (see Fig. 3.6). For simplicity, it is assumed that the operation steps in OS_1 can always be executed initially, and thus the event specified in OS_0 is omitted. The level-5 events $LHcon$ and $LLcon$ are used to represent the conditions that the sensor reading continues at the high and low levels for a long enough period, respectively. The events *openV1* and *closeV1* are the control actions to open and close valve V-1, while *openV2* and *closeV2* denote the corresponding control actions to manipulate V-2.
- Level 2: The model of valve V-1 is presented in Fig. 3.7. States $V1C$ and $V1O$ are used to represent the close and open positions, respectively, while the events $openV1$ and $closeV1$ denote the corresponding close-to-open and open-to-close actions. From Fig. 3.3, it is clear that these two events are triggered by the control signals of PLC, which is a level-1 component. On the other hand, the level-2 events $V1Ccon$ and $V1Ocon$ represent V-1 continues at close and open positions,

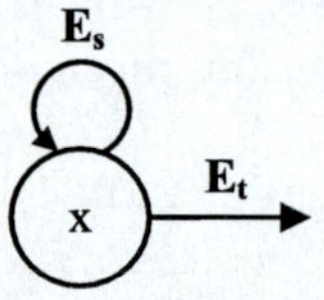

Fig. 3.5 Classification of active events on a component state (Reprinted with permission from Yeh 2013. Copyright 2013 National Cheng Kung University Library)

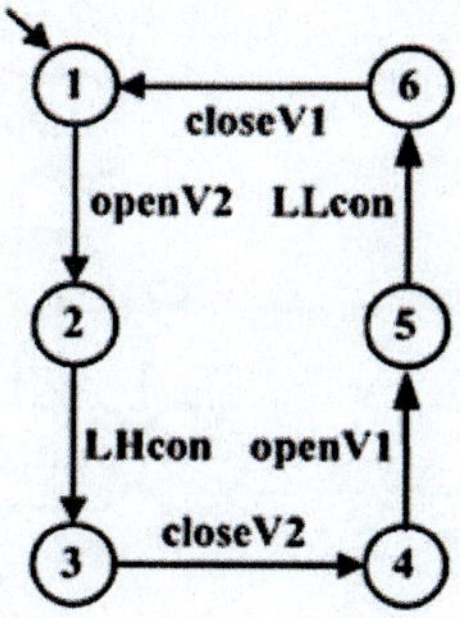

Fig. 3.6 Automaton model of controller in Example 3.1 (Reprinted with permission from Yeh and Chang 2011. Copyright 2011 Elsevier)

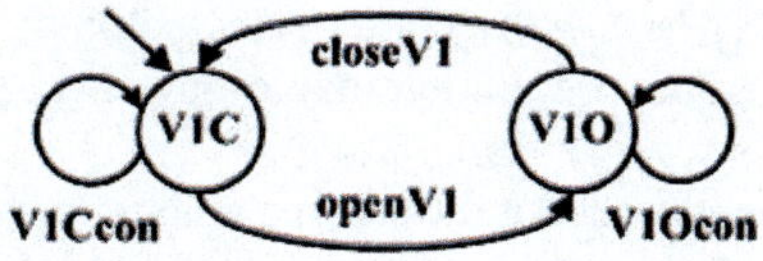

Fig. 3.7 Automaton model of valve V-1 in Example 3.1 (Reprinted with permission from Yeh and Chang 2011. Copyright 2011 Elsevier)

respectively, for a sufficiently long period of time. A similar model can be built for V-2 with the same approach.

- Level 3: The automaton model of outlet pipeline P-1 is presented in Fig. 3.8. There are two pipeline states, i.e., "flow" ($p1f$) and "no flow" ($p1nf$). It should be noted that the level-2 events $V1Ocon$ and $V1Ccon$ activate the "$p1nf$-to-$p1f$" and "$p1f$-to-$p1nf$" processes, respectively. For illustration simplicity, it is assumed in this example that the flow in outlet pipeline can be produced by opening V-1 even when the liquid level in the tank is low. This assumption can be removed in more realistic examples. A similar model for the inlet pipeline can be built with the same approach. The component model representing process configuration can then be obtained by performing a *parallel composition* on the above two pipeline models (see Fig. 3.9). A description of this operation can be found in the next subsection. Notice that the state-sustaining events $PC01con$ and $PC02con$, respectively, denote that the process configuration is maintained

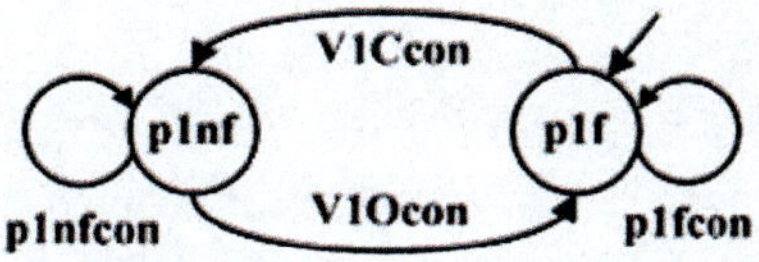

Fig. 3.8 Automaton model of outlet pipeline P-1 in Example 3.1 (Reprinted with permission from Yeh and Chang 2011. Copyright 2011 Elsevier)

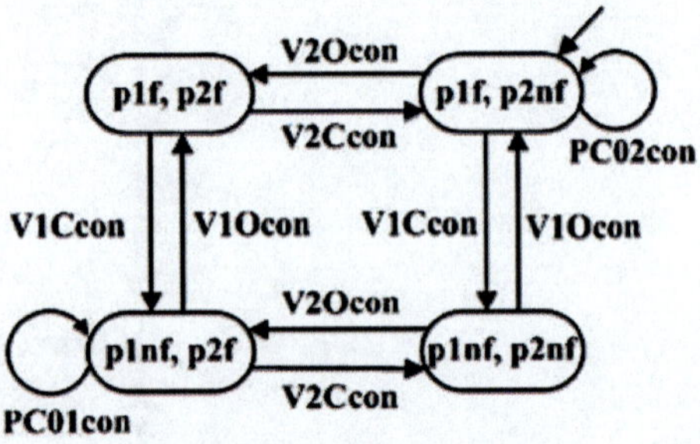

Fig. 3.9 Automaton model of process configurationin Example 3.1 (Reprinted with permission from Yeh and Chang 2011. Copyright 2011 Elsevier)

at $PC01$ (in which the pipeline states are $p1nf$ and $p2f$) and $PC02$ (in which the pipeline states are $p1f$ and $p2nf$) for a long enough period of time. Notice also that the state-sustaining events associated with the other two configurations, i.e., $(p1f, p2f)$ and $(p1nf, p2nf)$, are neglected on the ground that, in normal operation, these states may be unidentifiable or at best present for a very short period of time only.

- Level 4: The component model of the storage tank is presented in Fig. 3.10. It can be observed that two tank states are used in this model, i.e., "level high" (LH) and "level low" (LL). Notice that if the process configuration is kept at $PC01con$ (or the pipeline states continue at $p1nfcon$ and $p2fcon$), the LL-to-LH process should be realized. By the same token, the events $PC02con$ should cause state transition in the opposite direction. Finally, the level-4 events $LHcon$ and $LLcon$ represent the liquid level continues at high and low positions, respectively.
- Level 5: For the sake of brevity, it is assumed in the present example that only the liquid level is monitored online and the chance of sensor malfunctions is negligibly low. In other words, it is assumed that the desired level of sensing-system reliability can almost always be achieved by introducing hardware redundancy in design and also by adopting a proper maintenance policy (Liang and Chang 2008). As a result, the sensor model can be omitted here and the measurement readings are considered to be identical to the tank states.

After building automata to represent the normal behaviors of all components, additional mechanisms may be incorporated into each component model to describe failures. The general model structure is shown in Fig. 3.11. The top-layer states in this figure represent the normal states, i.e., NS_i $(i = 1, 2, \ldots, N)$, and the boxed automaton represents the normal operation cycle in which only routine events are allowed. Any failure event (i.e., FE_i and $i = 1, 2, \ldots, N$) could result in a

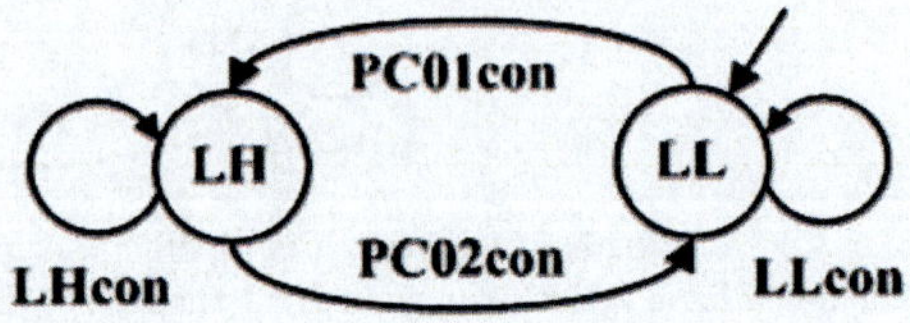

Fig. 3.10 Automaton model of tank in Example 3.1 (Reprinted with permission from Yeh and Chang 2011. Copyright 2011 Elsevier)

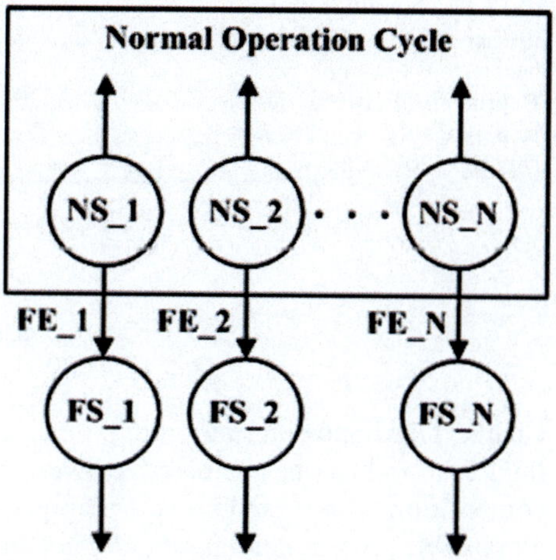

Fig. 3.11 General failure model (Reprinted with permission from Yeh and Chang 2011. Copyright 2011 Elsevier)

change from a normal state within the box to a failure state outside, i.e., FS_i and $i = 1, 2, \cdots, N$.

Let us revisit the automaton in Fig. 3.7 as an example to illustrate the aforementioned practice. From the modified automaton shown in Fig. 3.12, one can observe that the two extra abnormal valve states, i.e., "V-1 sticks at the close position" ($V1SC$) and "V-1 sticks at the open position" ($V1SO$), have been added. Notice also that, since the control actions $openV1$ and $closeV1$ cannot cause any state change in either case, they are treated as the state-sustaining events at $V1SC$ and $V1SO$. Finally, it should be noted that the same approach can be easily adopted to characterize the failures of V-2.

Since additional failure states and events are introduced into the normal valve models, it becomes necessary to modify the directly affected component models in the third level. Specifically, the normal outlet pipeline model in Fig. 3.8 should be modified accordingly and the resulting automaton is presented in Fig. 3.13. In the component model of outlet pipeline (P-1), failure $V1SC$ should cause the normal state $p1nf$ to be trapped in a new abnormal state $p1nf\_V1SC$, and failure $V1SO$ should

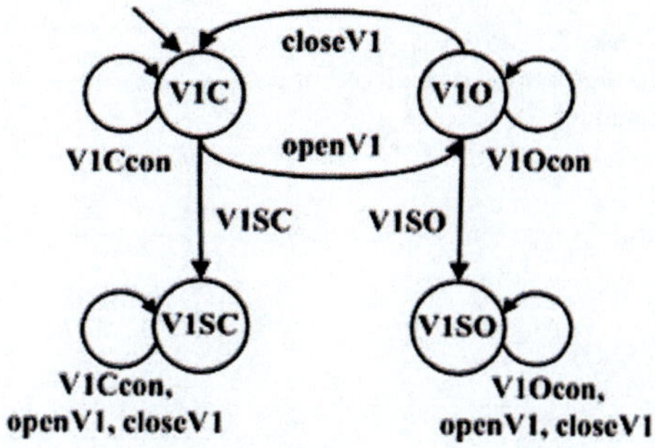

Fig. 3.12 Outlet valve model with failure mechanisms (Example 3.1) (Reprinted with permission from Yeh and Chang 2011. Copyright 2011 Elsevier)

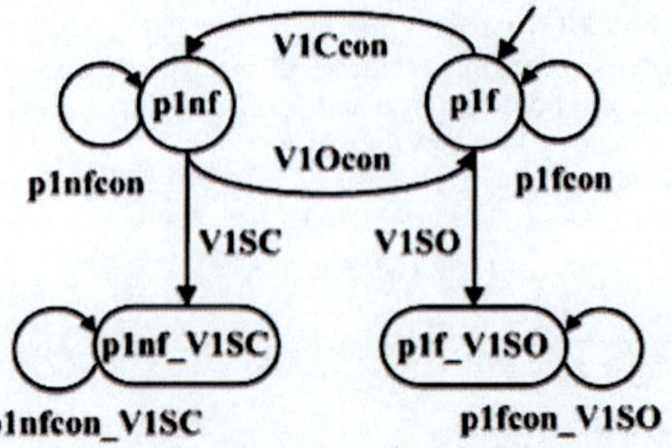

Fig. 3.13 Simplified outlet pipeline model with failure mechanisms (Example 3.1) (Reprinted with permission from Yeh and Chang 2011. Copyright 2011 Elsevier)

change the component state from $p1f$ to the new state $p1f\_V1SO$. Notice that the failures of P-2 can be characterized with the same approach. By applying a parallel composition with these two revised pipeline models, a complete representation of all possible process configurations can then be produced. As indicated previously in constructing the normal component model in Fig. 3.9, some of the configurations (states) can be judiciously ignored for simplicity. The same approach can be taken in this case and the kept states are listed in Table 3.1.

If an additional failure, i.e., the tank leakage, is to be considered in Example 3.1, then the corresponding failure mechanism should be introduced into the automaton in Fig. 3.10 (see Fig. 3.14). The abnormal tank states, i.e., LH_leak and LL_leak, represent "leakage occurs while level high" and "leakage occurs while level low", respectively. It should be noted that the event $LHcon$ is not allowed when tank leakage occurs. Notice also that the abnormal process configurations $PC04con$ and $PC05con$ may cause the normal tank state LH to be trapped in the abnormal states LH_V2SO and LH_V1SC, respectively. Similarly, the abnormal process configurations $PC03con$ and $PC07con$ may cause the normal tank state LL to be trapped in the abnormal states LL_V2SC and LL_V1SO, respectively. In addition, the abnormal process configuration $PC04con$ could also cause the abnormal tank state LH_leak to be trapped in another abnormal state LH_leak_V2SO, while $PC01con$, $PC03con$, $PC06con$, and $PC07con$ could cause the abnormal tank state LH_leak to be trapped in the abnormal states LL_leak_N, LL_leak_V2SC, LL_leak_V1SC, and LL_leak_V1SO, respectively.

Table. 3.1 Process configurations considered in Example 3.1

Valve states	Process configuration	Label
V1C, V2O	p1nf, p2f	PC01
V1O, V2C	p1f, p2nf	PC02
V1C, V2SC	p1nf, p2nf	PC03
V1O, V2SO	p1f, p2f	PC04
V1SC, V2C	p1nf, p2nf	PC05
V1SC, V2O	p1nf, p2f	PC06
V1SO, V2O	p1f, p2f	PC07

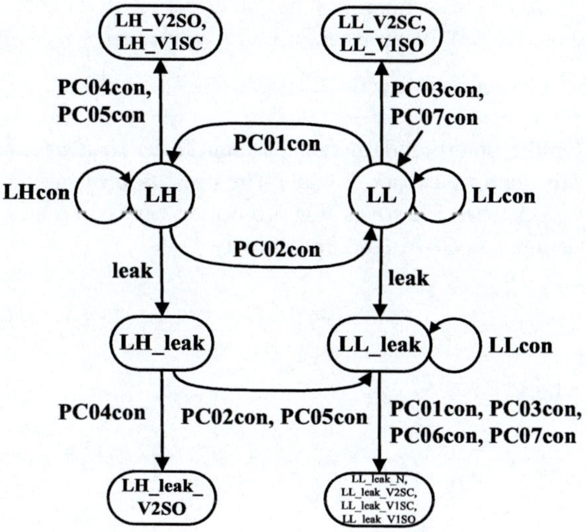

Fig. 3.14 Failure embedded tank model (Example 3.1) (Reprinted with permission from Yeh and Chang 2011. Copyright 2011 Elsevier)

3.2.2.2 Useful Operations on Automata

A wide variety of operations are available for manipulating automata. The useful ones are described as follows:

- **Accessible part:**

From the definitions of $\mathfrak{L}(A)$ and $\mathfrak{L}_m(A)$ in equations (3.9) and (3.10), it can be observed that any state that is not accessible (or reachable) from x_0 via a string in $\mathfrak{L}(A)$ can be deleted from A without affecting the languages generated or marked by A. Notice that if a state is deleted, all attached transitions should be deleted as well. This operation of taking the *accessible part* of an automaton is denoted as $Ac(A)$ and its formal definition is given as follows:

$$Ac(A) := \left(X_{ac}, E, f_{ac}, x_0, X_{ac,m}\right) \tag{3.16}$$

where

$$X_{ac} = \{x \in X : f(x_0, s) = x \text{and} s \in \mathfrak{L}(A)\} \tag{3.17}$$

$$X_{ac,m} = X_{ac} \cap X_m \tag{3.18}$$

$$f_{ac} = f|_{X_{ac} \times E \to X_{ac}}. \tag{3.19}$$

Since the Acoperation has no effect on $\mathfrak{L}(A)$ and $\mathfrak{L}_m(A)$, it is often assumed that an automaton is accessible, i.e., $A = Ac(A)$.

- **Co-accessible part**

On the other hand, a state x is said to be *co-accessible* if there is a path in A from this state to a marked state. The operation of taking the "co-accessible part" of$A$, i.e., deleting all states that are not co-accessible and all the attached transitions, is denoted as $CoAc(A)$. Specifically,

$$CoAc(A) := \left(X_{coac}, E, f_{coac}, x_{0,coac}, X_m\right) \tag{3.20}$$

where

$$X_{coac} = \left\{x \in X : \exists s \in E^* \Vdash f(x, s) \in X_m\right\} \tag{3.21}$$

$$f_{coac} = f|_{X_{coac} \times E \rightarrow X_{coac}} \tag{3.22}$$

$$x_{0,coac} = \begin{cases} x_0 & \text{if } x_0 \in X_{coac} \\ \text{undefined} & \text{otherwise.} \end{cases} \tag{3.23}$$

If $A = CoAc(A)$, then A is said to be *co-accessible* and $\mathfrak{L}(A) = \overline{(\mathfrak{L}_m(A)}$. Therefore, co-accessibility is closely related to the concept of blocking. An automaton is considered to be blocking if $\mathfrak{L}(A) \neq \overline{\mathfrak{L}_m(A)}$. In other words, blocking implies that $\overline{\mathfrak{L}_m(A)}$ is a subset of $\mathfrak{L}(A)$ and, thus, there are accessible states that are not co-accessible.

- **Trim**

If an automaton is both accessible and co-accessible, it is said to be *Trim*. The $Trim$operation can be expressed as follows:

$$Trim(A) = CoAc[Ac(A)] = Ac[CoAc(A)]. \tag{3.24}$$

- **Projection**

Let us consider an automaton A and also a subset of its event set, i.e., $E_s \subset E$. The *projections* of $\mathfrak{L}(A)$ and $\mathfrak{L}_m(A)$ from E^* to E_s^* are denoted, respectively, as $\mathcal{P}_s[\mathfrak{L}(A)]$ and $\mathcal{P}_s[\mathfrak{L}_m(A)]$, and they can be implemented by replacing all transition labels corresponding to the events in $E \backslash E_s$ by ε. Alternatively, let us consider the languages $\mathfrak{L}\left(A'\right) \subseteq E_s^*$ and $\mathfrak{L}_m\left(A'\right)$, and let $\mathcal{P}_s$ denotes the projection from E_l^* to E_s^* ($E_s^* \subset E_l^*$). The corresponding inverse projections $\mathfrak{L}_s^{-1}\left[\left(A'\right)\right]$ and $\mathfrak{L}_s^{-1}\left[_m\left(A'\right)\right]$ can be produced by adding self-loops for all events in $E_l \backslash E_s$ at all states of A'.

- Compositions

To build a system model, it is necessary to integrate all its components into a single one. Two operations may be performed for this purpose, i.e., the parallel composition (denoted by $\|$) and the product composition (denoted by $\times$). Specifically, let us consider two distinct automata expressed, respectively, as $A_1 = (X_1, E_1, f_1, \Sigma_1, x_{01}, X_{m1})$ and$A_2 = (X_2, E_2, f_2, \Sigma_2, x_{02}, X_{m2})$. The aforementioned operations yield two different types of joint behavior of A_1 and A_2 depending on how the private events, i.e., $e \in E_1/(E_1 \bigcap E_2)$ or$e \in E_2/(E_1 \bigcap E_2)$, are handled.

The *parallel composition* of two automata A_1 and A_2 is an automaton defined as follows:

$$A_1\|A_2 := Ac\big(X_1 \times X_2, E_1 \cup E_2, f_{1\|2}, \Gamma_{1\|2}, (x_{01}, x_{02}), X_{m1} \times X_{m2}\big) \tag{3.25}$$

where

$$f_{1\|2}((x_1, x_2), e) = \begin{cases} (f_1(x_1, e), f_2(x_2, e)) & \text{if } e \in \Gamma_1(x_1) \cap \Gamma_2(x_2) \\ (f_1(x_1, e), x_2) & \text{if } e \in \Gamma_1(x_1) \backslash E_2 \\ (x_1, f_2(x_2, e)) & \text{if } e \in \Gamma_2(x_2) \backslash E_1 \\ \text{undefined} & \text{otherwise} \end{cases} \tag{3.26}$$

$$\Gamma_{1\|2}(x_1, x_2) = (\Gamma_1(x_1) \cap \Gamma_2(x_2)) \cup (\Gamma_1(x_1)\backslash E_2) \cup (\Gamma_2(x_2)\backslash E_1). \tag{3.27}$$

A component in the parallel composition (i.e., A_1 orA_2) can execute its private events without the participation of the other component, while a common event can only take place if both components are able to execute it. If $E_1 = E_2$, then the parallel composition reduces to the *product composition* described below. On the other hand, if $E_1 \cap E_2 = \varnothing$, then there are no synchronized transitions and $A_1\|A_2$ is often referred to as the *shuffle* of A_1 andA_2.

Let's consider the fictitious example shown in Fig. 3.15. From Figs. 3.15a, b, the basic features of automata A_1 and A_2 can be summarized as follows:

- $A_1 : X_1 = \{x_1, x_2, x_3, x_4\}$; $E_1 = \{a, b, d, e\}$; $f_1(x_1, a) = x_2$, $f_1(x_2, b) = f_1(x_3, e) = x_3$, $f_1(x_2, d) = x_4$; $\Gamma_1(x_1) = \{a\}$, $\Gamma_1(x_2) = \{b, d\}$, $\Gamma_1(x_3) = \{e\}$; $x_{01} = x_1$; $X_{m1} = \varnothing$.
- $A_2 : X_2 = \{y_1, y_2, y_3, y_4, y_5\}$; $E_2 = \{a, b, c, d, e\}$; $f_2(y_1, a) = y_2$,$f_2(y_1, d) = y_5$, $f_2(y_2, c) = y_3$, $f_2(y_3, b) = f_2(y_4, e) = y_4$; $\Gamma_2(y_1) = \{a, d\}$, $\Gamma_2(y_2) = \{c\}$, $\Gamma_2(y_3) = \{b\}$, $\Gamma_2(y_4) = \{e\}$; $x_{02} = y_1$; $X_{m2} = \varnothing$.

According to the definition of$A_1\|A_2$, A_1 and A_2 can be combined according to equations (3.25)–(3.27) to form an automaton with the following basic features:

- $X_{A_1\|A_2} = \{(x_1, y_1), (x_2, y_2), (x_2, y_3), (x_3, y_4)\} \neq X_1 \times X_2$;
- $E_{A_1\|A_2} = \{a, b, c, e\} \neq E_1 \cup E_2$;
- $f_{1\|2}((x_1, y_1), a) = (x_2, y_2)$, $f_{1\|2}((x_2, y_2), c) = (x_2, y_3)$, $f_{1\|2}((x_2, y_3), b) = (x_3, y_4)$, $f_{1\|2}((x_3, y_4), e) = (x_3, y_4)$;

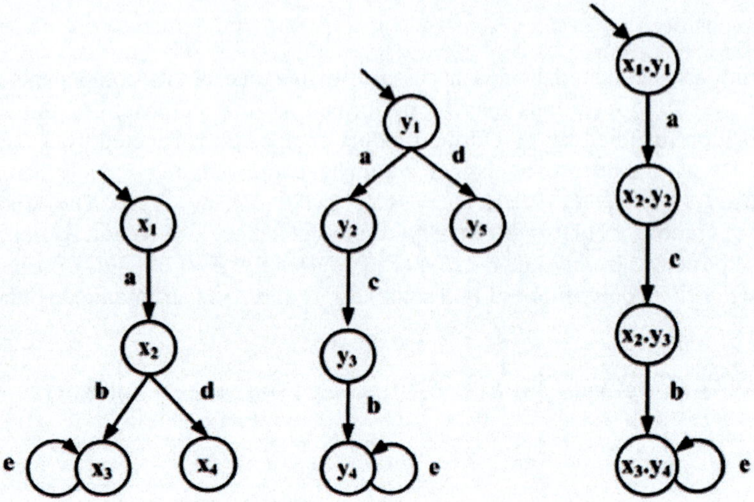

(a) (b) (c)

Fig. 3.15 An example of parallel composition: automaton A_1; **b** automaton A_2; **c** automaton $A_1 \| A_2$ (Reprinted with permission from Yeh 2013. Copyright 2013 National Cheng Kung University Library)

- $\Gamma_{1\|2}(x_1, y_1) = \{a\}$, $\Gamma_{1\|2}(x_2, y_2) = \{c\}$, $\Gamma_{1\|2}(x_2, y_3) = \{b\}$, $\Gamma_{1\|2}(x_3, y_4) = \{e\}$;
- $x_{A_1\|A_2,0} = (x_1, y_1)$;
- $X_{A_1\|A_2,m} = \varnothing$.

This automaton $A_1\|A_2$ can also be represented with the state-transition diagram shown in Fig. 3.15c. Notice that since event a is enabled by state x_1 in A_1 and state y_1 in A_2, respectively, event a should be enabled at state (x_1, y_1) in $A_1\|A_2$. Notice also that although event b is enabled at state x_2 in A_1, it is not enabled at y_2 in A_2. Clearly, event b should be postponed after completing event sequence ac to reach state y_3 in A_2.

Next, let us consider the definition of *product composition* of two automata A_1 and A_2:

$$A_1 \times A_2 := Ac(X_1 \times X_2, E_1 \cup E_2, f_{1\times 2}, \Gamma_{1\times 2}, (x_{01}, x_{02}), X_{m1} \times X_{m2}) \tag{3.28}$$

where

$$f_{1\times 2}((x_1, x_2), e) = \begin{cases} (f_1(x_1, e), f_2(x_2, e)) & \text{if } e \in \Gamma_1(x_1) \cap \Gamma_2(x_2) \\ \text{undefined} & \text{otherwise} \end{cases} \tag{3.29}$$

$$\Gamma_{1\times 2}(x_1, x_2) = \Gamma_1(x_1) \cap \Gamma_2(x_2). \tag{3.30}$$

The transitions of two components in the product (i.e., A_1 and A_2) are always synchronized on a common event in $E_1 \cap E_2$. It is also clear that

$$\mathfrak{L}(A_1 \times A_2) = \mathfrak{L}(A_1) \cap \mathfrak{L}(A_2) \tag{3.31}$$

$$\mathfrak{L}_m(A_1 \times A_2) = \mathfrak{L}_m(A_1) \cap \mathfrak{L}_m(A_2). \tag{3.32}$$

3.2.2.3 Assemblage of System Automata

Although both normal behaviors and failure mechanisms can be incorporated in a component model, there is still a need to impose additional *constraints* in the controller model so as to limit the scope of event evolution. These constraints are introduced mainly for the purposes of avoiding state explosion and producing a succinct system model. Specifically, it is assumed that.

- All state-sustaining events of the components in levels 2–4 (i.e., actuators, process configuration, and unit operations) must occur before the sensor state reaches the resulting activation condition in SFC.
- The failure event of any component in levels 2–5 (i.e., actuators, process configuration, unit operations and sensors) can only occur just before the controller triggers a subsequent actuator event. Furthermore, the aforementioned failure and actuator events are mutually exclusive.

These constraints can be incorporated into the *controller model* with additional self-looping transitions according to the following rules:

(1) Every state-sustaining event in level 2 should be incorporated with a self-looping transition at the controller state that enables the corresponding activation condition(s) in SFC.
(2) Every failure event in levels 2–5 should be incorporated with a self-looping transition at the controller state that enables the subsequent normal control action(s) in SFC.

In the liquid storage system mentioned above, let us assume that (a) the inlet valve may stick at close or open position, (b) the outlet valve may stick at close or open position, and (c) the tank may leak. The modified controller model for Example 3.1 is given in Fig. 3.16. Notice that.

- The state-sustaining events $V1Ocon$ and $V2Ccon$ of actuators V-1 and V-2 are both constrained at state 5, which is the controller state that enables the activation condition $LLcon$ in SFC. Similarly, the state-sustaining events $V1Ccon$ and $V2Ocon$ should both be constrained at state 2, which is the state that enables the activation condition $LHcon$ in SFC.

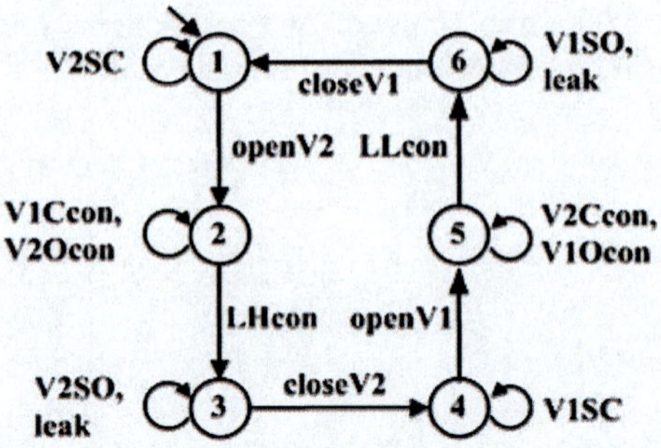

Fig. 3.16 The controller model that contains additional failure modules (Example 3.1) (Reprinted with permission from Yeh and Chang 2011. Copyright 2011 Elsevier)

- The failures of V-1, i.e., $V1SC$ and $V1SO$ in Fig. 3.13, should be constrained at state 4 (which enables the control action $openV1$) and state 6 (which enables the control action $closeV1$), respectively. Similarly, the failures of V-2, i.e., $V2SC$ and $V2SO$, are constrained at state 1 and state 3, respectively, for the same reasons. On the other hand, the tank failure, i.e., $leak$, should be constrained at the states 3 and 6 (which enable the control actions OS_2 and OS_1, respectively).

After introducing the aforementioned modifications, the *parallel composition* operation can then be performed to integrate all components into a system model. The resulting system model in Example 3.1 can be found in Fig. 3.17.

3.2.3 Petri Nets

The Petri nets (PNs) can be considered as an alternative to automata for modeling the discrete-event systems (DESs). They are related to automata in the sense that the PNs also explicitly characterize the transition functions of DESs. An automaton can always be expressed as a Petri net, while not all PNs can be interpreted with the finite-state automata.

Let us first consider the general structure of a Petri net model:

$$\mathcal{G} = (\mathbb{P}, \mathbb{T}, \mathbb{F}, \mathbb{W}, \boldsymbol{m}_0) \tag{3.33}$$

where $\mathbb{P} = \{p_1, p_2, \ldots, p_n\}$ and $\mathbb{T} = \{t_1, t_2, \ldots, t_m\}$ represent, respectively, the sets of places and transitions in a Petri net; $\mathbb{F}$ is the union of the sets of place-to-transition and transition-to-place arcs, i.e., $\mathbb{F} \subseteq (\mathbb{P} \times \mathbb{T}) \cup (\mathbb{T} \times \mathbb{P})$; $\mathbb{W}$ denotes the set of weighting functions associated with the arcs in $\mathbb{F}$; and $\boldsymbol{m}_0$ is the initial marking vector in which the initial token number in every place is stored. In addition to these usual definitions of Petri net components, it is assumed that not all process states, which are reflected with the numbers of tokens residing in the places, can be monitored online. In other words, the places in $\mathbb{P}$ can be classified as observable and unobservable ones and then collected in two separate subsets, i.e., $\mathbb{P} = \mathbb{P}_o \cup \mathbb{P}_u$. Notice

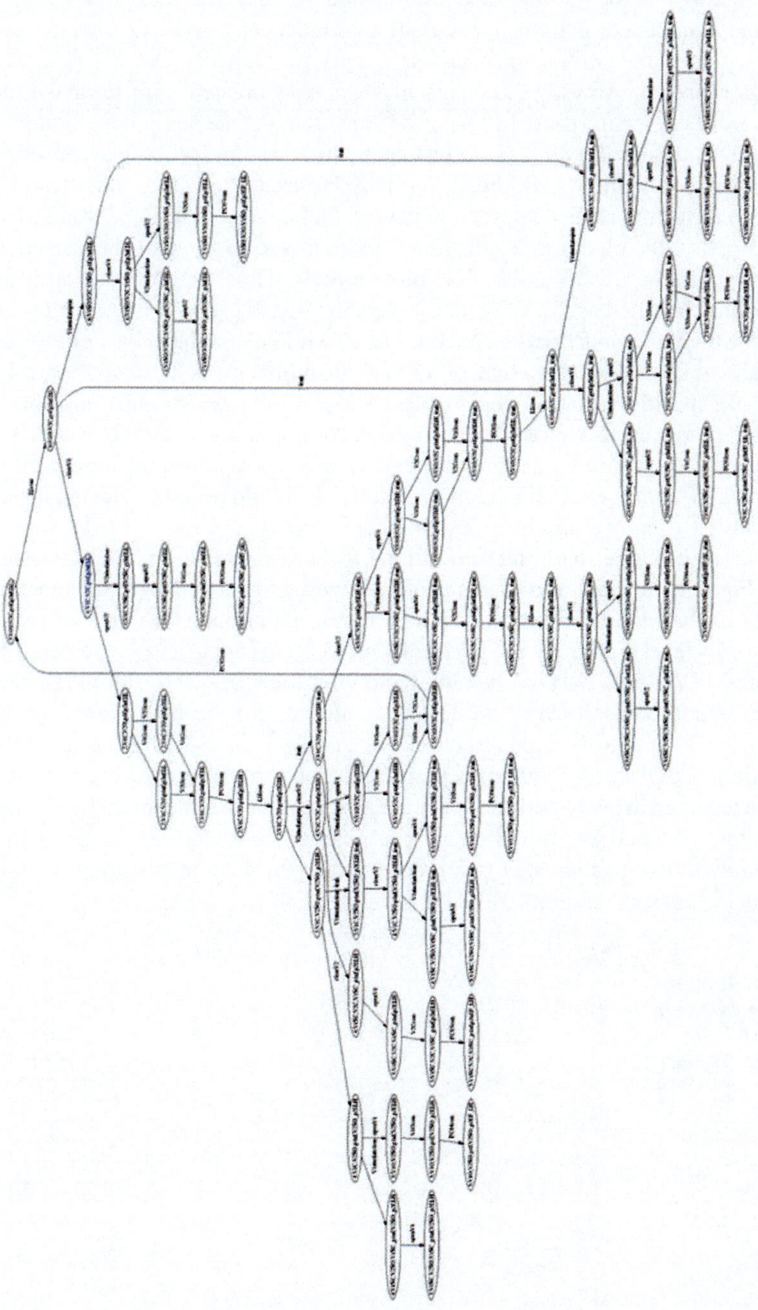

Fig. 3.17 The system model (Example 3.1) (Reprinted with permission from Yeh 2013. Copyright 2013 National Cheng Kung University Library)

that some of the unobservable states may be caused by failures. Thus, there is a need to further distinguish the unobservable normal and failed states, i.e., $\mathbb{P}_u = \mathbb{P}_n \cup \mathbb{P}_f$. On the other hand, the transitions in the Petri net model can also be divided into two groups to represent the normal and abnormal events, respectively, i.e., $\mathbb{T} = \mathbb{T}_N \cup \mathbb{T}_C$. The events represented by the elements in $\mathbb{T}_N$ are associated with normal state-transition processes, while those in $\mathbb{T}_C$ can be considered as the equipment failures. It is assumed that almost all events occur instantaneously except for some in the former case. In other words, the transitions in $\mathbb{T}_N$ may be fired after finite time delays to better characterize the realistic system behavior. Notice also that, in addition to the transitions in $\mathbb{T}_C$, the places in $\mathbb{P}_f$ may have to be linked to some of the transitions in $\mathbb{T}_N$ with inhibitor arcs to model the failure effects. Thus, the normal transitions can be further classified as$\mathbb{T}_N = \mathbb{T}_A \cup \mathbb{T}_B$. Specifically, $\mathbb{T}_A$ represents the subset of transitions which are unaffected by such failures, while $\mathbb{T}_B$ is the subset of affected ones. Finally, it should be noted that, other than the online measurements, controller execution of a specific operation step is considered as a *known* event. All transitions in the Petri net model can therefore be classified according to this alternative criterion, i.e.,$\mathbb{T} = \mathbb{T}_K \cup \mathbb{T}_{UK}$, where $\mathbb{T}_K$ denotes the set of transitions representing the controller actions and $\mathbb{T}_{UK}$ is the set of remaining transitions. It should be noted that the former transitions (in $\mathbb{T}_K$) may either be in $\mathbb{T}_N$ or in $\mathbb{T}_C$.

To illustrate the model-building conventions, let us revisit the liquid storage system shown in Fig. 3.2, in which the tank is equipped with an inlet and an outlet pipeline. The height of the liquid level in this tank is monitored online. Two distinct sensor signals, i.e., (1) level high (LH) and (2) level low (LL), are sent to a PLC to actuate the control valves (V-1 and V-2) on the outlet and inlet lines, respectively. In response to the LH signal, V-1 is opened while V-2 is closed. On the other hand, the LL signal triggers the control actions to close V-1 and to open V-2. It is assumed that the operation of this storage system is periodical, and the above two sets of control actions are repeated in every period. Under the assumptions that the initial liquid level in tank is low, V-1 and V-2 are at their close positions initially, the corresponding SFC is redrawn here to represent this procedure for the sake of illustration clarity (see Fig. 3.18, Tables 3.2, and 3.3).

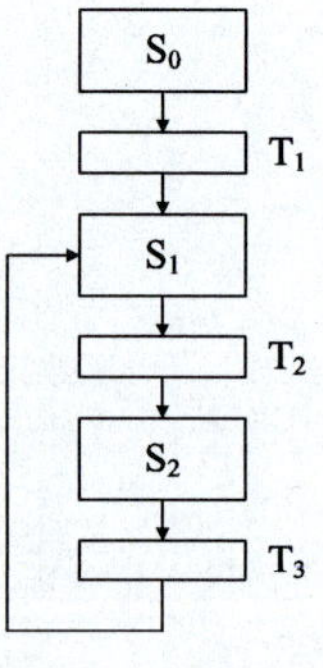

Fig. 3.18 SFC of the operating procedure in Example 3.1

Table. 3.2 Operation steps in Example 3.1

Operation step	Actions
S_0	Initialization
S_1	(1) Close V-1; (2) Open V-2
S_2	(1) Close V-2; (2) Open V-1

Table. 3.3 Activation conditions of the transitions in SFC of Example 3.1.

Symbol	Conditions
T_1	Start
T_2	LH
T_3	LL

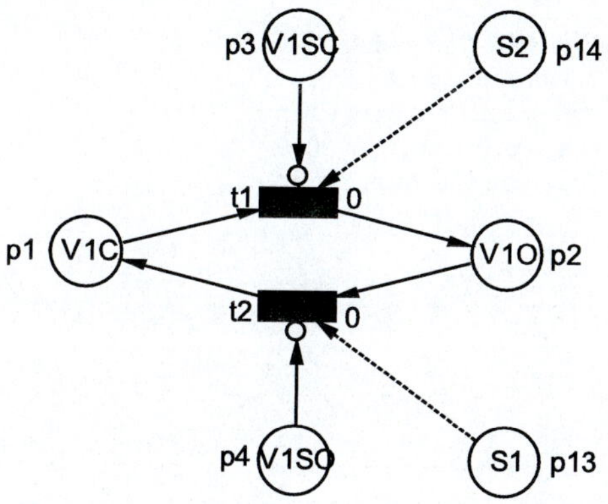

Fig. 3.19 The discharge valve model (Valve 1) (Reprinted with permission from Chen et al. 2010. Copyright 2010 American Chemical Society)

The Petri net representation of every component in this system under *normal* operating conditions is briefly described as follows:

- The Petri net model of valve V-1 is presented in Fig. 3.19. In this model, places p_1 (V1C) and p_2 (V1O) are used to represent the close and open positions, respectively, while the untimed transitions t_1 and t_2 denote the corresponding close-to-open and open-to-close processes. From Table 3.2, it is clear that these two events are triggered by the control actions in operation steps S_2 and S_1, respectively. These two cause-and-effect relations are represented with the input places of t_1 and t_2 (i.e., p_{14} and p_{13}). A similar Petri net model for V-2 can be found in Fig. 3.20.
- The Petri net model of the outlet pipeline is presented in Fig. 3.21. There are two pipeline states, i.e., "flow" (p_7) and "no flow" (p_8). It should be noted that, other than the open state of V-1 (p_2), an additional precondition, i.e., "tank is full" (p_{12}), is needed to trigger the untimed transition from the OPNF state, i.e., "no flow in outlet pipeline", to the OPF state, i.e., "flow in outlet pipeline". On the

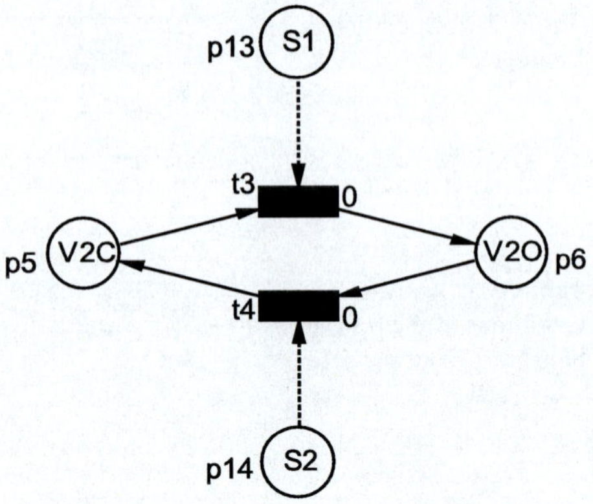

Fig. 3.20 The inlet valve model (Valve 2) (Reprinted with permission from Chen et al. 2010. Copyright 2010 American Chemical Society)

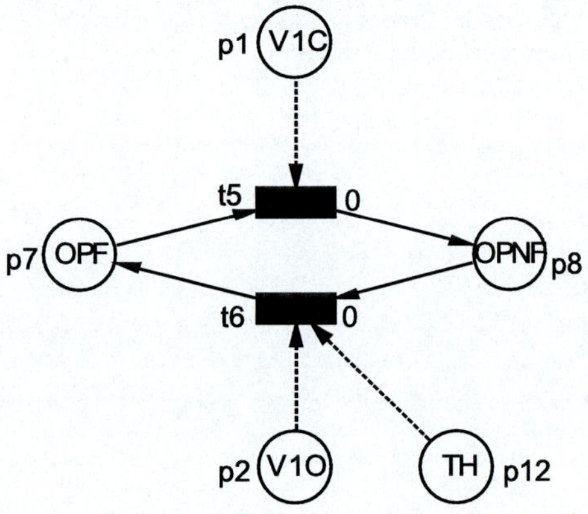

Fig. 3.21 The outlet pipeline model (Reprinted with permission from Chen et al. 2010. Copyright 2010 American Chemical Society)

other hand, this precondition is not needed in the Petri net model for inlet pipeline (see Fig. 3.22).

- The tank model is given in Fig. 3.23. Two tank states are considered here, i.e., "empty" (p_{11}) and "full" (p_{12}), and these two states should be both observable online. In addition, to better describe the transition processes between these two states, a time delay of 1 is assigned to t_9 and also to t_{10}, respectively, and these delays are shown next to the transitions.
- The PLC model can be constructed in a straightforward fashion with the aforementioned places $p_{11} - p_{14}$ according to Tables 3.2 and 3.3 (see Fig. 3.24). For simplicity, it is assumed that the operation steps in S_1 can always be executed initially and thus the place representing S_0 is omitted in this model.

After building the above models to represent normal behaviors, additional mechanisms should then be incorporated into each component Petri net to characterize failures. The general model structure in Fig. 3.25 is used to represent *all* possible fault scenarios. In this model, the direct outcome of a failure is viewed as a change

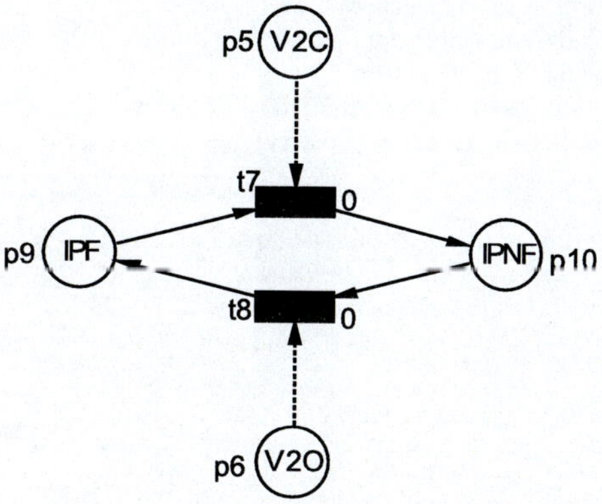

Fig. 3.22 The inlet pipeline (Reprinted with permission from Chen et al. 2010. Copyright 2010 American Chemical Society)

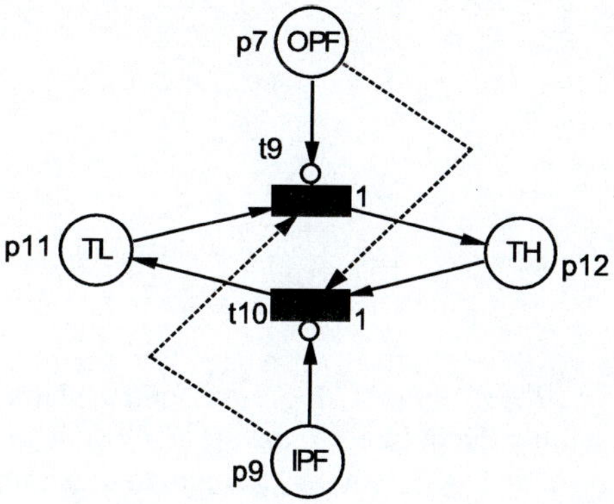

Fig. 3.23 The tank model (Reprinted with permission from Chen et al. 2010. Copyright 2010 American Chemical Society)

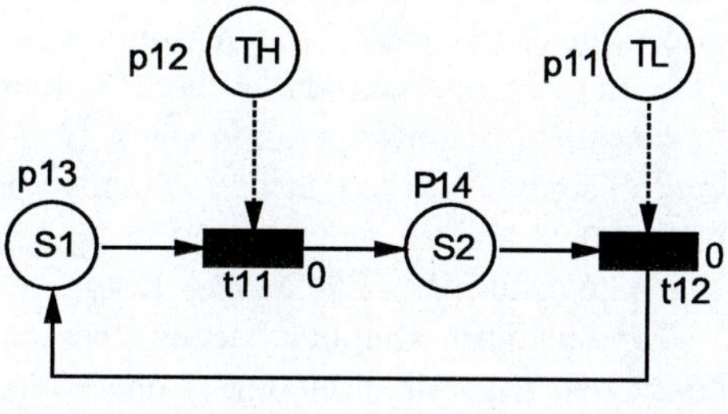

Fig. 3.24 The controller model (Reprinted with permission from Chen et al. 2010. Copyright 2010 American Chemical Society)

in the equipment state. The equipment state caused by its $i$thfailure mode is represented by the place $PFS(i)$ $(i = 1, 2, \cdots)$. The effects of a failure are regarded as the outcomes created by replacing a set of routine events occurred during normal operation with an alternative set of abnormal events. These effects can be readily modeled with a **combination** of the inhibitor arcs and test arcs (see Fig. 3.25). The former arcs are used to disable the transitions corresponding to the routine events,

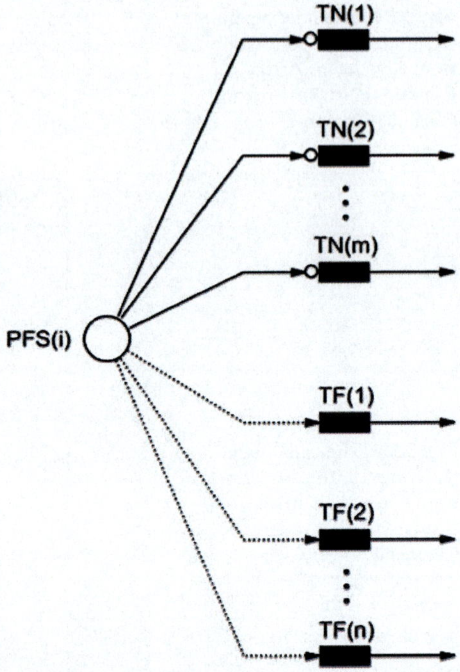

Fig. 3.25 The generalized failure model (Reprinted with permission from Chen et al. 2010. Copyright 2010 American Chemical Society)

i.e., $TN(j)$ $(j = 1, 2, \ldots, m)$, and the latter activate the alternative transitions representing the failure events, i.e., $TF(k)$ $(k = 1, 2, \ldots, n)$. Let us first use the Petri net given in Fig. 3.19 as an example to illustrate this model-building approach. In particular, the abnormal valve states, i.e., "V-1 sticks at the close position" (V1SC) and "V-1 sticks at the open position" (V1SO), are represented with p_3 and p_4, and their effects are characterized with inhibitor arcs. Notice that no test arcs are needed in this case. Let us next assume that the PLC may occasionally send out spurious signals to execute erroneous operation steps. If an additional fault origin, i.e., the spurious control signal, is to be considered in the operation of V-2, then the controller model in Fig. 3.24 and the valve model in Fig. 3.20 could be changed to those shown in Fig. 3.26a and Fig. 3.26b, respectively.

For illustration simplicity, let us assume in the present example that V-1 is the only component that might fail during operation. As a result, the places and transitions in this simplified operation can be classified as follows:

$$\mathbb{P}_o = \{p_{11}, p_{12}\}$$

$$\mathbb{P}_n = \{p_1, p_2, p_5, p_6, p_7, p_8, p_9, p_{10}, p_{13}, p_{14}\}$$

$$\mathbb{P}_f = \{p_3, p_4\}$$

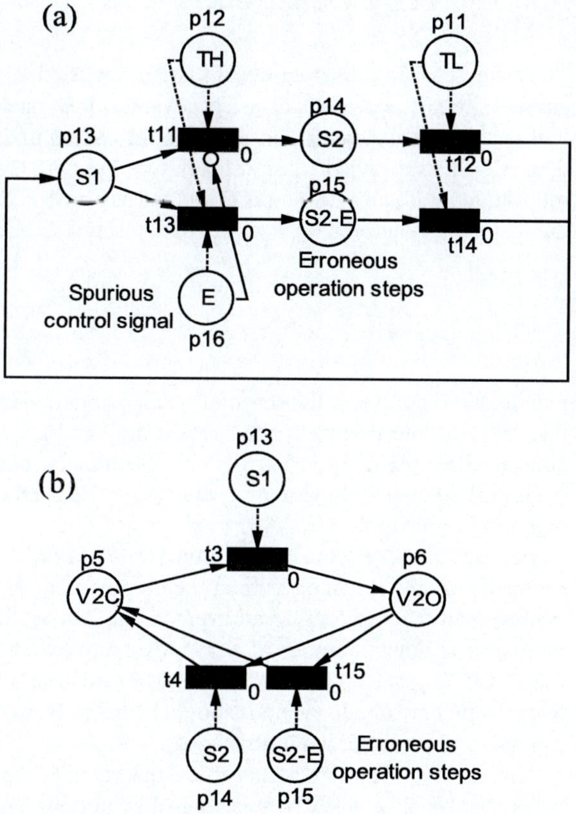

Fig. 3.26 **a** The controller model that contains an additional failure module; **b** The inlet valve model that contains an additional failure module (Reprinted with permission from Chen et al. 2010. Copyright 2010 American Chemical Society)

$$\mathbb{T}_A = \{t_3, t_4, t_5, t_6, t_7, t_8, t_9, t_{10}, t_{11}, t_{12}\}$$

$$\mathbb{T}_B = \{t_1, t_2\}$$

$$\mathbb{T}_C = \emptyset$$

$$\mathbb{T}_K = \{t_{11}, t_{12}\}$$

$$\mathbb{T}_{UK} = \{t_1, t_2, t_3, t_4, t_5, t_6, t_7, t_8, t_9, t_{10}\}.$$

3.3 Supervisory Controllers

The event set E in a discrete-event system can usually be partitioned into two disjoint subsets, i.e., $E = E_c \cup E_{uc}$. The events in E_c are those that can be *forbidden* with a supervisory controller, whereas the events in E_{uc} are bound to occur in due course. In the supervisory control paradigm (see Fig. 3.27), the plant to be operated is represented with an automaton P and the supervisor S is viewed as a *mapping* from the language generated by P to the power set of E if all events in E are observable, i.e.,

$$S : \mathfrak{L}(P) \rightarrow 2^E \tag{3.34}$$

where $\mathfrak{L}(P)$ represents the set of all strings obtained from automaton$P$. If$t \in \mathfrak{L}(P)$, then $S(t)$ is interpreted in the present applications as the set of actuator actions allowed after realizing string$t$. More specifically, automaton P is a model of the uncontrolled plant behavior, and supervisor S is used to represent the corresponding operating procedure.

For each $t \in \mathfrak{L}(P)$, the intersection $S(t) \cap \Gamma(f(x_0, t))$ represents the set of enabled events that plant P can execute at the current state $f(x_0, t)$. To put it differently, P is prevented from executing an event in the current active event set $\Gamma(f(x_0, t))$ as long as it is not a member of $S(t)$. Also, supervisor S is said to be *admissible* if $E_{uc} \cap \Gamma(f(x_0, t)) \subset S(t)$ and $\forall t \in \mathfrak{L}(P)$, that is, S is not allowed to disable any feasible uncontrollable event. It should be noted from the outset that only admissible supervisors are discussed in this book.

Given plant P and an admissible supervisor S, the resulting closed-loop system is denoted by S/P. Clearly, this controlled system S/P is also a DES which can be characterized with its generated and marked languages. The language generated by S/P can be defined recursively as follows:

(1) $\varepsilon \in \mathfrak{L}(S/P)$;
(2) $[(t \in \mathfrak{L}(S/P))\text{and}(t\sigma \in \mathfrak{L}(P)) \text{ and } (\sigma \in S(t))] \iff [t\sigma \in \mathfrak{L}(S/P)]$.

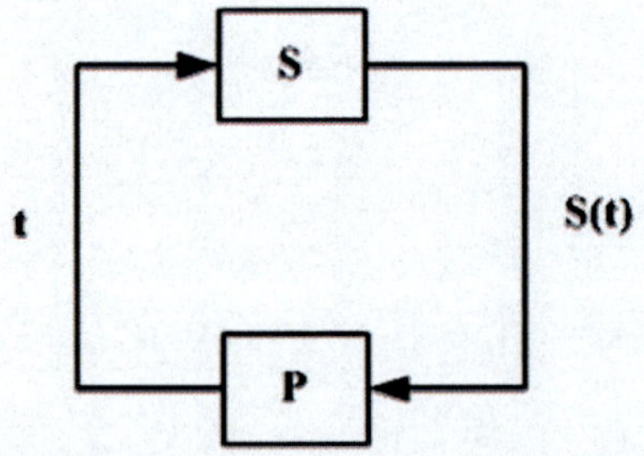

Fig. 3.27 The feedback loop of supervisory control (Reprinted with permission from Kang and Chang 2014. Copyright 2014 Elsevier)

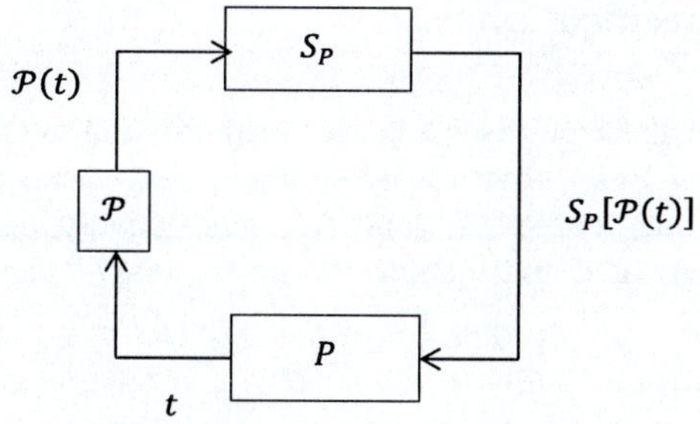

Fig. 3.28 Feedback loop of partial-observation supervisory control

Furthermore, the language marked by S/P can be obtained by taking the intersection of $\mathfrak{L}(S/P)$ and $\mathfrak{L}_m(P)$, i.e., $\mathfrak{L}_m(S/P) = \mathfrak{L}(S/P) \cap \mathfrak{L}_m(P)$.

In practical applications, usually not all events in plant P can be monitored online. Thus, the event set E of P can be partitioned accordingly into two disjoint subsets, i.e., $E = E_o \cup E_{uo}$. The events in E_o are observable, while those in E_{uo} are hidden from the supervisor. The corresponding feedback loop in Fig. 3.28 is obtained by inserting a language projector $\mathcal{P}$ into the loop in Fig. 3.27 and then replacing the full-observation supervisor S with a partial-observation counterpartS_P. On the basis of the definitions given in equations (3.3)–(3.5), one can see that the above projection is the mapping $\mathcal{P} : E^* \to E_o^*$. Due to the presence of projection$\mathcal{P}$, the partial-observation supervisor S_P cannot distinguish strings (say $t_1, t_2, \ldots, t_n$) that have the same projection and, consequently, issue the same control action, e.g., $S_P[\mathcal{P}(t_1)]$. To be more specific, the partial-observation supervisor can be treated as a function, i.e., $S_P : \mathcal{P}[\mathfrak{L}(P)] \to 2^E$, and referred to as a *P-supervisor*. Thus, a new control action can be issued only after the emergence of an observable event, that is, when string $\mathcal{P}(t)$ grows. It is also assumed that when an enabling observable event occurs, the control action is immediately updated before any unobservable event takes place.

The closed-loop behavior of S_P/P can be characterized with the same approach as that in the full-observation scenario. In particular, the language generated by S_P/P is also defined recursively as follows:

(1) $\varepsilon \in \mathfrak{L}(S_P/P)$;
(2) $[(t \in \mathfrak{L}(S_P/P))\text{and}(t\sigma \in \mathfrak{L}(P))\text{and}(\sigma \in S_P[\mathcal{P}(t)])] \iff [t\sigma \in \mathfrak{L}(S_P/P)]$.

The language marked by S_P/P can be constructed in the same way as that for $\mathfrak{L}_m(S/P)$ by taking the intersection of $\mathfrak{L}(S_P/P)$ and $\mathfrak{L}_m(P)$, i.e., $\mathfrak{L}_m(S_P/P) = \mathfrak{L}(S_P/P) \cap \mathfrak{L}_m(P)$. Finally, it should be emphasized that the languages $\mathfrak{L}(S_P/P)$ and $\mathfrak{L}_m(S_P/P)$ should be defined over the entire event set E (not its subset E_o) of plant automaton P.

3.4 Control Specifications

As mentioned above, the supervisor S is used primarily to limit the behavior of plant P within a subset of $\mathfrak{L}(P)$, that is, $\mathfrak{L}(S/P) \subseteq L_a \subset \mathfrak{L}(P)$ and/or $\mathfrak{L}_m(S/P) \subseteq L_{am} \subset \mathfrak{L}_m(P)$. In this notation, L_a and L_{am}, respectively, denote the admissible languages of the generated and marked languages of S/P, and L_a is assumed to be prefix-closed, i.e., $L_a = \overline{L_a}$. The constraint $\mathfrak{L}(S/P) \subseteq L_a$ should be satisfied if blocking is not of concern, while $\mathfrak{L}_m(S/P) \subseteq L_{am}$ if otherwise. Finally, notice that S should be replaced by S_P in the above constraints in the partial-observation case.

In practical applications, it is usually necessary to construct the automata to characterize the control specifications which are usually expressed in natural languages. For examples, such specifications may be used to avoid a list of illegal states, to enforce a first-come first-serve policy, to alternate two distinct events, to execute specific events in a given precedence order, etc. Therefore, the construction of the automaton that generates/marks the admissible language L_a should be preceded by the construction of one or more simple automaton that captures the essence of the natural language specification(s). Let us denote these automata as $H_{spec,1}, H_{spec,2}, \cdots$. By using the parallel composition, one can combine $H_{spec,i}$ $(i = 1, 2, \cdots)$ with P to form a final automaton H_a for generating the admissible language, i.e., $\mathfrak{L}(H_a) = L_a$. Note that, in most cases, all states of $H_{spec,i} (i = 1, 2, \ldots)$ should be marked so that the markings in H_a are solely determined by P.

Following are a few examples of the above modeling practices:

- Illegal states. If certain states in P are considered to be illegal, it is necessary to remove them and all the attached transitions from P. The Acoperation should then be performed on the resulting automaton to produce H_a and, as mentioned before, $\mathfrak{L}(H_a) = L_a$. However, if nonblocking is required after deleting the illegal states, then a $Trim$ operation should be used to replace Ac for the same purpose of creating H_a. In this second scenario, $\mathfrak{L}_m(H_a) = L_{am}$ and$\mathfrak{L}(H_a) = \overline{L_{am}}$.
- State splitting. If two or more paths leading to a particular state, respectively, activate different sets of feasible events, then this state in P should be split into as many states to form H_a.
- Eventalternation. If two events must occur alternately, then a two-state automaton H_{spec} can be built to capture this alternation (see Fig. 3.29). The desired automaton H_a should be generated by taking parallel composition $H_a = H_{spec} \| P$.

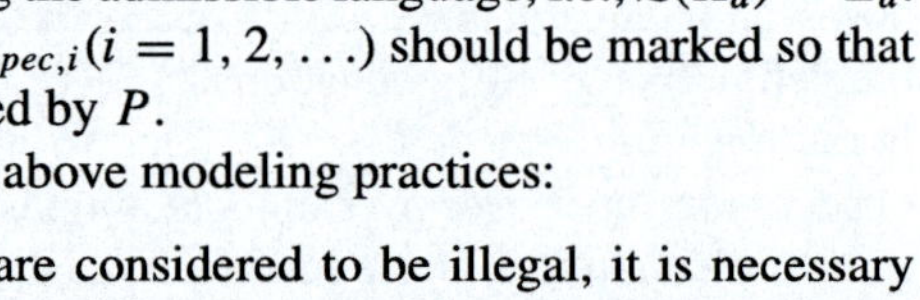

Fig. 3.29 Two-state automaton that yields event alternation

3.5 Systematic Synthesis of Operating Procedures

The automaton-based method presented here can be used for synthesizing the *normal* operating procedures in chemical processes automatically. By following the methodical model-building procedures presented previously in Sect. 3.2.2, a set of automata can be first constructed to characterize the component behaviors in any given *uncontrolled* plant, i.e., levels 2–5 in Fig. 3.1. A number of synthesis rules are given in the present section for conjecturing and modeling the control requirements (or specifications). The plant supervisor can then be produced by applying the parallel composition operation with these and maybe some additional auxiliary automata for setting the operation target(s) and bounds.

Specifically, the operating procedures can usually be produced in four steps:

- Build the automaton models of all components in the uncontrolled plant (levels 2–5);
- Construct automata to represent all control specifications;
- Combine the automata created in steps (1) and (2) to produce an *admissible* supervisor;
- Produce an *implementable* supervisor by augmenting the admissible supervisor with auxiliary automata and then identify the suitable operating procedures accordingly.

A simple example (Example 3.2) is used in the sequel to illustrate such a procedure.

3.5.1 Process Description

Let us consider the heated vessel given in Fig. 3.30. The height of the liquid level in the tank is monitored with a level sensor and two distinct sensor signals, i.e., (1) level high (LH) and (2) level low (LL), may be issued to trigger the corresponding material-transfer operations. In addition, the liquid temperature is also measured, and two critical states, i.e., temperature high (TH) and temperature low (TL), may

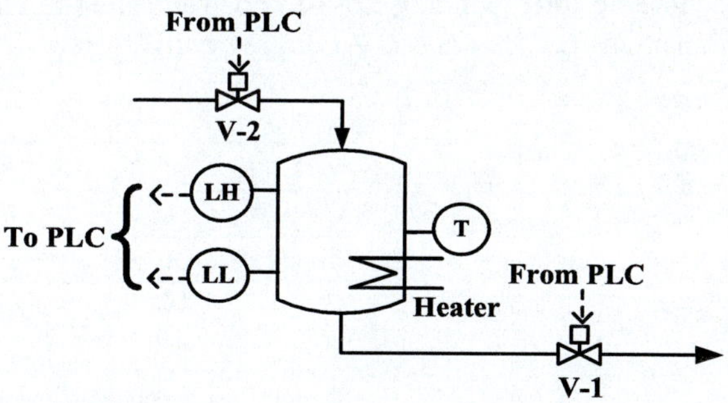

Fig. 3.30 A storage tank equipped with heating equipment (Example 3.2) (Reprinted with permission from Yeh 2013. Copyright 2013 National Cheng Kung University Library)

be reported to, respectively, deactivate and activate the heating operation. The inlet valve V-2 should obviously be open in the tank-filling operation under the condition that the outlet valve V-1 is kept closed. On the other hand, V-1 should be open while keeping V-2 closed if tank draining is required. In order to facilitate safe operations, liquid should be transported into or from the tank only when the corresponding sensor signal, i.e., LL or LH, can be confirmed. Notice also that the heater should be turned off after reaching the target temperature, and it should not be turned on before the tank is LH. Initially, the liquid level and the temperature in the tank are at LL and TL, respectively, both valves are closed, and the heater is off.

3.5.2 Components

For the sake of illustration simplicity, all equipment failures are ignored in this example. The component models in levels 2–5 are outlined as follows:

Level 2 and 3:

The process configuration is governed by the collective states of actuators. Although there are three actuators in Fig. 3.30 and, therefore, eight possible configurations, four of them are disallowed to ensure operability and safety. Specifically, the justifications of this practice are listed as follows:

(1) To avoid opening both valves simultaneously, the configurations $(V1O, V2O, Hon)$ and $(V1O, V2O, Hoff)$ are excluded;
(2) To avoid heating before V-1 is closed, the configuration $(V1O, V2C, Hon)$ is not considered;
(3) To avoid opening V-1 before the heater is switched off, the configuration $(V1C, V2O, Hon)$ is not included in the model.

And the permitted configurations are referred to in this example as GV_{01},GV_{02}, GV_{03}, and GV_{04}, respectively (see Table 3.4). The actuator models can be built according to these allowed configurations. The automaton representation of V-1 can be found in Fig. 3.31a. States $V1C$ and $V1O$ in this model are used to represent the close and open positions, respectively, while the events $openV1$ and $closeV1$ denote the corresponding close-to-open and open-to-close processes. In addition, the transitions GV_acon and GV_bcon, respectively, represent V-1 continues at close and

Table. 3.4 Actuator combinations in Example 3.2

V-1*	V-2*	Heater[a]	Symbol
C	C	C	GV_{01}
C	C	O	GV_{02}
C	O	C	GV_{03}
O	C	C	GV_{04}

[a]O and C denote open/on and close/off positions, respectively

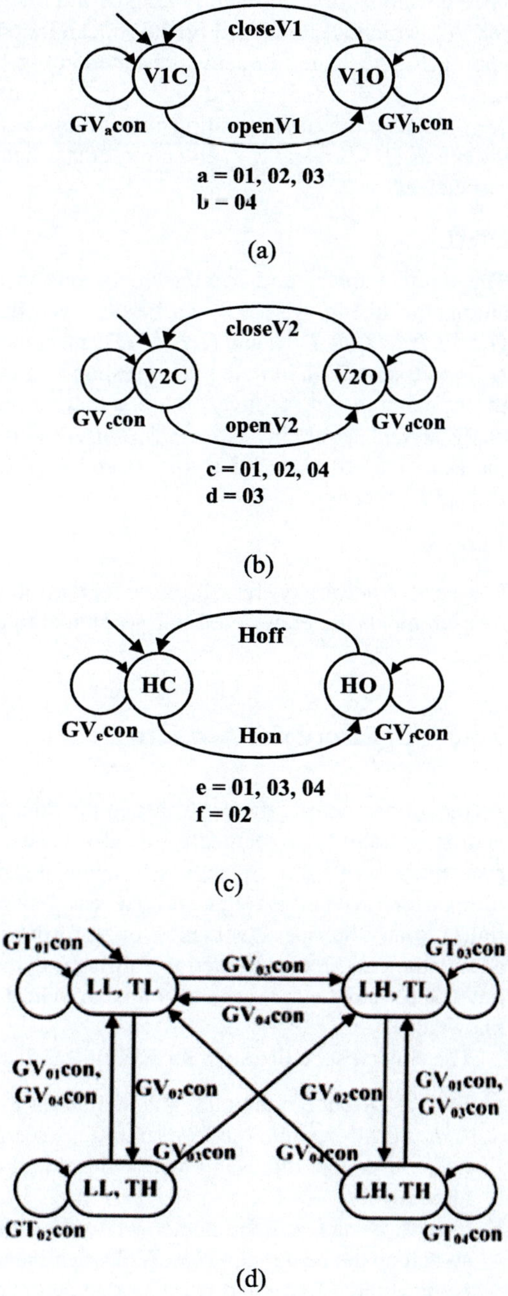

Fig. 3.31 Component models in Example 3.2: **a** valve V-1 model; **b** valve V-2 model; **c** heater model; **d** tank model associated with level and temperature (Reprinted with permission from Yeh 2013. Copyright 2013 National Cheng Kung University Library)

open positions for a sufficiently long period of time. Notice also that a similar model for V-2, which is shown in Fig. 3.31b, can be established with essentially the same approach. The heater model is presented in Fig. 3.31c. States HF and HN here are used to represent the off and on positions, respectively, while the events Hon and $Hoff$ denote the corresponding close-to-open and open-to-close processes. Finally, the events GV_econ and GV_fcon represent the heater continues at off and on positions, respectively.

Level 4:

The liquid transfer and heating operations in the tank can be modeled with the automaton in Fig. 3.31d. It can be observed that four tank states, i.e., (LL, TL), (LL, TH), (LH, TL), and (LH, TH), are considered here, and $GT_{01}con$, $GT_{02}con$, $GT_{03}con$, and $GT_{04}con$ are the corresponding events denoting the tank continues at these four states, respectively. Notice that all process configurations defined previously should result in state transitions. For instance, $GV_{03}con$ should facilitate the (LL, TL)-to-(LH, TL) process and $GV_{04}con$ should cause the (LH, TL)-to-(LL, TL) process.

Level 5:

For the sake of brevity, the sensor model is omitted in the present study and the online measurements are considered to be identical to the tank states.

3.5.3 Control Specifications

As mentioned before, the control specifications are usually adopted to ensure safety and/or operability. Specifically, they are used to achieve or forbid a variety of prescribed event/state sequences to avoid physically inadmissible behaviors, e.g., filling a tank when it is full, heating a vessel when it is empty, and transferring material(s) to an improper destination or to form a hazardous mixture. More than one automaton can be constructed to fulfill these requirements (Cassandras and Lafortune 2008), and they are usually conjectured in an ad hoc fashion according to process knowledge.

The control specifications for Example 3.2 can be summarized as follows:

- *Spec 1*: Discharge the tank and switch on the heater while the liquid level is at LH and fill it while the level is at LL (see Fig. 3.32). Notice that GT_gcon and GT_hcon denote the events where the liquid level is maintained at LL and LH, respectively.
- *Spec 2*: Switch off the heater when the liquid temperature is at TH whereas switch on the heater and close V-2 when the temperature is at TL (see Fig. 3.32). Notice that GT_icon and GT_jcon denote the events where the liquid temperature is maintained at TH and TL, respectively.
- *Spec 3*: Avoid opening both valves simultaneously (see Fig. 3.32c).

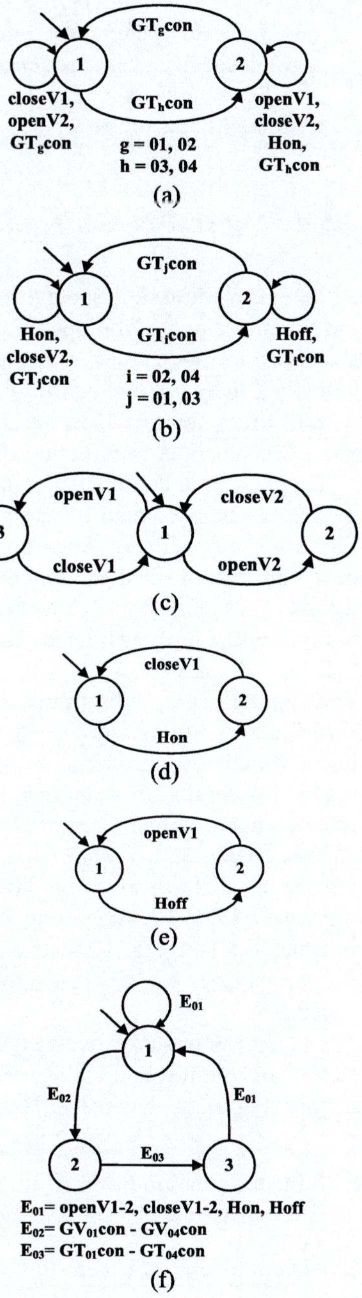

Fig. 3.32 Specification models in Example 3.2: **a** Spec 1; **b** Spec 2; **c** Spec 3; **d** Spec 4; **e** Spec 5; **f** Spec 6 (Reprinted with permission from Yeh 2013. Copyright 2013 National Cheng Kung University Library)

- *Spec 4*: Avoid heating before V-1 is closed (see Fig. 3.32d).
- *Spec 5*: Avoid opening V-1 before heater is switched off (see Fig. 3.32e).
- *Spec 6*: Specify event sequence, i.e., $E_{01}E_{02}E_{03}$. Notice that event E_{01} denotes actuator actions, whereas E_{02} and E_{03} represent process configurations and combined states of processing units, respectively (see Fig. 3.32f).

3.5.4 Supervisor Generation

Although an admissible supervisor (e.g., see Fig. 3.33) can be obtained by applying the *parallel composition* operation on the aforementioned component models and specification models, an implementable supervisor must be further identified by extracting the *supremal controllable sublanguage* from this admissible supervisor with auxiliary automata. A fictitious example is adopted in the sequel to illustrate these two supervisor generation steps.

Let us assume that there are only two components in the given system, and their model can be presented by the automata C_1 and C_2 presented in Fig. 3.34a and Fig. 3.34b, respectively. According to the automata building steps described previously, the uncontrolled plant model D in Fig. 3.34c can be produced by performing a parallel composition on C_1 and C_2. In addition, the specification models E_1 and E_2 in Fig. 3.34d, e are constructed primarily to enforce alternations of two sets of events, i.e., $(a1, b1)$ and$(a2, b2)$. After combining D, E_1, and E_2 with parallel composition, one can then produce the admissible supervisor G in Fig. 3.34f. Let us further assumed that automaton H in Fig. 3.34g is the auxiliary automaton and let the set of uncontrollable events be $E_{uo} = \{a2, b2\}$, that is, events $a2$ and $b2$ cannot be disabled. It can be observed from automaton G that the event sequence $a1a2$ can be extended with the uncontrollable event $b2$, but the event sequence $a1a2b2$ is not present in automaton H. On the other hand, a prefix $a1$ which can be extended within G by the uncontrollable event$a2$, and the event sequence $a1a2$ is not in H after removing the trace $a1a2b1b2$. According to Cassandras and Lafortune (2008), the resulting automaton I (see Fig. 3.34h) can be obtained after removing all traces with prefix $a1$ or prefix $a1a2$ in H. The automaton I can be considered as the implementable supervisor.

In the present study, two auxiliary automata are constructed to define the target state(s) or event(s) and also to set the upper limit on the total number of actuator actions. Let us again consider Example 3.2 here for illustration convenience:

- The auxiliary automaton in Fig. 3.35a is adopted for the purpose of specifying termination mechanism, i.e., stopping operation after events $GV_{03}con$ and $GT_{03}con$ twice. Thus, state 5 should be marked as the operation target and disallow any events to occur later.
- The auxiliary automaton in Fig. 3.35b is used to limit the total number of operation steps. Since the initial state 0 is driven to state n ($n = 1, 2, \ldots, N$) after n operation steps, the maximum number of operation steps, i.e., N, in the operation procedure(s) can be imposed by augmenting the admissible supervisor with this

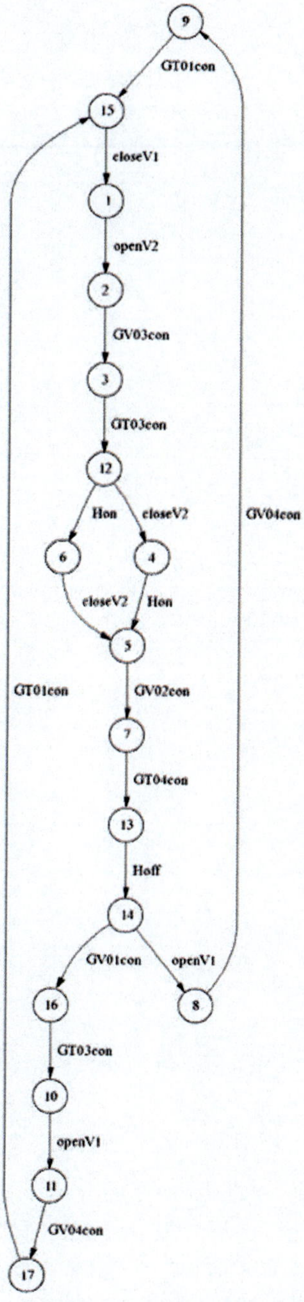

Fig. 3.33 Admissible supervisor for Example 3.2 (Reprinted with permission from Yeh 2013. Copyright 2013 National Cheng Kung University Library)

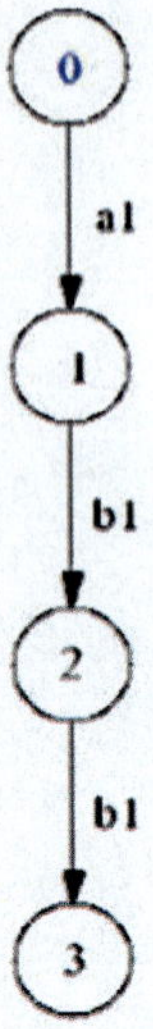

(a)

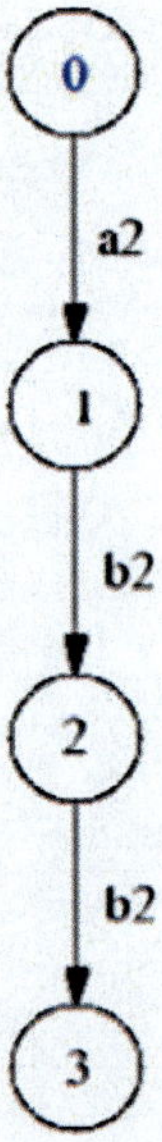

(b)

Fig. 3.34 A fictitious example: **a** Automaton C_1; **b** Automaton C_2; **c** Automaton D; **d** Automaton E_1; **e** Automaton E_2; **f** Automaton G; **g** Automaton H; **h** Automaton I (Reprinted with permission from Yeh 2013. Copyright 2013 National Cheng Kung University Library)

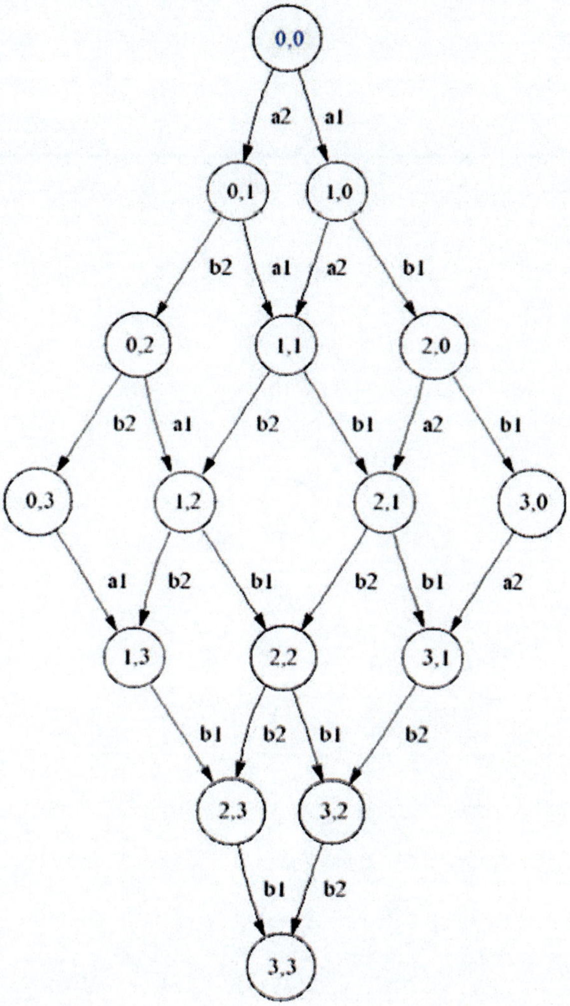

(c)

Fig. 3.34 (continued)

automaton. Notice also that, in order to allow fewer operation steps to be taken in the operation procedure(s), all states are marked in this model. Consequently, this auxiliary automaton facilitates easy identification of all feasible procedures with $n \leq N$ operation steps and also the most efficient one(s) among them.

The above two automata have been incorporated into the admissible supervisor in Fig. 3.33 to generate the implementable supervisor in Fig. 3.36.

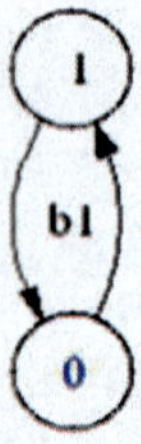

(d)

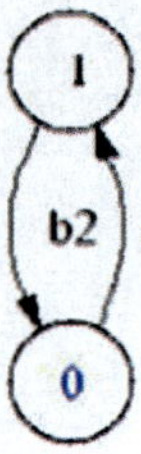

(e)

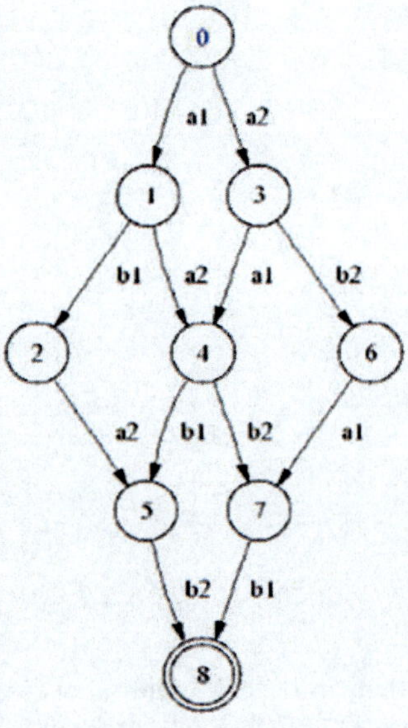

(f)

Fig. 3.34 (continued)

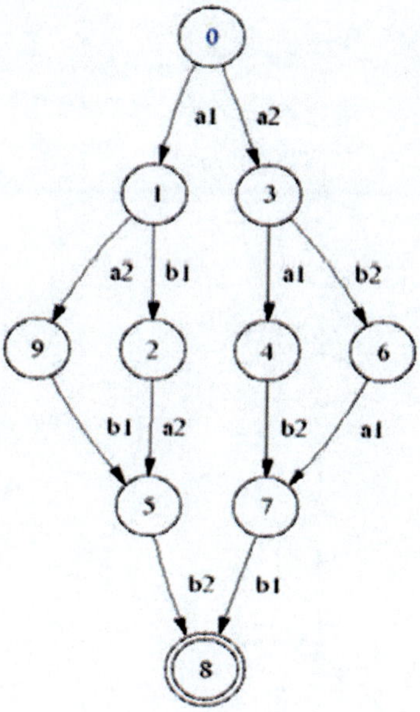

(g)

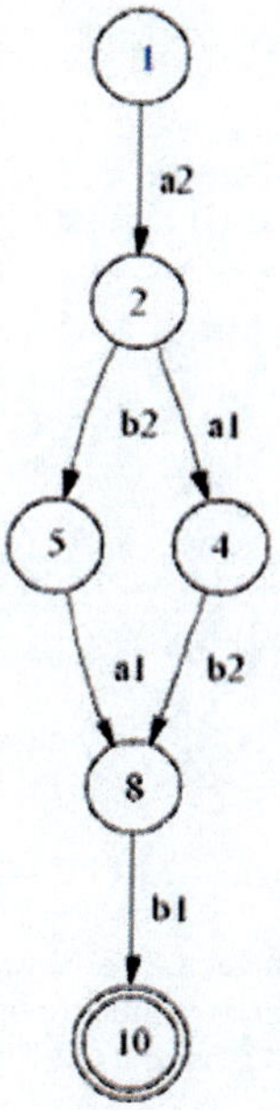

(h)

Fig. 3.34 (continued)

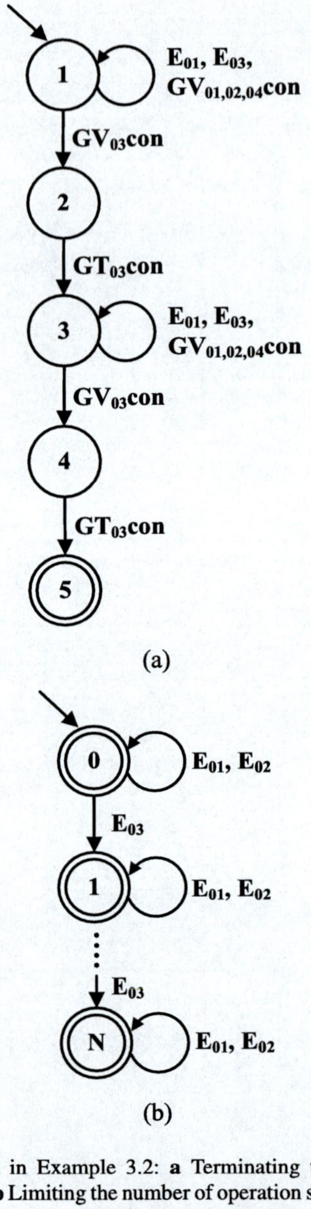

Fig. 3.35 Auxiliary automata in Example 3.2: **a** Terminating the admissible supervisor after $GV_{03}con$ and $GT_{03}con$ twice; **b** Limiting the number of operation steps (Reprinted with permission from Yeh 2013. Copyright 2013 National Cheng Kung University Library)

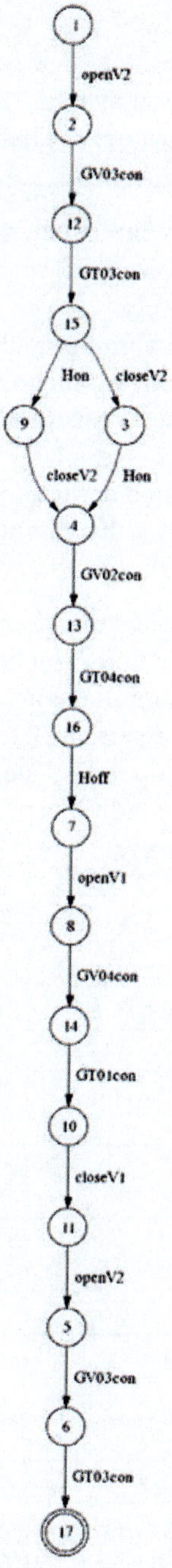

Fig. 3.36 Implementable supervisor for Example 3.2 (Reprinted with permission from Yeh 2013. Copyright 2013 National Cheng Kung University Library)

3.5.5 Procedure Synthesis

Having created the implementable supervisor, the corresponding operating procedures can then be identified accordingly. This task should be performed in the following steps:

1. Classify the supervisor events into three sets, namely, E_{AC}, E_{PC}, and E_{AA}, which are associated with the *activation conditions, process configurations*, and *actuator actions*, respectively.
2. Produce a *reduced* supervisor by grouping the consecutive actuator actions (i.e., the events in E_{AA}) between activation conditions (i.e., the events in E_{AC}) and *process configurations* (i.e., the events in E_{PC}). It is assumed that the grouped events occur simultaneously.
3. Remove all events in E_{PC}, which are treated as unobservable events.
4. Identify all possible procedures by enumerating traces generated from the reduced supervisor.

Let us consider the implementable supervisor shown in Fig. 3.36 as an example. Notice that actuator actions $openV2$ and Hon between events E_{AC} and E_{PC} should be grouped together and treated as simultaneous events. The reduced implementable supervisor can then be obtained for Example 3.2. The most efficient operating procedure, i.e., the ones with the fewest operation steps, can be enumerated accordingly. The operation steps and activation conditions of the corresponding SFC are listed in Table 3.5a and Table 3.5b, respectively.

Table 3.5 Identified SFC in Example 3.2: **a** Operation steps; **b** Activation conditions

(a)

Operation step	Control actions
S_0	Initialization
S_1	(1) Close V-1 (2) Open V-2
S_2	(1) Switch on Heater (2) Close V-2
S_3	(1) Switch off Heater (2) Open V-1

(b)

Symbol	Conditions
T_1	Start
T_2	LH
T_3	TH
T_4	TL and LL

References

Blanke M, Kinnaert M, Lunze J, Staroswiecki M (2003) Diagnosis and Fault-tolerant Control. Springer, Berlin

Brandin BA, Wonham WM (1994) Supervisory control of timed discrete-event systems. IEEE Trans Automat Control 39:329–342

Cassandras CG, Lafortune S (2008) Introduction to discrete event systems, 2nd edn. Springer Science+Business Media, LLC, New York

Chen CL, Chen WC (1994) Fuzzy controller-design by using neural-network techniques. IEEE Trans Fuzzy Syst 2:235–244

Chen YC, Yeh ML, Hong CL, Chang CT (2010) Petri-net based approach to configure online fault diagnosis systems for batch processes. Ind Eng Chem Res 49(9):4249–4268

Chou HH, Chang CT (2005) Petri-net-based strategy to synthesize the operating procedures for cleaning pipeline networks. Ind Eng Chem Res 44:114–123

Crooks CA, Macchietto SA (1992) Combined MILP and logic-based approach to the synthesis of operating procedures for batch plants. Chem Eng Commun 114:117–144

Dietrich P, Malik R, Wonham WM, Brandin BA (2002) Implementation considerations in supervisory control. In: Caillaud B, Darondeau P, Lavagno L, Xie X (eds) Synthesis and control of discrete event systems. Kluwer, pp 185–201

Falkman P, Lennartson B, Tittus M (2009) Specification of a batch plant using process algebra and petri nets. Control Eng Pract 17:1004–1015

Fravolini ML, Campa G (2011) Design of a neural network adaptive controller via a constrained invariant ellipsoids technique rai. IEEE Trans Neural Netw 22:627–638

Fusillo RH, Powers GJ (1987) A synthesis method for chemical plant operating procedures. Comput Chem Eng 11:369–382

Galán S, Barton PI (1997) Dynamic optimization formulations for operating procedure synthesis. Paper presented at the annual meeting of the American Institute of Chemical Engineers

Hamid MKA, Sin G, Gani R (2010) Integration of process design and controller design for chemical processes using model-based methodology. Comput Chem Eng 34:683–699

Hashizume S, Yajima T, Ito T, Onogi K (2004) Synthesis of operating procedures and procedural controllers for batch processes based on Petri nets. J Chin Inst Chem Eng, 35:363–369

Hashizume S, Yajima T, Kuwashita Y, Onogi K (2008) Integration of fault analysis and interlock controller synthesis for batch processes. Chin J Chem Eng 16(1):57–61

Hoshi K, Nagasawa K, Yamashita Y, Suzuki M (2002) Automatic generation of operating procedures for batch production plants by using graph representations. J Chem Eng Jpn 35:377–383

Ivanov VA, Kafarov VV, Perov VL, Reznichenko AA (1980) On algorithmization of the start-up of chemical productions. Eng Cybern 18:104–110

Kang A, Chang CT (2014) Automata generated test plans for fault diagnosis in sequential material- and energy-transfer operations. Chem Eng Sci 113:101–115

Kaspar MH, Ray WH (1992) Chemometric methods for process monitoring and high-performance controller-design. AIChE J 38:1593–1608

Kinoshita A, Umeda T, O'Shima E (1982) An approach for determination of operational procedure of chemical plants. In: Proceedings of the international symposium on process systems engineering, pp 114–120

Kim J, Kim J, Moon I (2009) Error-free scheduling for batch processes using symbolic model verifier. J Loss Prev Process Ind 22:367–372

Kim J, Moon I (2009) Automatic verification of control logics in safety instrumented system design for chemical process industry. J Loss Prev Process Ind 22:975–980

Koutsoukos XD, Antsaklis PJ, Stiver JA, Lemmon MD (2000) Supervisory control of hybrid systems. Proc IEEE 88:1026–1049

Lai JW, Chang CT, Hwang SH (2007) Petri-net based binary integer programs for automatic synthesis of batch operating procedures. Ind Eng Chem Res 46(9):2797–2813

Lakshmanan R, Stephanopoulos G (1988) Synthesis of operating procedures for complete chemical plants—I. Hierarchical, structured modeling for nonlinear planning. Comput Chem Eng 12:985–1002

Liang KH, Chang CT (2008) A simultaneous optimization approach to generate design specifications and maintenance policies for the multi-layer protective systems in chemical processes. Ind Eng Chem Res 47(15):5543–5555

Li HS, Lu ML, Naka Y (1997) A two-tier methodology for synthesis of operating procedures. Comput Chem Eng 21S:S899

Malik P, Malik R (2006) Modular control-loop detection. In: Proceedings of the 8th international workshop on discrete event systems, Ann Arbor, Michigan, USA, July 10–12

Moon I, Powers GJ, Burch JR, Clarke EM (1992) Automatic verification of sequential control systems using temporal logic. AIChE J 38(1):67–75

Naka Y, Lu ML, Takiyama H (1997) Operational design for start-up of chemical processes. Comput Chem Eng 21:997–1007

Ouedraogo L, Khoumsi A, Nourelfath M (2010) A new method for centralised and modular supervisory control of real-time discrete event systems. Int J Control 83:1–39

Panjapornpon C, Soroush M, Seider WD (2006) Model-based controller design for unstable, non-minimum-phase, nonlinear processes. Ind Eng Chem Res 45:2758–2768

Patton RJ (1997) Fault-tolerant control, the 1997 situation. In: Proceedings of safeprocess, Hull,UK, pp 1033–1055

Ramadge PJ, Wonham WM (1987) Supervisory control of a class of discrete event processes. SIAM J Control Optim 25:206–230

Ramadge PJ, Wonham WM (1989) The control of discrete event systems. Proc IEEE 77:81–98

Rivas JR, Rudd DF (1974) Synthesis of failure-safe operation. AIChE J 20:320–325

Sanchez A, Macchietto S (1995) Design of procedural controllers for chemical processes. Comput Chem Eng 19:S381–S386

Tan KS, Yamashita Y (2010) Design of a dependable process control system. In: Proceedings of the 5th international symposium on design, operation and control of chemical processes, pp 446–453

Viswanathan S, Johnsson C, Venkatasubramanian V, Arzen KE (1998a) Automating operating procedure synthesis for batch processes: Part I. Knowledge representation and planning framework. Comput Chem Eng 22(11):1673–1685

Viswanathan S, Johnsson C, Venkatasubramanian V, Arzen KE (1998b) Automating operating procedure synthesis for batch processes: Part II. Implementation and application. Comput Chem Eng 22 (11):1687–1698

Wang YF, Chou HH, Chang CT (2005) Generation of batch operating procedures for multiple material-transfer tasks with petri nets. Comput Chem Eng 29:1822–1836

Wonham WM (2000) Supervisory control of discrete-event systems: an introduction. In: Proceedings of the IEEE international conference on Industrial Technology 2000, Goa, India, pp 474–479, January 19–22

Yamalidou EC, Kantor JC (1991) Modeling and optimal control of discrete-event chemical processes using petri nets. Comput Chem Eng 15:503–519

Yamashita Y (2007) Toward dependable process control systems: integration of fault diagnosis and controller redesign. In: PSE Asia 2007, Xi'an

Yang H, Li N, Li SY (2011) A data-driven bilinear predictive controller design based on subspace method. Asian J Control 13(2):345–349

Yeh ML (2013) Automata based methods for abnormal situation management in batch processes. Doctoral dissertation, Department of Chemical Engineering, National Cheng Kung University, Tainan

Yeh ML, Chang CT (2011) An automaton-based approach to evaluate and improve online diagnosis schemes for multi-failure scenarios in batch chemical processes. Chem Eng Res Des 89:2652–2666

Zhang Y, Jiang J (2003) Bibliographical review on reconfigurable fault-tolerant control systems. In: Proceedings of safeprocess, Washington, USA, pp 265–276

Chapter 4
Normal Operating Procedures Based on Dynamic Simulations

Startup Operations of Simple and Extractive Distillation Columns (DCs)

Startup procedure of DC is one of the most complicated dynamic operations in the chemical industry. This procedure is energy and time consuming, and the process is unproductive during this transient period before reaching steady state.

Besides, the startup operations of DCs are also a challenging act because most of the manipulated variables such as heating power, reflux ratio, and feed flow rate change rapidly and are modified more than one time.

According to the analysis of Ruiz et al. (1988), there are three characteristic steps during the startup operations. Those three stages are shown as below.

1. Discontinuous stage: The shortest period of three stages, including initial charge and the starting of heating. In this stage, the variables of fluid dynamics alter violently. At the end of this stage, there is full of liquid in each tray of DC.
2. Semi-continuous stage: This duration is the most important part of startup operation and takes the longest time to finish. In this duration, the variables of thermodynamics change rapidly and non-linearly. However, the variables of fluid approach nominal values, and the specifications of the product are reached gradually yet would be easily affected by the disturbance caused by the other variables.
3. Continuous stage: In this stage, all of the variables response linearly and would reach nominal values at the end of this stage.

4.1 Startup Operations of Simple DCs

The base case of the startup operations of simple DC are elaborated in Chap. 2. According to the previous results, the concentrations of $CHCl_3$ and CCl_4 tend to steady state as operating for 3.7 h. The time for finishing the startup operation is about 4.6 h. The off-specification products during the startup operation are 23,316 kg. The total energy consumption is 28.1 GJ before settling steady.

C.-T. Chang et al., *Process Plant Operating Procedures*, Advances in Industrial Control,
https://doi.org/10.1007/978-3-030-70978-5_4

However, this example has a lot of action which the objective or occasion is not clear, and some of the products which the specifications have not been satisfied are produced. Thus, the startup procedure should be modified.

The optimal startup procedure would be divided into four parts: feed control strategy, bottom flow control strategy, reflux and distillate flow control strategy, and the control strategy of heating.

In the startup operation, the height of sump level would change after the column was started heating. If the height of sump level is too law, the reboiler would dry and lead to dangerous situation. Besides, in order to maintain the steady of throughput, the DC needs to feed more than one time during the startup operation to avoid the above condition. Hence, the optimal startup operation would use the height of sump level to control the opportunity of feeding. For this case, we set the highest point is the 90% position of sump which the feed to the DC would be stopped when the liquid height overs this value, and the lowest is the 75% position of sump which the feed to the DC would be fed again when the height is below this height.

In the cases offered by Aspen Plus Dynamics®, the bottom flow that starts to produce under the specification is not reached. To this end, the optimal strategy would also focus on reducing the production of off-specification bottom product. The strategy of this part is switch the LC_B form manual to auto mode when the concentration achieves the specification which is under the circumstance that the composition of the bottom flow is measurable, hypothetically.

In the normal operation of startup of DC, the DC is under the total reflux status at the beginning of startup. Till the specification of top product is satisfied, the top product is started to be exhausted. However, the top product is produced before the specification is reached would decrease the time needed for startup and energy consumption. The only drawback is some of the top product would be wasted.

For reflux control strategy, there are two methods would be used to determine the switch point. The one is the minimum of the sum of temperature difference (MT) algorithm, the other is the dynamic composition trend of top product.

MT algorithm is proposed by Yasuoka et al. (1987). It is a characteristic function used to determine the optimal switch point of reflux flow control. This function records the differences of all the tray temperatures during the whole startup operation with the nominal values. The optimal switch point is the minimum solution solved from the followed equation (Eq. 4.1).

$$\mathrm{MT} = \sum_{j=1}^{NT} \left| T_j - T_j^{sp} \right|. \tag{4.1}$$

T_j is the temperature of sample instant of the jth tray, and T_j^{sp} is the nominal value of the jth tray.

The result of MT is illustrated in Fig. 4.1a. According to the result, the value of MT is decrease to minimum value first, then settle steady which means the temperature of DC had already achieved nominal values. The optimal switch point for reflux control is at 1.33 h.

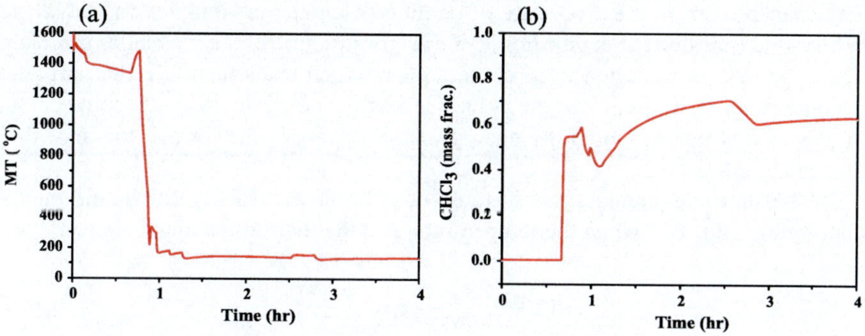

Fig. 4.1 Results of **a** MT algorithm, **b** dynamic composition trend of top product

The dynamic composition trend of top product method also assumes the concentration of top product is measurable. The result is demonstrated in Fig. 4.1b, and the optimal switch point is at the time that the concentration of $CHCl_3$ reached 50 wt.%.

According to the Woinaroschy (2008), there are two methods for heating medium temperature control method. One is the Piecewise Constant Control, the other is Piecewise Linear Control. The schematic illustrations of those two method are shown in Fig. 4.2. Besides, there are also two path of raising the temperature of heating medium which is proposed by Wendt et al. (2003). One is raising the temperature to nominal value gradually, the other is increasing the temperature to the maximum operation temperature (usually 1.2 time of nominal value) then back to nominal value.

The optimal heating method would use simulated annealing approach (SA) to figure out. SA is proposed by Metropolis et al. (1953). However, this algorithm is started to be noticed when the Kirkpatrick et al. (1983) found out the similarity between annealing process and combination optimization problem. Because it has the easy procedure and could avoid solving a local optimum, SA is used widely recently.

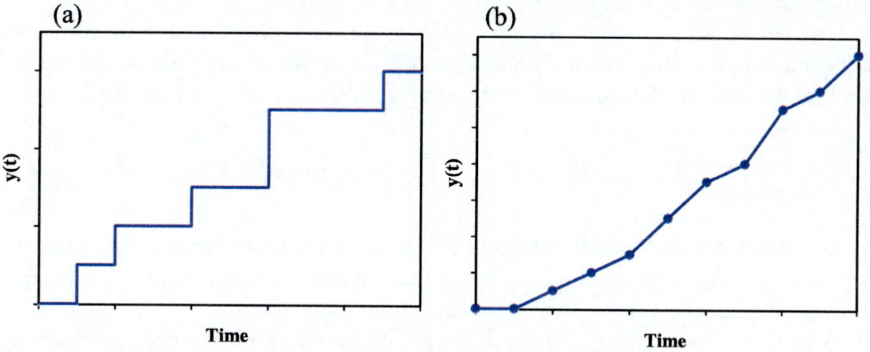

Fig. 4.2 **a** Piecewise constant control. **b** Piecewise linear control

The concept is from the procedure of cooling down and crystalizing in real world. When solid is heated to become liquid, the origin structure of the molecules is broken. Then, the system is started to be cooled down which leads the molecular structure becomes in order again. Until the system is totally cooled down and become crystal in expected structure which the energy of the system is the lowest, the annealing procedure is finished.

In the thermodynamics, the system could be described by Boltzmann energy distribution (Eq. 4.2) when the temperature is at the thermodynamic balance.

$$\mathrm{P}\{E\} = \frac{1}{Z(t)}\exp\left(\frac{-E}{K_bT}\right). \tag{4.2}$$

E is the system energy in J, T is system temperature in K, K_b is Boltzmann's constant (1.3806×10^{23} J/K) and $1/Z(t)$ is the normalization factor.

Using SA in solving optimum solution, we define the objective function as the energy E and all of the design variables as a set of d which would referred to an E_d, then randomly generate a new set call d^{new} by mimicking the disturbance behavior of molecules which would also referred to an $E_{d^{new}}$. Later, through the Metropolis method (Eq. 4.3) to decide the new status would be accepted yet or not.

$$P = \begin{cases} 1, & E_{d^{new}} - E_d \le 0 \\ r = \dfrac{\frac{1}{Z(t)}\exp\left(-\frac{E_{d^{new}}}{K_bT}\right)}{\frac{1}{Z(t)}\exp\left(-\frac{E_d}{K_bT}\right)} = \exp\left[-\frac{(E_{d^{new}} - E_d)}{K_bT}\right], & E_{d^{new}} - E_d > 0 \end{cases} \tag{4.3}$$

where P is the probability of accepting d^{new}, if $E_{d^{new}} - E_d > 0$, the probability of acceptance r is between 0 and 1. At this situation, there is a random number ξ is generated and if $r > \xi$, the d^{new} would be accepted, otherwise not.

The optimal startup strategy with SA should define the initial temperature (T_0), final temperature (T_f), criteria for thermodynamic equilibrium (EqN_T) and temperature decrement factor (α) first in order to simulate the real annealing schedule. Temperature decrement factor affects the final solution is global optimum or not. In this case, α is set between 0.8 and 0.994.

The objective function of startup (J, Eq. 4.4) is the sum of square of difference between sample instant and nominal values of concentration of $CHCl_3$ at the top and CCl_4 at the bottom. The algorithm of startup with SA is illustrated in Fig. 4.3a.

$$\mathrm{J} = \int_0^{t_f} \left\{\left[x_D(t) - x_D^{sp}\right]^2 + \left[x_B(t) - x_B^{sp}\right]^2\right\} dt. \tag{4.4}$$

Besides, there are two different control schemes used for the startup operation and nominal operation which are illustrate in Fig. 4.3b and c, respectively. The controller setting for startup control scheme is demonstrated in Table 4.1.

Adopting above strategy, there are eight different startup strategies which are demonstrated in Table 4.2. According to the results of all strategies, strategy **A** is the

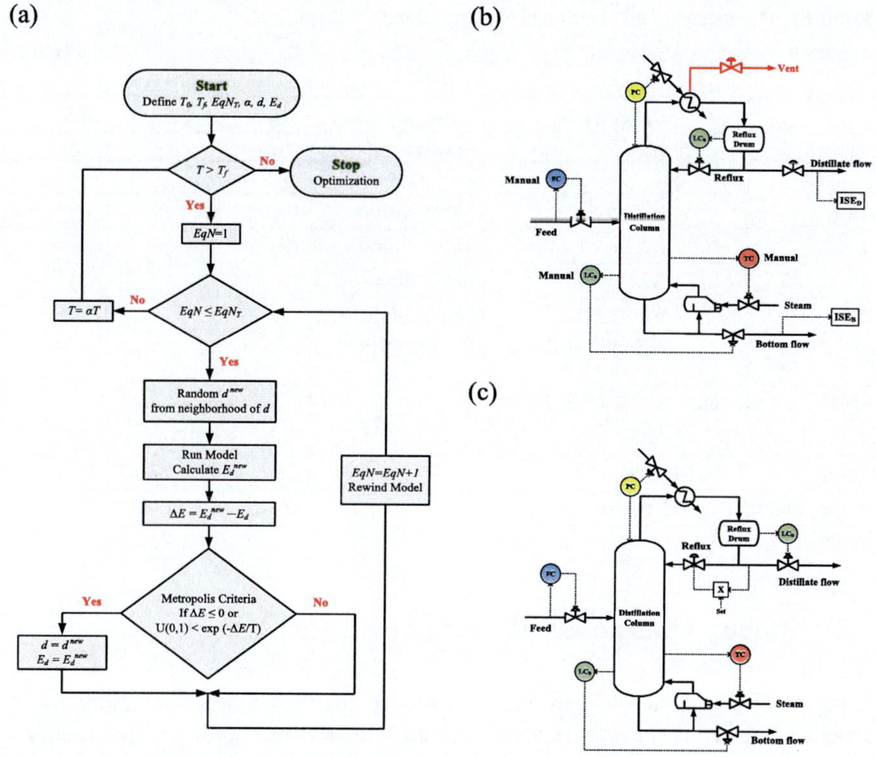

Fig. 4.3 **a** The algorithm of startup with SA. **b** The control scheme of startup operation. **c** The control scheme of nominal operation

optimal one in which the switch point for reflux control is used the MT algorithm and the heating method is piecewise constant control, raising the temperature to nominal value gradually and with SA. The results are shown in Figs. 4.4, 4.5 and 4.6.

For the switch point for reflux control is used the dynamic composition trend of top product method, the strategy **F** is the optimal one which the heating method is piecewise constant control, raising the temperature to 1.2 time nominal value then back to nominal value and with SA. The results are shown in Figs. 4.7, 4.8 and 4.9.

Based on the results of strategy A, though the time of startup increase 0.4 h, the off-specification product and the energy consumption have reduced almost 39% off and 3% off compared with the base cases, respectively. Besides, according to the results of strategy F, not only the time of startup is decreased to 0.3 h, but also the off-specification product and the energy consumption have been reduced almost 41% off and 17% off compared with the base cases, respectively.

Table 4.1 Parameters of all controllers of startup control scheme

Controller	Set point	Manipulated variables	K_c	τ_I (min)
FC	10,000 kg/h	Valve position	1	0.5
PC	2 bar	Flowrate of cooling water	10	12
LC_B	1.25 m	Valve position	10	60
LC_D	1.25 m	Valve position	10	60
LC_R	1.25 m	Reflux flowrate	10	60
TC	102.75 °C (the 16th stage temperature)	Temperature of heating medium	1	20
Objective **function**	*Reference point*		*Measured variable*	
ISE_D	0.831		Conc. of $CHCl_3$ at the top	
ISE_B	0.99		Conc. of CCl_4 at the bottom	
Ratio controller	*Set point*		*Measured variable*	
Reflux ratio	Nominal value: 5		Flowrate of top product	

4.2 Startup Operations of Extractive DCs

Compared with the simple distillation, extractive distillation which is mainly used to separate the azeotropes is more often adopted in the chemical industry. The theory of the extractive distillation is adding the third component as the entrainer of the system to change the relative volatility of azeotropes. Usually, the solvent which has high boil point, low relative volatility and would not generate new azeotropes with the other component is selected as the entrainer. However, the study of startup policies for the azeotropic mixture is quite rare in the open literature. The essential difficulty in modeling column startup lies in the fact that it is a quite complicated dynamic process. Further, most of the research use the customized formulation and did the simulation case by case in conventional distillation.

Isopropyl alcohol (IPA) is widely used in the semiconductor industry as a cleaning agent, thus the recovery of IPA from the waste solvent is very important. The waste solvent contains mainly IPA and water which exists a minimum-boiling azeotrope. Hence, it is hard to use the simple distillation to purify the IPA. Arifin and Chein (2008) have proposed an extractive distillation process via dimethyl sulfoxide (DMSO) as entrainer for dehydration of IPA process. The stream result and steady-state flowsheet are illustrated in Fig. 4.10. In this process, the extractive column (C1) has 41 stages. Feed location of IPA-water mixture and entrainer are at the 35th and 7th trays, respectively. Feed flow rate of IPA mixture and overall entrainer into extractive distillation are 100 and 102.5 kmol/h. The top product of C1 is IPA. The bottom stream of C1 is feeding into the 9th stage of entrainer recovery column (C2).

Table 4.2 All startup strategies

Strategy	Feed control	Bottom flow control	Reflux control	Heating method (T_m: heating medium temperature, $T_{m,ss}$: nominal heating medium temperature)	
A	Height of sump over its 90%, stop feeding; height of sump lower its 70%, start feeding	Wait for the specification of bottom product is reached, then start to produce	MT algorithm (switch point at 1.33 h	Constant	$T_m = 20$ °C → $T_{m,SS} = 140$ °C
B				Constant	$T_m = 20$ °C → 1.2 $T_{m,SS} =$ 168 °C → $T_{m,SS} = 140$ °C
C				Linear	$T_m = 20$ °C → $T_{m,SS} = 140$ °C
D				Linear	$T_m = 20$ °C → 1.2 $T_{m,SS} =$ 168 °C → $T_{m,SS} = 140$ °C
E			When $x_{CHCl3} = 50$ wt.%	Constant	$T_m = 20$ °C → $T_{m,SS} = 140$ °C
F				Constant	$T_m = 20$ °C → 1.2 $T_{m,SS} =$ 168 °C → $T_{m,SS} = 140$ °C
G				Linear	$T_m = 20$ °C → $T_{m,SS} = 140$ °C
H				Linear	$T_m = 20$ °C → 1.2 $T_{m,SS} =$ 168 °C → $T_{m,SS} = 140$ °C

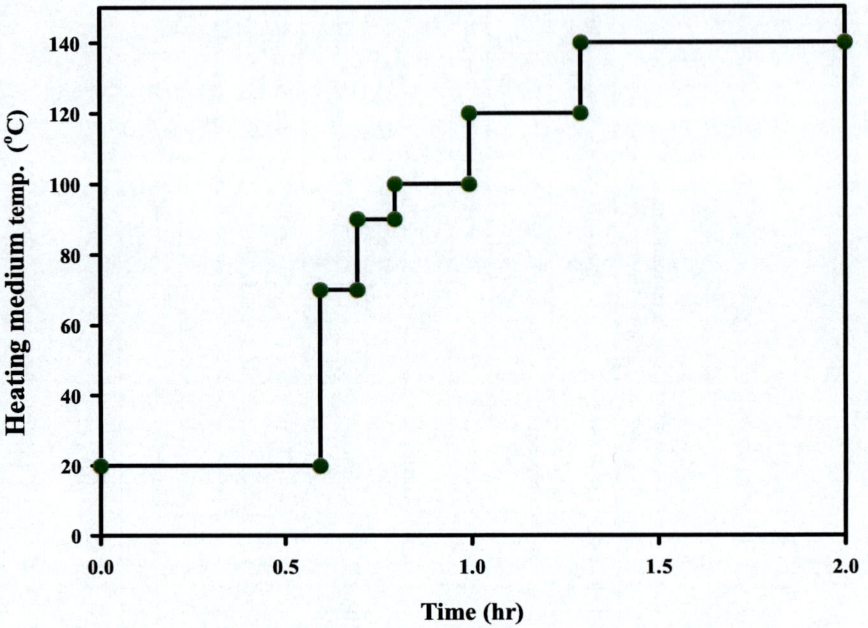

Fig. 4.4 The heating path for strategy A

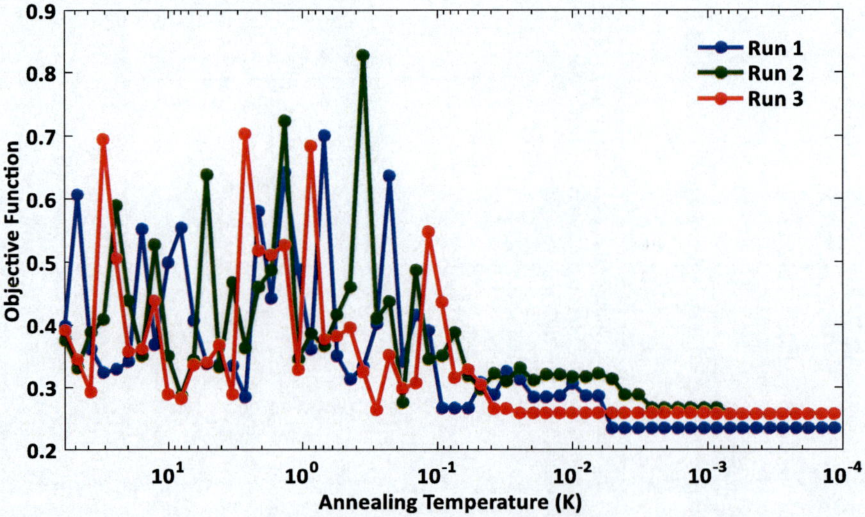

Fig. 4.5 The comparison of objective function by SA for strategy A

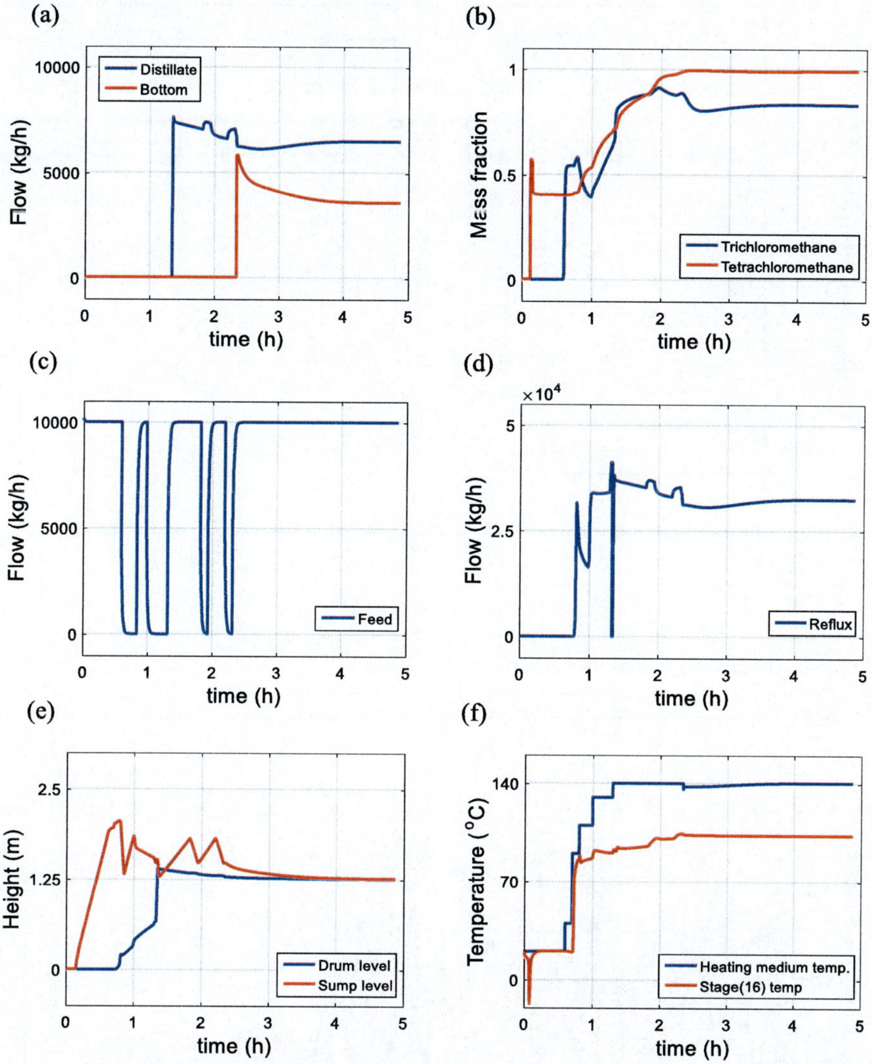

Fig. 4.6 The results of strategy A, **a** product flows, **b** product purities, **c** feed flowrate, **d** reflux flowrate, **e** reflux drum and sump levels, **f** temperatures of heating medium and the 16th stage

DMSO is recovered from the bottom of C2. Besides, the control scheme is shown in Fig. 4.11 which is used to operate in nominal condition.

Compared with the simple distillation, there are two different operations while start-up the DC. Thus, the above control scheme is not suitable. One of the differences is initial feed flow rate of DMSO. Because the make-up stream is too small to maintain the feed ratio between azeotrope mixture and entrainer, additional DMSO feed is

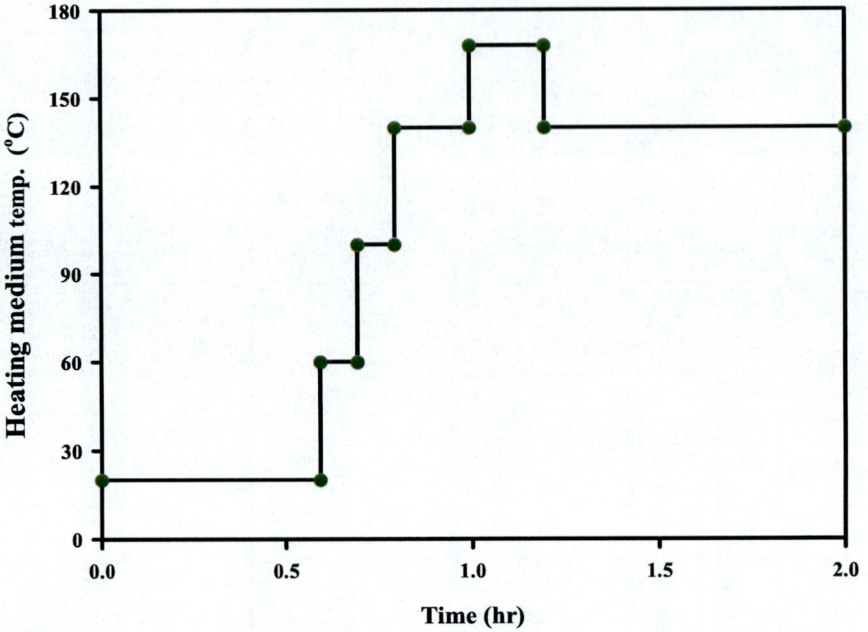

Fig. 4.7 The heating path for strategy F

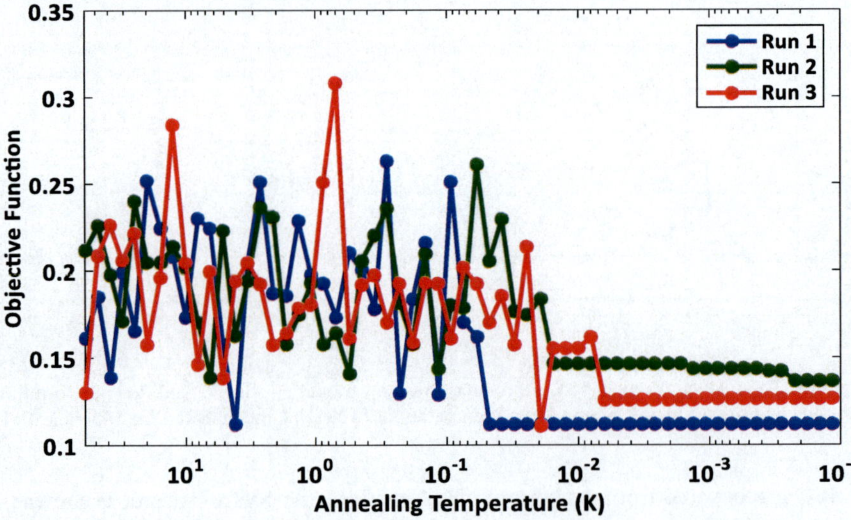

Fig. 4.8 The comparison of objective function by SA for strategy F

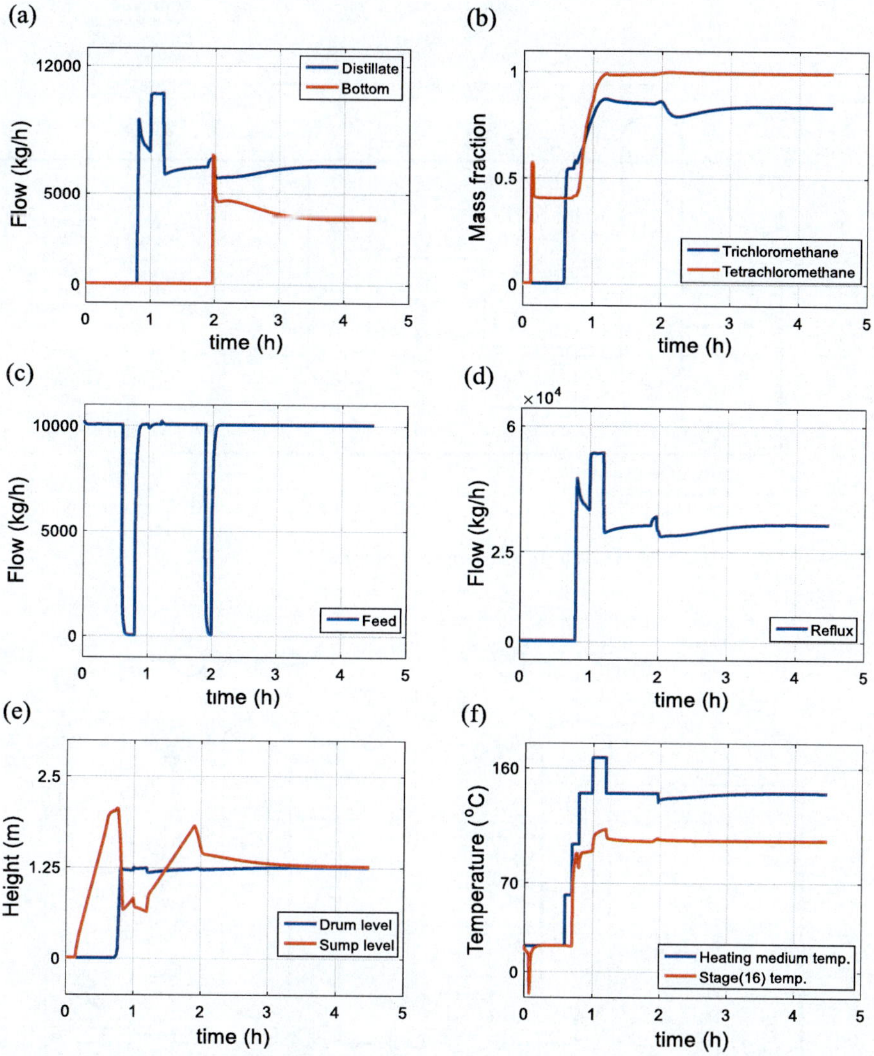

Fig. 4.9 The results of strategy F, **a** product flows **b** product purities, **c** feed flowrate, **d** reflux flowrate, **e** reflux drum and sump levels, **f** temperatures of heating medium and the 16th stage

required for the startup duration. The other difference is how the excess DMSO would be purged during the startup operation. The modified control scheme is shown in Fig. 4.12. There are five additional streams required for this process startup. Streams FN1 and FN2 are used to remove nitrogen during the C1 and C2 startup. The entrainer DMSO is fed by stream FEI at the beginning of startup. Because the total amount of entrainer cannot be estimated easily in the startup period, the excess entrainer DMSO needs partial drawn out to prevent accumulation in the whole process. Therefore, the

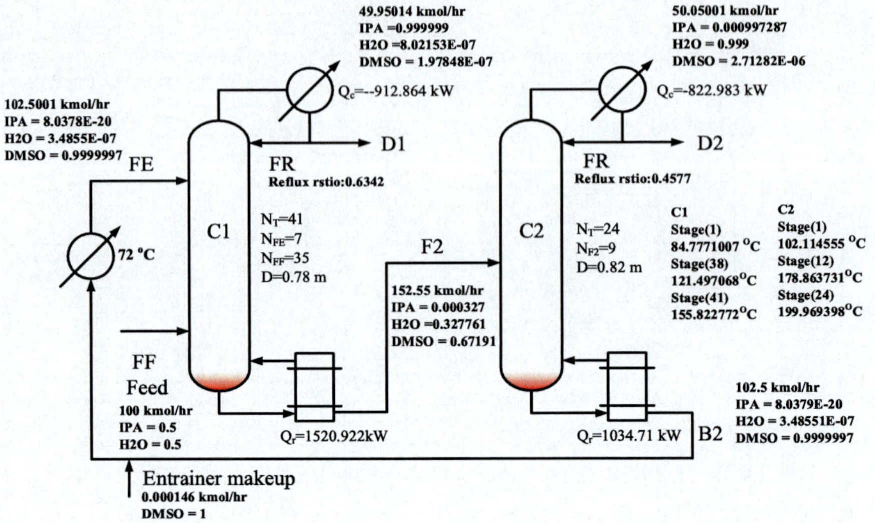

Fig. 4.10 Steady-state flowsheet of IPA dehydration process

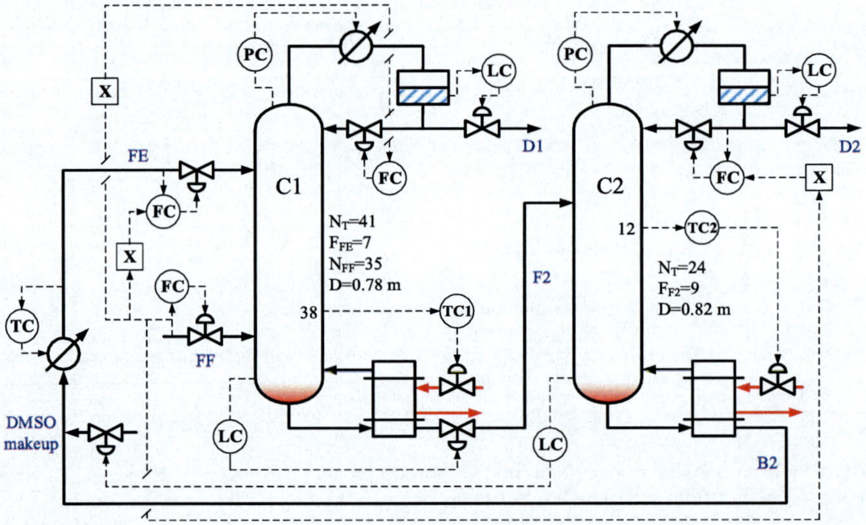

Fig. 4.11 Overall control scheme of IPA dehydration process

excess DMSO is designed to purge out by stream FEO at the bottom of C2. Unlike the nominal condition, the sump level of C2 is maintained by bottom stream flow rate. The reflux drum level of each column is controlled by its reflux flow rate in the startup period.

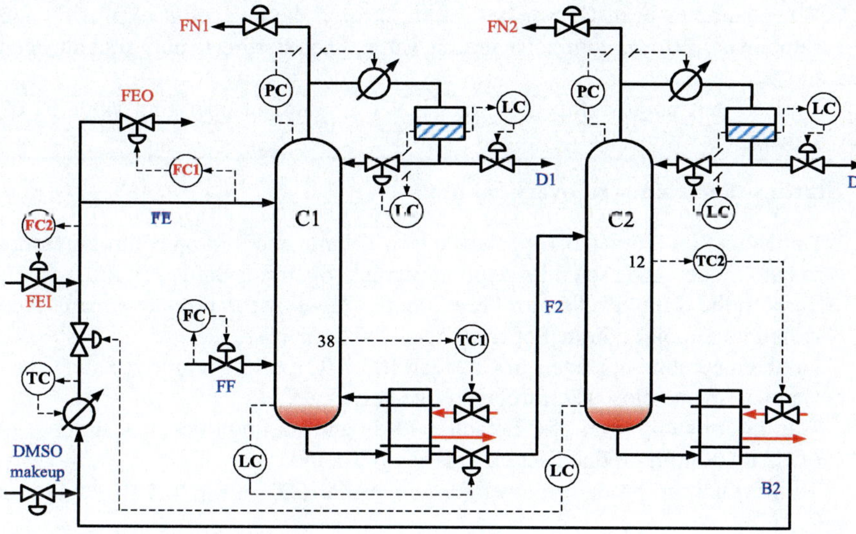

Fig. 4.12 Control scheme of IPA dehydration process for startup

Because the purities of two products are very high, the overall startup operation will be finished when both of the product concentrations are satisfied the following equation (Eq. 4.5).

$$\left|X_{p,i} - X_{p,iss}\right| \leq 0.001. \tag{4.5}$$

The total distillate strategy (Kruse et al. 1996) would be adopted in the following section first. The concept of total distillate policy is operating column without any reflux flow initially. After reflux flow switch time, the reflux flow is set as nominal operating value. The steps of startup are described as followed:

- **Startup of extractive column**

1. At t = 0.05 h, IPA/water mixture initial charge is started until sump level of C1 reach 1.5 m. Starting to heat column.
2. Nitrogen is started to purge by FN1 stream and turn on the pressure controller of C1. Setpoint is 1.1 atm.
3. Using top product stream of C1 to maintain reflux drum level, setpoint is 0.44 m.
4. Until temperature of stage 1 of C1 reach 82 °C (temp. of azeotrope), Starting to feed DMSO in 102.5 kmol/h.
5. Waiting for 0.05 h and closing nitrogen purge valve.
6. Until temperature of stage 1 of C1 reach nominal operating value 84.5 °C, starting to feed fresh IPA/water mixture continuously.

7. When sump level of C1 reaches 1.9 m, turn on the level control in auto, and setpoint is 1.26 m as nominal operating value. Then starting to purge out nitrogen of C2.
8. Until switch point, reflux flow rate is set as nominal operating value 31.67 kmol/h.

- **Startup of entrainer recovery column**

1. Until sump level reach 0.1 m, start to heat column and turn on control pressure in auto mode. And assign its setpoint nominal operating value 1.1 atm.
2. Turn on the drum level controller to control level by top product stream. And assign its setpoint nominal operating value 0.2663 m.
3. Until temperature of stage 1 of C2 reach 102.7 °C (boiling point. of water), then wait 0.05 h and close C2 nitrogen purge valve.
4. Waiting for sump level of C2 reaches 1.9 m and starting to control sump level of C2 by bottom product, set point is set as 1.2 m.
5. Turn on bottom product temperature controller (TC in Fig. 4.12), setpoint is 72 °C.
6. Turn on overall entrainer flow controller (FC1 in Fig. 4.12), setpoint is 102.5 kmol/h.
7. Until reflux switch point, reflux flow rate is set as nominal operating value 22.31 kmol/h.

The heating strategy is an important key factor for startup operation and could be divided into two different methods, open-loop operation and close loop operation. Energy supply is gradually increased in manual during all startup period is called open-loop operation. Some works of literature have studied how to optimize the heating steps of this strategy. The condition of the column may be easier to achieve nominal operating condition if temperature controller of the column is applied in some moment during the startup procedure is named as close loop operation. These two kinds of operations for reboiler heating are also considered. The steps of these two operations are shown as follows:

1. Reboilier duty of C1 and C2 is adjusted from 0, 50, 100 to 150% of its nominal operating values, and the interval of each change is set equal to 0.1 h.

TD-Case 1 Open-loop operation

2. Until temperature of sensitive stage equal to the nominal value, then reboilier duty is set as its nominal operating value directly.

TD-Case 2 Close loop operation

3. Until temperature of sensitive stage equal to the nominal value, then turn on column temperature controller in auto mode.

Because the total distillate policy does not have reflux flow at the beginning of startup, a switch time is required to turn on the reflux flow rate to its nominal operating value. Therefore, the reflux switch time selection of total distillate policy is determined by MT function which is presented in Eq. 4.1. The reflux switch time selection result of TD-Case 1 and TD-Case 2 are shown in Fig. 4.13. The switch point of C1 and C2 are 0.593 and 1.1 h for TD-Case 1, respectively. For TD-Case 2, the switch point of C1 is 0.596 h, and C2 is 1.06 h.

The overhead compositions of C1 for both TD-Case 1 and 2 are shown in Fig. 4.14. The red dash line represents the reflux switch point. It is obvious that the IPA composition of TD-Case 1 increase faster than TD-Case 2 at the beginning of startup. After 1.5 h, the trend of IPA composition become mildly increase for both cases.

At the beginning of both operations, it is found that the overheads contain a little DMSO. After the reflux switch time, DMSO composition is monotonic decrease. Until $t = 1.2$ h, DMSO becomes quite tiny in the overhead of C1.

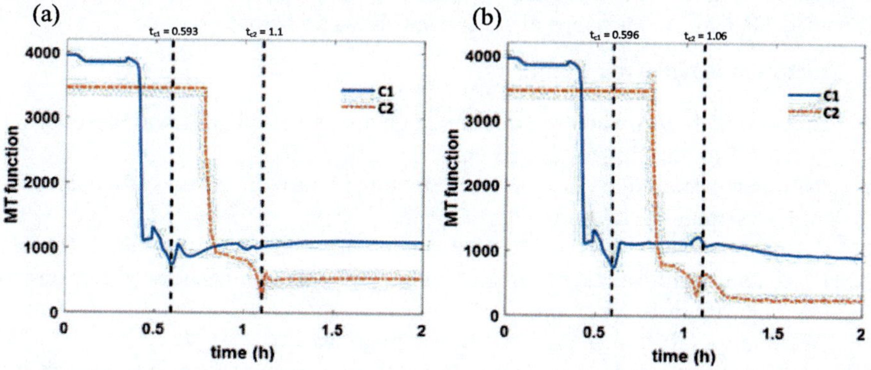

Fig. 4.13 Switch point selection of **a** TD-Case 1 open-loop strategy. **b** TD-Case 2 close loop strategy

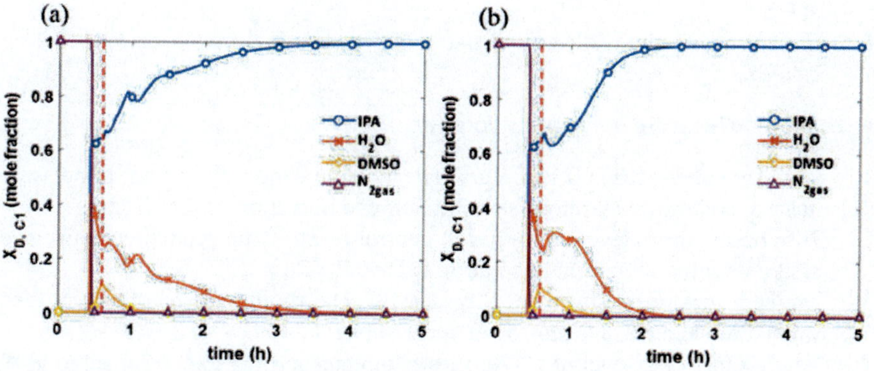

Fig. 4.14 Startup results of IPA composition of C1 overhead for **a** TD-Case 1 open-loop strategy. **b** TD-Case 2 close loop strategy

From the Fig. 4.14, the total startup period spends about 4.8 h for TD-Case 1. Comparing with TD-Case 2, it is interesting that the startup period is only half of TD-Case 1. It can be found that the behavior of TD-Case 2 is quite similar to TD-Case 1 before the reflux switch time. After switch point, Unlike IPA composition of TD-Case 1, the response goes down first and then up later. The response of TD-Case 2 can quickly increase to the nominal value by close loop operation. Comparing the result of TD-Case 1 with TD-Case 2, it is shown that the close loop operation has shorter startup period than open-loop operation.

Based on the results of the previous section, close loop operation is applied to total reflux policy (Ruiz et al. 1988). Total reflux policy is also a common approach for conventional column startup. Unlike total distillate policy, the process operated in total reflux at the beginning of startup before reflux switch time. After this switch point, the reflux flow is set as the nominal operating value, and the distillate is used to control reflux drum level. For reflux switch time selection, the MT-function method is named TR-Case 1. The steps of startup are shown in below.

- **Startup of extractive column**

1. At t = 0.05 h, IPA/water mixture initial charge is started until sump level of C1 reach 1.5 m. Starting to heat column.
2. Nitrogen is started to purge by FN1 stream and turn on the pressure controller of C1. Setpoint is 1.1 atm.
3. Using top product stream of C1 to maintain reflux drum level, setpoint is 0.44 m.
4. Until temperature of stage 1 of C1 reach 82 °C (temp. of azeotrope), Starting to feed DMSO in 102.5 kmol/h.
5. Waiting for 0.05 h and closing nitrogen purge valve.
6. Until temperature of stage 1 of C1 reach nominal operating value 84.5 °C, starting to feed fresh IPA/water mixture continuously.
7. When sump level of C1 reaches 1.9 m, turn on the level control in auto, and setpoint is 1.26 m as nominal operating value. Then starting to purge out nitrogen of C2.
8. Until switch point, reflux flow rate is set as nominal operating value 31.67 kmol/h.

- **Start-up of entrainer recovery column**

1. Until sump level reach 0.1 m, start to heat column and turn on control pressure in auto mode. And assign its setpoint nominal operating value 1.1 atm.
2. Turn on the drum level controller to control level by top product stream. And assign its setpoint nominal operating value 0.2663 m.
3. Until temperature of stage 1 of C2 reach 102.7 °C (boiling point. of water), then wait 0.05 h and close C2 nitrogen purge valve.
4. Waiting for sump level of C2 reaches 1.9 m and starting to control sump level of C2 by bottom product, set point is set as 1.2 m.
5. Turn on bottom product temperature controller (TC in Fig. 4.12), setpoint is 72 °C.

6. Turn on overall entrainer flow controller (FC1 in Fig. 4.12), setpoint is 102.5 kmol/h.
7. Until reflux switch point, reflux flow rate is set as nominal operating value 22.31 kmol/h.

The result of switch time and startup of TR-Case 1 are shown in Fig. 4.15. The switch point of C1 is t = 0.58 h and C2 is t = 0.98 h for TR-Case 1. The response shows that the composition of IPA would have faster raise up response at the beginning. However, the response of IPA composition goes down and becomes oscillatory when the composition of IPA is near 0.8. The startup time of TR-Case 1 is 2.54 h.

The comparison of startup behavior by total distillate and total reflux policy is shown in Fig. 4.16. It shows IPA composition increases slowly under the total distil-

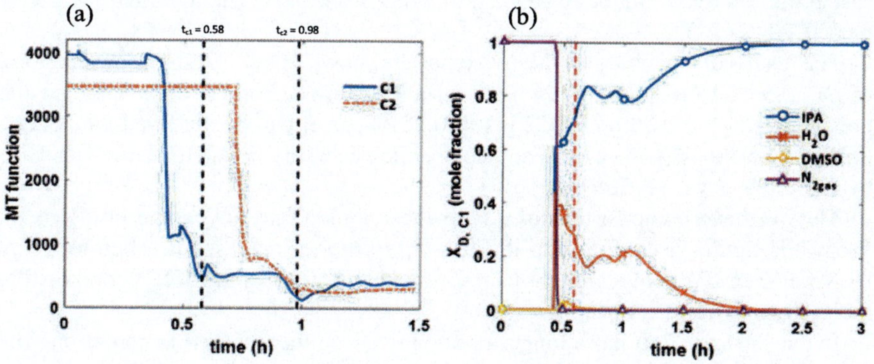

Fig. 4.15 Results of TR-Case1 **a** Switch point selection. **b** Startup result of IPA composition of C1 overhead

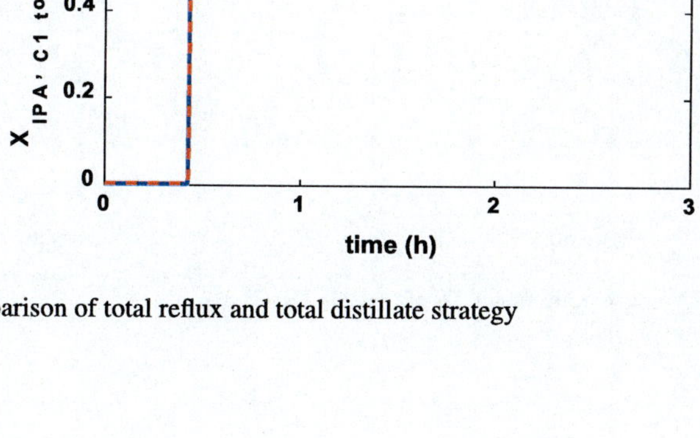

Fig. 4.16 Comparison of total reflux and total distillate strategy

late strategy. The response of IPA composition goes down and becomes oscillatory with total reflux policy during t = 0.7 to 1.1 h. Accordingly, the startup time of total reflux policy is shorter than total distillate policy.

Based on the above results, total reflux policy is more suitable than total distillate policy for the extractive distillation process. However, the reflux switch time selecting method, MT function, is not easy to be applied in the industrial column because temperatures of all stages need to be measured. Wendt et al. (2003) proposed another simple method to select the switch point. They observed the bottom temperature of the column and set the reflux switch time at the point while the temperature is not increasing. Another idea is modified method proposed by Wendt et al. (2003). The nominal value of C1 bottom temperature is selected as switch point. Since C1 bottom mainly contains water and DMSO mixture, the bottom temperature would not keep at a constant value except the composition is quite close to pure DMSO. Hence, the case proposed by Wendt et al. (2003) is named TR-Case 2 The modified method is named TR-Case 3.

The results of switch time with two cases are shown in Fig. 4.17. The switch point of C1 is t = 1.15 h and C2 is t = 0.97 h for TR-Case 2. For TR-Case 3, the switch point of C1 is t = 0.58 h and C2 is t = 0.98 TR hr. It can be observed that switch time of C2 for these two cases is quite close. However, the switch time of C1 in these two methods is very different.

The overhead compositions of C1 for these three cases are shown in Fig. 4.18. From this figure, it can be seen the IPA composition increases fast before 0.5 h. However after IPA composition is over 0.7 mol fraction, the trend of IPA composition increasing has various raring rates.

In the TR-Case 2, it has a longer period for C1 on the total reflux condition. The response of IPA composition in overhead becomes flatter than others. It is because the overhead composition is quite close to the IPA/water azeotrope, and this composition is always refluxed before the switch point. After switch point, the IPA composition in the overhead of C1 increases like an open-loop response of first-order transfer function to the product specification. The total startup time of TR-Case 2 is 2.58 h.

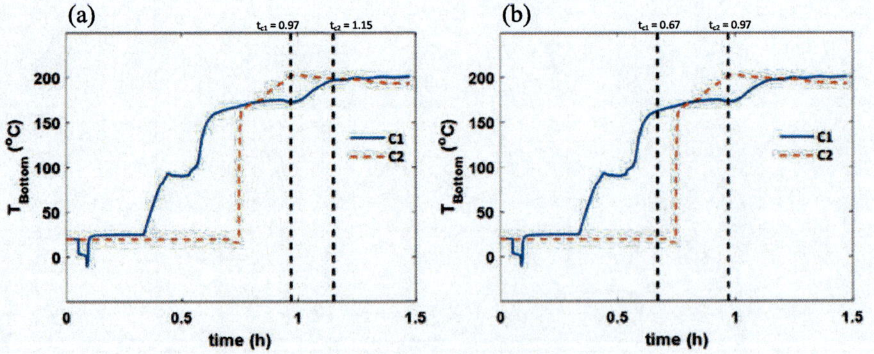

Fig. 4.17 Switch point selection of **a** TR-Case 2. **b** TR-Case 3

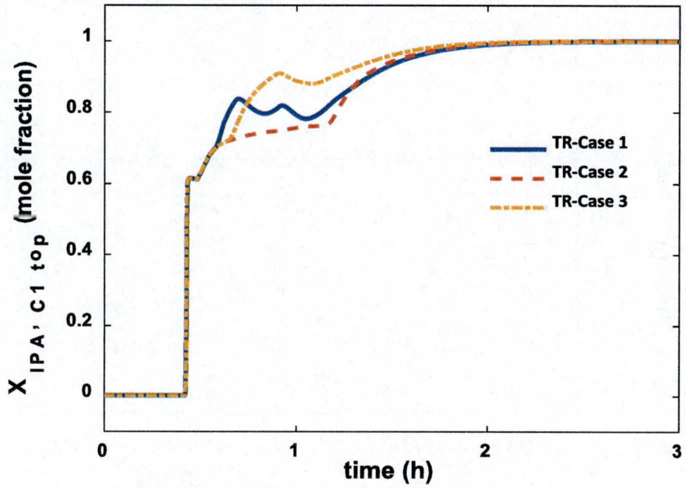

Fig. 4.18 Comparison of three cases of total reflux strategy

From Fig. 4.18, the responses show the startup period of TR-Case 1 goes down and becomes oscillatory. And the response of TR-Case 2 spends too long time at total reflux condition. The process studied by Wendt et al. (2003) was a simple binary mixture process. It means that the bottom temperature is quite close to the boiling point of the pure component when the bottom temperature does not increase in their process. However, for this process, there is still a mixture composited by water and DMSO in the bottom of the extractive DC. The TR-Case 3 provides a faster response and IPA composition does not decrease much than TR-Case 1. Thus, the reflux switch strategy of TR-Case 3 should be chosen.

The overall startup time and energy consumption during the startup period are demonstrated in Tables 4.3 and 4.4, respectively. In Table 4.4, the lower startup time

Table 4.3 Startup time of all tests

	Total distillate		Total reflux		
	TD-Case 1	*TD-Case 2*	*TR-Case 1*	*TR-Case 2*	*TR-Case 3*
t	4.8 h	2.63 h	2.54 h	2.58 h	2.41 h
%	99.2	9.1	5.4	7.1	–

Table 4.4 Energy consumption of all tests

	Total distillate		Total reflux		
	TD-Case 1	*TD-Case 2*	*TR-Case 1*	*TR-Case 2*	*TR-Case 3*
Q	36.7 GJ	19.4 GJ	14.58 GJ	17.97 GJ	13.61 GJ
%	169.6	42.5	7.1	32	–

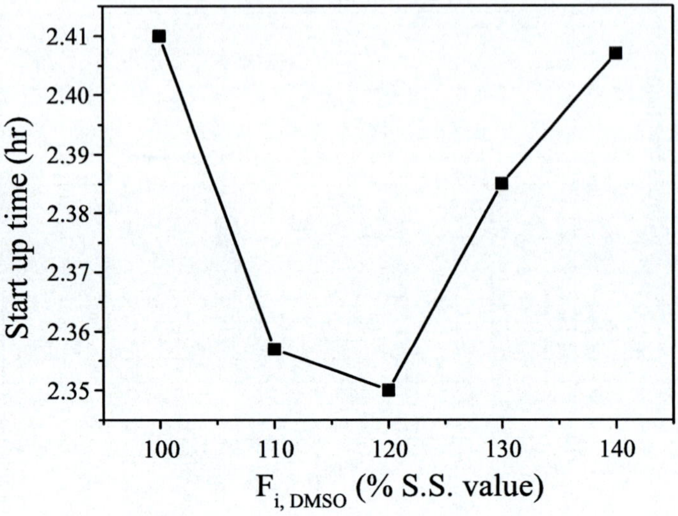

Fig. 4.19 Test result of initial entrainer flow rate

is with lower energy consumption. The startup time of the worst case is twice than the best one. Furthermore, the energy consumption of the worst case is triple than the best one.

For extractive distillation, the initial feed flow rate of entrainer is an important variable at the beginning of startup. In this section, various entrainer feed flow rates are investigated. Notice that the overall entrainer flow into C1 will be set as nominal operating value while entrainer from C2 recycle back to C1. At this condition, FEI stream will be closed, and excess DMSO would be purged by FEO stream. All simulations are based on the TR-Case 3 to observe the effects. The test result is shown in Fig. 4.19. The optimal entrainer flow rate is 120% of the nominal operating value at the beginning. If operating initial entrainer feed flow rate on this value, the startup time of TR-Case 3 could reduce from 2.41 to 2.35 h.

It is found that TR-Case 3 illustrates a better dynamic behavior and faster response to complete whole startup period than other cases. This startup policy could reduce total startup time near 99% than the total distillate policy in open loop operating. Furthermore, the energy consumption during the startup period can be reduced to 13.61 GJ. From the result, the entrainer initial flow rate should be set at 1.2 times of nominal operating value.

References

Arifin S, Chien IL (2008) Design and control of an isopropyl alcohol dehydration process via extractive distillation using dimethyl sulfoxide as an entrainer. Ind Eng Chem Res 47(3):790–803

Kirkpatrick S, Gelatt CD Jr, Vecchi MP (1983) Optimization by simulated annealing. Science 220:671–680

Kruse Ch, Fieg G, Wozny G (1996) A new time-optimal strategy for column startup and product changeover. J Proc Cont 6:187–193. https://doi.org/10.1016/0959-1524(95)00044-5

Metropolis N, Rosenbluth AW, Rosenbluth MN, Teller AH, Teller E (1953) Equation of state calculations by fast computing machines. J Chem Phys 21:1087–1092

Ruiz CA, Cameron IT, Gani R (1988) A generalized dynamic model for distillation columns III. Study of startup operations. Comput Chem Eng 12:1–14

Wendt M, Königseder R, Li P, Wozny G (2003) Theoretical and experimental studies on startup strategies for a heat-integrated distillation column system. Chem Eng Res Des 81:153–161

Woinaroschy A (2008) Time-optimal control of startup distillation columns by iterative dynamic programming. Ind Eng Chem Res 47:4158–4169

Yasuoka H, Nakanishi E, Kunigita E (1987) Design of an on-line startup system for a distillation column based on a simple algorithm. Inter Chem Eng 27:466–472

Chapter 5
Petri Net-Based Operating Procedures

The mathematical representation of the ordinary Petri net is provided by Peterson (1981). As initially planned, it is just made out of three kinds of components, i.e., a set of discrete places P, a set of discrete transitions T, and a set of interconnected normal arcs A. A discrete place is graphically communicated in the Petri net with a circle and a discrete transition is with a bar. A normal arc is represented with a directed solid line. It interfaces either a place to a transition or vice versa. To encourage proper characterization of the material-transfer patterns in pipeline networks, additional extensions are also used in this work, i.e., the weighted arcs, the inhibitor arcs, and the static test arcs. The execution of a Petri net is controlled according to the numbers and dissemination of tokens within the places of Petri net. The vector of all token numbers at a particular instance is alluded to as a marking. A more point by point review of these and other Petri net elements and also the transition enabling and firing rules can be found elsewhere (David and Alla 1994; Wang and Chang 2004). On the premise of these crucial advancements, a systematic approach has been proposed by Wang and Chang (2003, 2004) to build Petri nets for modeling the batch operations. In specific, a system model can be amassed with a progression of four different levels of components. In any existing batch process, the first-level component (which is usually a programmable logic controller or a human operator) is utilized to execute the operating steps indicated in a recipe on the basis of a predetermined time schedule or a set of sensor measurements. Its actions modify the states of valves, pumps, and compressors in the second level. The states of these components in turn decide the process configuration and, thus, the operation mode and equipment condition of each process unit in the third level. Finally, these process states are monitored via sensors within the final level, which may or may not be utilized as the premise for further controller actions.

C.-T. Chang et al., *Process Plant Operating Procedures*, Advances in Industrial Control,
https://doi.org/10.1007/978-3-030-70978-5_5

5.1 Cleaning Operation of a Pipeline Network

Cleaning pipelines is one of the normal tasks that must be acted in any chemical plant. In continuous operations, the removal of scales and/or sediments collected during regular activity is essentially a periodic housekeeping procedure. The cleaning routines, on the other hand, may play a more crucial role in keeping a batch process operable. Notice that the current units are mostly shared by several different processes in a multi-product batch plant and the raw materials, intermediates and products of these processes must be transported from one unit to another via a common pipeline network. In order to prevent contamination of foreign substances, the pipeline network needs to be washed, disinfected, or purged prior to each batch cycle. The "cleaning" method can basically be seen as the operation steps to move a detergent, a disinfectant or an inert substance from the inlets (sources) to the outlets (sinks) of a pipeline network, and any part of the system should also be protected by the transfer routes. The tasks of identifying all possible cleaning routes and then synthesizing the corresponding operating procedure are usually performed on an ad hoc basis manually. For a complex chemical process, the demand for time and effort for these tasks may be daunting, and the resulting formula is often prone to error. Therefore, it is highly desirable to establish a systematic strategy to correctly and efficiently synthesize the necessary operating procedures in order to alleviate the workload and also to improve the cleaning efficiency.

In the literature, you can find many similar studies. Rivas and Rudd (1974) suggested a system for the synthesis of failure-safe procedures in a seminal study to assist operators in emergency situations to make effective decisions. To meet the specified operation target, a valve operation sequence can be calculated quickly. With a more successful solution technique, O'Shima (1978) managed this problem. The author developed algorithms for finding the routes between a material stream's defined starting and ending points and also for evaluating the flow state along the stream in each unit. On the basis of those algorithms, the operating procedures were then synthesized. The state of fragments in a plant structure with a set of condition lists was described by Foulkes et al. (1988). A mix of artificial intelligence techniques, pattern matching and path search algorithms were used to classify all possible routes for the transfer of a given material from one storage.

It should be noted that the above-mentioned approaches in the present application are not directly applicable. First of all the purpose of a cleaning procedure is in general not the same as that of simple material transfer. In the former case, it is necessary to ensure that all parts of the pipeline network are included in the routes of material transfer, while this requirement is not enforced in the latter. Second, with the available methods for achieving a multi-route cleaning schedule, it is difficult to generate valve-sequencing steps. Thus, the design of operating procedures to conduct multiple material-transfer tasks for the cleaning of the entire network can be applied with the Petri net approach.

5.1.1 Representation of Material-Transfer Paths

The division of the system into distinct components should be the first critical issue in modeling any network. To this end, the idea of piping fragments (Foulkes et al. 1988) is adopted in this work. A fragment is defined, in particular, as a set of pipeline branches and/or processing units separated by valves, pumps, and other means of flow blockages in the pipeline network from other fragments (or the environment). As an example, let us consider Fig. 5.1. According to this description, eight fragments can be identified, i.e., FR1–FR8. In this case, each pump and its insulation valves are seen as one lumped power-generating system, and if it is switched off this system is regarded as a flow blockage. Also, note that the pipeline networks contain dead branches in several industrial plants. Blanks, slip plates and/or closed and locked valves typically separate these divisions from the external atmosphere.

Let us first discuss the most fundamental fragment structure, i.e. a pipe branch separated by an inlet valve and an outlet valve, for illustrative convenience (see Fig. 5.2a). Note that in this case, only in one direction is the flow permitted in either valve. In Fig. 5.2b, the corresponding Petri net model is presented. The FR location in this model is used to represent the state of the fragment. More precisely, a token entering such a position denotes the condition that the detergent from an upstream source fragment is delivered to the subsequent fragment. PK1 is used to record the connection status of FR with its upstream fragment, and PK2 is used for the same purpose with respect to the downstream fragment of FR. The T1 and T2 transitions can be regarded as the operator/controller acts to open the V1 and V2 valves, respectively. On the other hand, the fragment model shown in Fig. 5.2b can

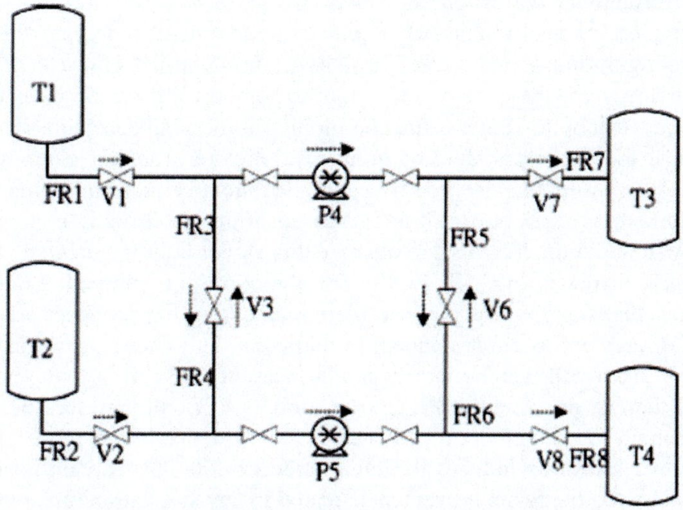

Fig. 5.1 A typical example of pipeline network (reprinted with permission from Chou and Chang 2005. Copyright 2005 American Chemical Society)

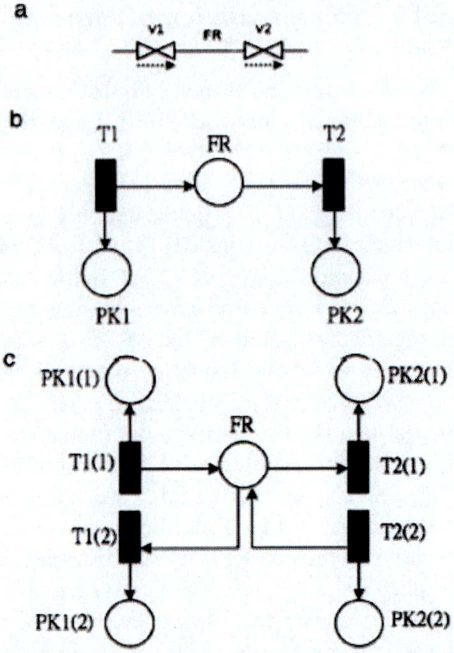

Fig. 5.2 **a** Basic structure of a piping fragment with two single-direction valves, **b** Petri net Model of a basic fragment with two single-direction valves, **c** Basic structure of a basic fragment with two bidirection valves (reprinted with permission from Chou and Chang 2005. Copyright 2005 American Chemical Society)

be modified to the one shown in Fig. 5.2c if both valves permit bidirectional material transfer. Note that any transition in the former Petri net is now replaced by two transitions in order to signify the behavior of material transfer to and from the FR fragments through the corresponding valve.

In theory, in a Petri net assembled by connecting fragment models according to the network configuration, all mass-transfer paths can be identified. However, if there are bidirectional valves in the system in question, one or more infinite loops can be found. Obviously, these looping routes cannot be adopted as the candidate paths for cleaning operations. In the Petri net model, additional restrictions are thus put to remove certain possibilities. In particular, any record-keeping location, i.e., a location marked with "PK", is linked to the input transitions of the location representing its downstream fragment with inhibitor arcs. To demonstrate this model-building practice, let us use the pipeline network in Fig. 5.1. In Fig. 5.3, the corresponding path model can be found. Note that for the sole purpose of implementing the proposed restrictions, the inhibitor arcs in this net are added. In particular, let us take transition T1 as an example. A token will then be introduced in location FR3 if it is shot, and another token will then be deposited in PK1. Transition T 3(2) will therefore be inhibited and will remain so even after a token enters FR4. Note that in this model, the firing of every other transition induces the same effects on its downstream positions. In this Petri network, the token movements created in any simulation run are therefore bound to follow a loop-free path. If the inhibitor arcs in Fig. 5.3 are all removed, a

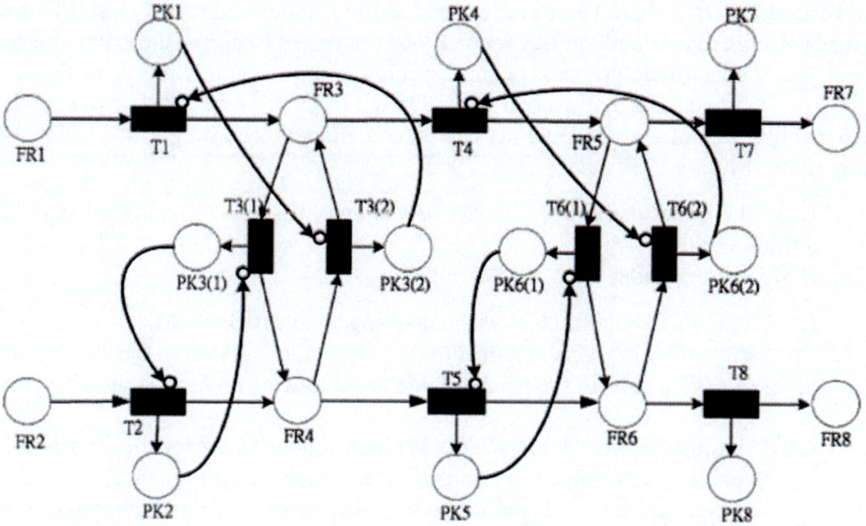

Fig. 5.3 Path model of the example network in Fig. 5.1 (reprinted with permission from Chou and Chang 2005. Copyright 2005 American Chemical Society)

token may travel endlessly in one of the following two loops: (1) FR3 → FR4 → FR3 → FR4 … and (2) FR5 → FR6 → FR5 → FR6 ….

5.1.2 *Enumeration of Possible Routes*

Since there may be more than one route emanating from a particular source fragment to the sink fragments of a pipeline network, it is desirable to first identify all of them to ensure thoroughness of the cleaning operation. This task can be achieved by constructing a reachability tree from a given initial condition on the basis of the Petri net model. Specifically, by firing the transitions enabled initially, one can first obtain as many "new" markings as the number of the fired transitions. From each new marking, one can then again generate more markings with the same approach. Repeating this procedure over and over results in a reachability tree. This tree consists of nodes and arcs. Other than the node representing the initial state, each node is associated with a generated marking and its input arc denotes the corresponding fired transition. All nodes in the tree can be classified into four different types: (1) frontier nodes, (2) interior nodes, (3) duplicate nodes, and (4) terminal nodes. The frontier nodes are nodes that have not yet been developed by the tree-building algorithm, whereas the interior nodes are processed nodes. The duplicate nodes are the ones that have appeared more than once in the tree. The terminal nodes are nodes that cannot lead to any enabled transition. It should be noted that the construction of reachability tree should continue as long as the frontier nodes still exist. In other words, every frontier node must eventually be converted to one of the other nodes.

The tree construction process begins by defining the initial marking to be the root node of the tree and also a frontier node. On the basis of breadth-first strategy, the reachability tree of a given Petri net can be constructed according to the following algorithm (Murata 1989; Wang et al. 2002):

1. Label the initial marking $\mathbf{M}_0$ as the root node of the tree and, initially, tag it as a frontier node.
2. If the frontier nodes exist, do the following:
   (a) Select a frontier node. Let the marking of this node be **M**.
   (b) If the marking **M** is identical to that of an existing node in the constructed tree, then convert the frontier node to a duplicate node and then go to step 2(a).
   (c) Use the revised Murata's (1989) state equation to obtain all possible enabled transitions. If no transitions are enabled for the marking M, then convert the selected frontier node to a terminal node and then go to step 2(a). If the enabled transitions can be identified, then select each enabled transition t_j as a firing transition t_f and carry out the following tasks repeatedly:
3. Obtain the marking $\mathbf{M}_k$ by firing t_f.
4. Include $\mathbf{M}_k$ as a node, draw a directed arc with label t_f from **M** to $\mathbf{M}_k$, and tag $\mathbf{M}_k$ as a frontier node.
5. Remove the original tag from **M** and tag it as an interior node.

Let us now turn to the Petri net model presented in Fig. 5.3. By assuming that the detergent is stored in tank T1 initially, the corresponding reachability tree (shown in Fig. 5.4) can be generated according to the above algorithm. The markings associated with the nodes in this tree can be found in Table 5.1. In order to conveniently identify the elements in a marking, the token numbers are into 2 subsets and arranged sequentially in a classified vector, i.e.,

$$\mathbf{M}_k = [\mathbf{FR}_k | \mathbf{PK}_k]$$

and $k = 0, 1, 2, \ldots, 10$. Here, each subset label is identical to the place labels of its elements. For example, the elements stored in $\mathbf{FR}_k$ are the token numbers of places representing the fragment states. Based on this convention, useful information can be directly acquired from the marking of each terminal node in the reachability tree. Specifically, the sink fragment of a material-transfer route should be associated with 1 in subset $\mathbf{FR}_k$ of the terminal node. The corresponding connection status among various fragments in the network can be identified from $\mathbf{PK}_k$. From the reachability tree given in Fig. 5.4 and its markings in Table 5.1, it can be seen that there are four terminal nodes, i.e., $\mathbf{M}_4$, $\mathbf{M}_7$, $\mathbf{M}_9$, and $\mathbf{M}_{10}$. Four material-transfer routes can be identified accordingly, i.e.,

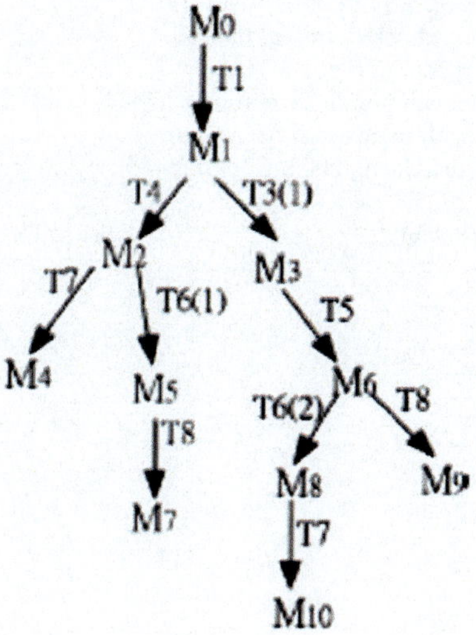

Fig. 5.4 Reachability tree of the path model in Fig. 5.3: starting nonempty FR1 (reprinted with permission from Chou and Chang 2005. Copyright 2005 American Chemical Society)

Table 5.1 Markings $\mathbf{M}k$s (= [$\mathbf{FR}_k$ I$\mathbf{PK}_k$]) of the reachability tree in Fig. 5.4

k	FR_k	PK_k
0	{10000000}	{0000000000}
1	{00100000}	{1000000000}
2	{00001000}	{1000100000}
3	{00010000}	{1010000000}
4	{00000010}	{1000100010}
5	{00000100}	{1000101000}

$$FR1 \rightarrow FR3 \rightarrow FR5 \rightarrow FR7 \tag{5.1}$$

$$FR1 \rightarrow FR3 \rightarrow FR5 \rightarrow FR6 \rightarrow FR8 \tag{5.2}$$

$$FR1 \rightarrow FR3 \rightarrow FR4 \rightarrow FR6 \rightarrow FR5 \rightarrow FR7 \tag{5.3}$$

$$FR1 \rightarrow FR3 \rightarrow FR4 \rightarrow FR6 \rightarrow FR8. \tag{5.4}$$

Similarly, another reachability tree can be built from the second source fragment FR2 of the Petri net given in Fig. 5.3. This tree is presented in Fig. 5.5 and the markings of its nodes can be found in Table 5.2. Four more material-transfer routes can be found in this tree, i.e.,

$$FR2 \rightarrow FR4 \rightarrow FR6 \rightarrow FR8 \tag{5.5}$$

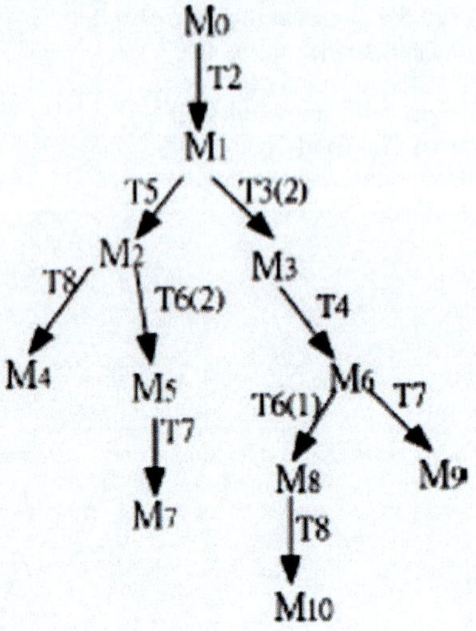

Fig. 5.5 Another reachability tree of the path model in Fig. 5.3: starting nonempty FR2 (reprinted with permission from Chou and Chang 2005. Copyright 2005 American Chemical Society)

Table 5.2 Markings $\mathbf{M}_k$ s (= [$\mathbf{FR}_k$ |$\mathbf{PK}_k$]) of the reachability tree in Fig. 5.5

k	FR_k	PK_k
0	{0 1 0 0 0 0 0 0}	{0 0 0 0 0 0 0 0 0 0}
1	{0 0 0 1 0 0 0 0}	{0 1 0 0 0 0 0 0 0 0}
2	{0 0 0 0 0 1 0 0}	{0 1 0 0 0 1 0 0 0 0}
3	{0 0 1 0 0 0 0 0}	{0 1 0 1 0 0 0 0 0 0}
4	{0 0 0 0 0 0 0 1}	{0 1 0 0 0 1 0 0 0 1}
5	{0 0 0 0 1 0 0 0}	{0 1 0 0 0 1 0 1 0 0}

$$\mathrm{FR2} \to \mathrm{FR4} \to \mathrm{FR6} \to \mathrm{FR5} \to \mathrm{FR7} \tag{5.6}$$

$$\mathrm{FR2} \to \mathrm{FR4} \to \mathrm{FR3} \to \mathrm{FR5} \to \mathrm{FR6} \to \mathrm{FR8} \tag{5.7}$$

$$\mathrm{FR2} \to \mathrm{FR4} \to \mathrm{FR3} \to \mathrm{FR5} \to \mathrm{FR7}. \tag{5.8}$$

5.1.3 Route Selection Procedures

Although every material-transfer route identified from the reachability trees of a Petri net model can be adopted to clean a portion of the given pipeline network, it may not be necessary to include all of them to achieve the operation objective. The task of cleaning a pipeline network is considered to be accomplished if the detergent is transported either (1) through every fragment at least once or (2) across every blockage at least once. Notice that the former criterion may result in a less rigorous operating procedure than the latter. This is because the requirement of moving material through a pipeline fragment only guarantees continuous fluid flows in its inlet and outlet branches. On the other hand, since by definition there is always a blockage on every branch of a fragment, the quality of the cleaning operation can be better ensured with the second criterion. However, it should be noted that the dead branches in the pipeline network are always not cleanable according to this blockage-based criterion. In this case, the blanks, slip plates, or valves on all dead branches may have to be removed/opened to allow the needed detergent flows.

5.1.3.1 Sequential Operations

A systematic procedure has been developed in this study to select the appropriate material-transfer routes so that either one of the above two criteria can be satisfied with sequential cleaning operations. For illustration convenience, let us consider all possible cleaning routes of the pipeline network given in Fig. 5.1, i.e., routes (5.1)–(5.8). These routes can be arranged in a matrix form as shown in Table 5.3. Notice that its columns are associated with the fragments in a pipeline network and each row represents a route. This matrix is referred to as a path matrix in this paper. The route selection procedure can be viewed conceptually as the symbolic manipulation steps to identify the independent vectors that can "span" the row space of path matrix. In order to reduce the number of cleaning routes as much as possible, a simple heuristic rule is used in the proposed selection procedure, i.e., the route having the largest number of fragments should be chosen first. Although the resulting routes may not be optimal, this approach is taken because of the relative easiness in implementation. The specific selection steps are presented in the sequel:

1. Select a row in the path matrix with the largest number of elements.
2. Delete the row identified in step 1.

Table 5.3 A fragment-based path matrix

Route no	FR1	FR2	FR3	FR4	FR5	FR6	FR7	FR8
1	0		0		0		0	
2	0		0		0	0		0
3	0		0	0	0	0	0	
4	0		0	0		0		0
5		0		0		0		0
6		0		0	0	0	0	
7		0	0	0	0	0		0
8		0	0	0	0		0	

3. Delete all columns in which the elements of the row identified in step 1 are located.
4. Repeat steps 1 to 3 until all columns are deleted.

Let us now apply this procedure to Table 5.3. The detailed implementation steps are marked in Fig. 5.6. A brief description of these steps is provided below (1) Select and then delete the 3rd row since it contains the maximum number of elements; (2) Delete columns 1 and 3–7 since row 3 contains elements in these columns; (3) Select and then delete row 5 since it contains the most elements in the remaining matrix obtained after carrying out the previous two steps; (4) Delete column 2 and column 8 since row 5 contains elements in these two columns and all other columns have already been deleted before. Notice that the selection process is terminated after all columns have been eliminated in step (4). From Fig. 5.6, it is clear that the first cleaning criterion can be satisfied by selecting routes (5.3) and (5.5).

It should be noted that a similar path matrix can be constructed if routes (5.1)–(5.8) are expressed in terms of the removed blockages. As mentioned before, a feasible route can be identified from the marking of a terminal node in the reachability tree. Specifically, the elements of subset PKk in this markings reflect the states of valves, pumps, and compressors in the pipeline network to facilitate material transfer via the corresponding route. For example, route (5.1) can also be written as

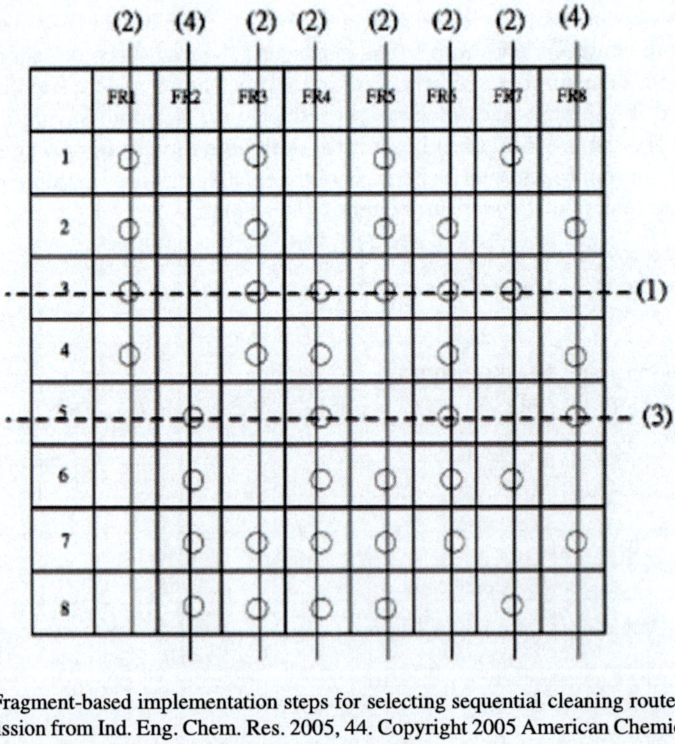

Fig. 5.6 Fragment-based implementation steps for selecting sequential cleaning routes (reprinted with permission from Ind. Eng. Chem. Res. 2005, 44. Copyright 2005 American Chemical Society)

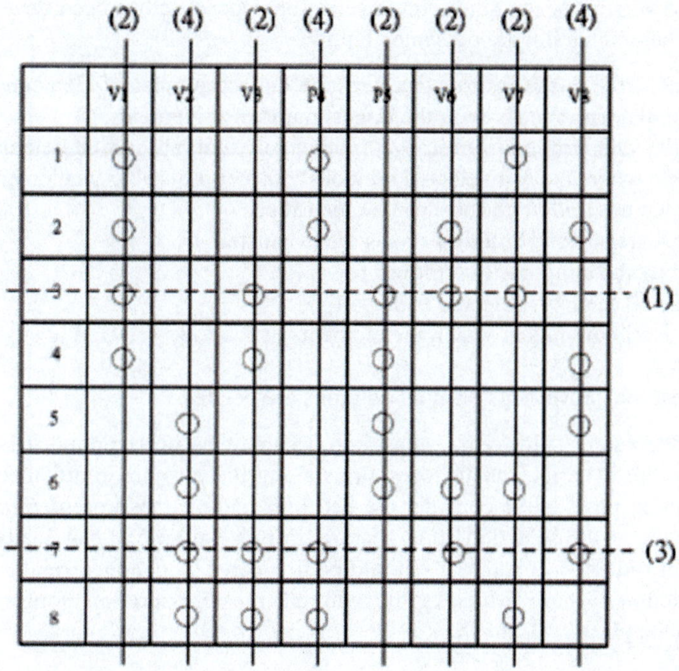

Fig. 5.7 Blockage-based implementation steps for selecting sequential cleaning routes (reprinted with permission from Chou and Chang 2005. Copyright 2005 American Chemical Society)

$$V1 \rightarrow P4 \rightarrow V7. \tag{5.9}$$

Consequently, the same procedure can be followed to select the routes that satisfies the second criterion. From the implementation steps presented in Fig. 5.7, one can see that routes (5.3) and (5.7) should be chosen in this case.

5.1.3.2 Concurrent Operations

Note that the chosen material-transfer routes may be partially overlapped by the above method. For example, if the fragment-based criterion is used the pipeline network routes chosen (5.3) and (5.5) in Fig. 5.1 share two common elements, i.e., FR4 and FR6. As another example, note that valves V3 and V5 are both needed to be open in order to facilitate material transfer through route (5.3) and route (5.7) in the same pipeline network. Due to this possibility of overlapping routes, it is important to sequentially carry out the cleaning operations of different routes. Therefore, there are incentives to define non-overlapping routes in order to save operating time so that the concurrent cleaning methods can be formulated accordingly. Thus, for this reason, a modified version of the above route selection procedure has been developed. Its steps of implementation are presented below.

1. Select a row in the path matrix with the largest number of elements. Select a row in the path matrix with the largest number of elements.
2. Identify and temporarily remove another row with at least one element in the column where the row selected previously in step 1 also has an element. Repeat this step until no more rows can be identified.
3. Repeat steps 1 and 2 until all rows are exhausted.
4. Recover the temporarily removed rows.
5. Delete all the rows selected in steps 1–3.
6. Delete all columns in which the elements of the rows selected in steps 1–3 are located.
7. Repeat the above steps until all columns are deleted.

Since the routes selected in steps 1 to 3 cannot be overlapping, it is possible to execute the material-transfer operations along these routes simultaneously. For illustration purpose, let us consider the fictitious path matrix presented in Fig. 5.8. On the basis of the selection steps marked in this figure, one can easily see that cleaning of routes (5.1) and (5.2) should be first carried out concurrently and, after the completion of these two tasks, the material-transfer operation along route (5.4) can then take place.

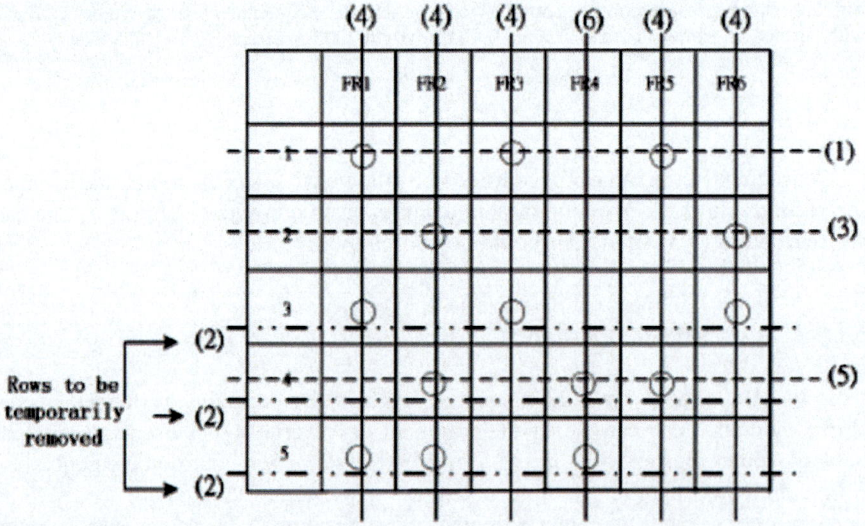

Fig. 5.8 Implementation steps for selecting concurrent cleaning routes (reprinted with permission from Chou and Chang 2005. Copyright 2005 American Chemical Society)

5.1.4 Equipment Models

The Petri net model of a pipeline network must include not only the component models of the fragments, but also those of the mounted devices, such as valves, pumps, and compressors in order to generate the specific operational steps to carry out the cleaning tasks of the chosen material-transfer routes. In Fig. 5.9, the valve model is presented. The PV(O) and PV(C) locations here show two opposite positions of the valve, i.e., open and close. The TV(O) and TV(C) transitions reflect the valve-switching behavior between PV(C) and PV(O) and vice versa. Note that the input position PC(O) of the transitions TV(O) can be interpreted as the valve-opening requirement once a particular material-transfer action is selected in a route. Similarly, the place PC(C) can be considered as the demand for valve-closing operation. The output places PR(O) and PR(C) of the two transitions TV(O) and TV(C) can be used to record the actual number of times that the corresponding valve-switching actions have been carried out. Since it is possible to call for a material-transfer action when the corresponding valve is already open, transition TR(O) is introduced as the output of both PC(O) and PV(O) in this model. A normal arc is adopted in the former case to avoid a token being permanently kept in place PC(O), while a test arc is used in the latter to prevent loss of the tokens in PV(O). Finally, note that the transition TR(C) is adopted for the same reason.

Since the operating procedures of pumps, compressors and their isolation valves can be considered as well-established industrial practices, e.g., see Karassik and McGuire (1998), their detailed steps are not described in the equipment models for the sake of simplicity. Specifically, the Petri net presented in Fig. 5.9 is also used to model a power-generating system in this study. In this case, the places PV(O) and PV(C) represent two opposite states, i.e., on and off, of the system, respectively.

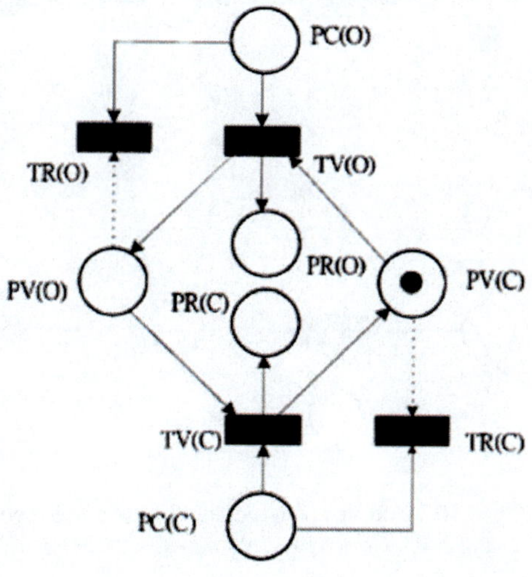

Fig. 5.9 Standard valve model (reprinted with permission from Chou and Chang 2005. Copyright 2005 American Chemical Society)

The transitions TV(O) and TV(C) can be regarded as a series of standard operation actions to turn on and off the pump/compressor system.

5.1.5 Generation of Operation Steps

Based on a system model, the operating procedure for cleaning a chosen route in the pipeline network can be obtained. The equipment models should be one-by-one connected to an updated Petri net representation of the material-transfer paths in order to construct such a model. By eliminating all inhibitor arcs and then reversing the paths of all connecting arcs between the record-keeping places, (i.e., the places labeled with "PK") and their input transitions, this updated route model can be converted from its original version. It is now possible to interpret the above-mentioned positions in the updated net as the requirements to link the corresponding fragments upstream and downstream. Let us consider the device as an example in Fig. 5.1 again. The path model in Fig. 5.3 can be converted to the Petri net presented in Fig. 5.10 by introducing the proposed modifications. This net is then expanded by connecting its transitions, respectively, to the places PC(O)s in the corresponding equipment models with normal arcs. A detailed listing of these connections can be found in Table 5.4. This practice is meant to reflect the relationship between each material-transfer action and the need to open the corresponding valve or to turn on the corresponding pump/compressor.

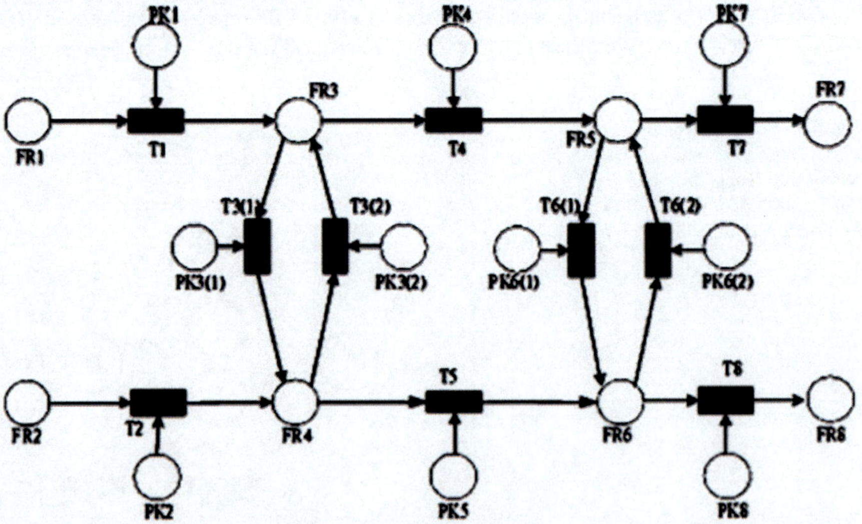

Fig. 5.10 Modified path model of the example network (reprinted with permission from Chou and Chang 2005. Copyright 2005 American Chemical Society)

Table 5.4 Normal-arc connections between transitions in modified path model (Fig. 5.10) and places *PC*(*O*)s in different equipment models (Fig. 5.9)

Transition	Equipment
$T1$	$V1$
$T2$	$V2$
$T3(1)$	$V3$
$T3(2)$	$V3$
$T4$	$P4$
$T5$	$P5$
$T6(1)$	$V6$
$T6(2)$	$V6$
$T7$	$V7$
$T8$	$V8$

Given a route and a set of initial valve states, the required operation steps can be synthesized by performing simulation with the system model. Under the condition that all valves are closed and all pumps are switched off initially, it can be determined easily by inspecting the pipeline network in Fig. 5.1 that valves V2 and V8 should be open and also pump P5 must be turned on for transporting material through route (5.5). On the other hand, notice that the initial system condition of the Petri net model can be set by introducing a token in FR2 and also in the place representing the close position of every valve in the system. Notice also that the selected route (5.5) can be stipulated by providing a token in every place representing the demand to connect a pair of fragments in route (5.5), i.e., PK2, PK5, and PK8. A collection of operation steps can then be identified by executing the Petri net-based simulation accordingly. It has been verified that the simulation results are the same as those obtained by inspection.

However, if a different set of initial equipment states is adopted in the simulation run, it can be shown that the resulting operating procedure still remains unchanged for route (5.5). In the network presented in Fig. 5.1, if the cleaning of route (5.3) is carried out before that of route (5.5) and only pump P5 and valve V1 are switched off to terminate the former task, the operating procedure identified above is clearly insufficient for accomplishing the latter. Thus, in order to guarantee the feasibility and safety of the material-transfer steps through a selected route, it is necessary to impose additional auxiliary control rules in operating the related valves, pumps and/or compressors. These *equipment operation rules* are summarized below.

Equipment Operation Rules: *Given a specific material-transfer action, all valves and/or pumps surrounding its downstream fragment (except the one used for facilitating the present material transfer) should be closed/switched off.*

To realize this requirement of blocking all the entrances and exits not located on the selected route, additional normal arcs should be introduced to connect the transition representing the given action in the modified path model to the places PC(C)s in Petri net models of the equipments surrounding its downstream fragment. In the case of our example system in Fig. 5.1, these additional connections are shown in Table 5.5.

Table 5.5 Normal-arc connections between transitions in modified path model (Fig. 5.10) and places $PC(C)$s in different equipment models (Fig. 5.9)

Transition	Equipments
$T1$	$V3, P4$
$T2$	$V3, P5$
$T3(1)$	$V2, P5$
$T3(2)$	$V1, P4$
$T4$	$V6, V7$
$T5$	$V6, V8$
$T6(1)$	$P5, V8$
$T6(2)$	$P4, V7$
$T7$	–
$T8$	–

As a result, the cleaning procedure of route (5.5) can be correctly generated from any given initial condition with the proposed simulation approach. To be specific, let us assume that the valves V3, V6, and V7 are left open after cleaning route (5.3) in our previous example, while in the meantime the other valves are closed and both pumps are off. The operating procedure to clean route (5.5) in this situation can be found to be close V3 and V6, open V2 and V8, and then turn on P5.

5.1.6 Execution of Multiple Tasks

In this case, it is assumed that two separate cleaning tasks can be scheduled sequentially according to the Gantt chart shown in Figs. 5.11a and concurrently according to Fig. 5.11b. It is required in the former case that:

$$t_0 < t_1 \leq t_2 < t_3 \tag{5.10}$$

while, in the latter case, the constraint is either,

$$t_0 \leq t_1 < t_2 \leq t_3 \tag{5.11}$$

or,

$$t_0 \leq t_1 < t_3 \leq t_2. \tag{5.12}$$

The time needed to accomplished a particular task should be determined on a case-by-case basis.

In the previous parts, the systematic method for generating the operation steps of a single-route cleaning task has already been presented. Here to synthesize the

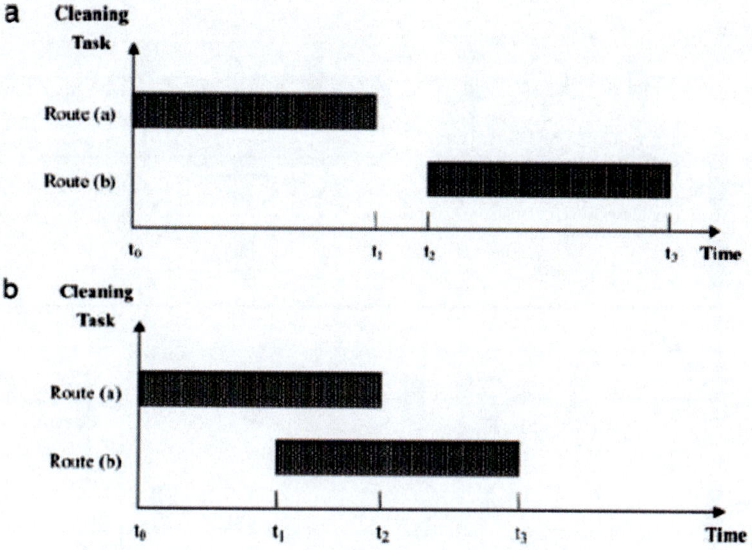

Fig. 5.11 **a** Typical sequential schedule. **b** Typical concurrent schedule (reprinted with permission from Chou and Chang 2005. Copyright 2005 American Chemical Society)

operating procedure for performing a multi-task schedule, the suggested Petri net-based techniques are extended. The Petri net model of a so-called schedule manager is attached to the system model in order to organize the execution times of various different tasks according to the defined schedule. If the route cleaning tasks (5.3) and (5.5) in the example framework are to be performed in accordance with the schedule set out in Fig. 5.11a, this schedule planner can be modeled on the basis of the Petri network set out in Fig. 5.12. Notice first that, instead of a single event, transition TX1 should be interpreted as a collection of operation steps (events) to establish route (5.3) for cleaning purpose and, similarly, TX6 represents another set of operation steps to establish route (5.5). To activate the operation steps associated with route (5.3), transition TX1 is connected to the place representing source fragment, i.e., FR1, and also to those representing the operation demands for establishing fragment connections, i.e., PK1, PK3(1), PK5, PK6(2), and PK7. Similarly, TX6 is connected to FR2, PK2, PK5, and PK8 to trigger the operation actions for cleaning route (5.5). On the other hand, transitions TX4 and TX9 represent the actions to terminate the cleaning operations of routes (5.3) and (5.5), respectively. It is assumed in the present study that each termination procedure consists of only two steps, i.e., switching off the pump on the cleaning route and then closing the exit valve of source fragment. The connections from TX4 and TX9 to the places in system model reflect the demands for these actions. Notice also that the delay times of TX1, TX4, TX6, and TX9 are assigned to meet the given schedule exactly. Another type of connections between the Petri net model of the schedule manager and the system model is concerned with places P2, P4, P6, and P9. They are used simply for maintenance purposes. Places

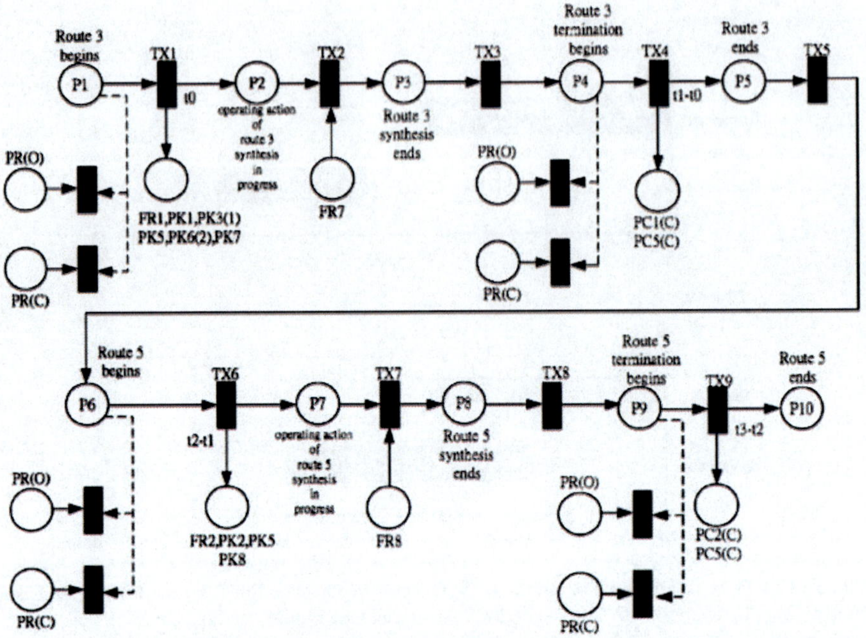

Fig. 5.12 Petri net model of a sequential schedule manager (reprinted with permission from Chou and Chang 2005. Copyright 2005 American Chemical Society)

P2 and P6 mark the initialization phases of the operations to establish routes (5.3) and (5.5), respectively, while P4 and P9 represent the preparation stages prior to the termination steps for these two routes, respectively. It should be noted that every such place and all the places labeled with "PR(O)" and "PR(C)" in system model are connected to a common output transition. Since the operation records of pumps and valves are stored in the latter places, this practice is in essence to reset these records before carrying out each of the above four distinct sets of operation steps.

Let us presume, for illustrative convenience, that all valves are closed and that all pumps are initially off in the example method. An operating procedure can be created to realize the schedule in Fig. 5.11a by placing a token in the P1 position and then executing the simulation run. Table 5.6 presents the resulting procedure phases.

Table 5.6 Fragment-based operation steps for cleaning the pipeline network in Fig. 5.1

Time	Operation steps
t_0	Open valves $V1$, $V3$, $V6$ and $V7$
t_1	Switch off pump $P5$
t_2	Close valves $V3$ and $V6$
t_3	Open valves $V2$ and $V8$ Switch off pump $P5$

Table 5.7 Blockage-based operation steps for cleaning the pipeline network in Fig. 5.1

Time	Operation Steps
t_0	Open valves $V1$, $V3$, $V6$ and $V7$
t_1	Switch off pump $P5$
t_2	Close valve $V7$
t_3	Open valves $V2$ and $V8$ Switch off pump $P4$

Notice that it is also possible to generate an operating procedure obtained on the basis of a blocking criterion using the same method (see Table 5.7). To clean routes (5.3) and (5.7) in sequence, these activity steps can be followed. Finally, while it is not possible to provide a concrete example of concurrent operations at this point, it should be stressed that the Petri net model of the corresponding schedule manager can be easily constructed in a similar way.

5.2 Material Transportation in a Batch Plant

Many high-value low-volume commercial products, such as pharmaceuticals, foods, specialty chemicals and certain forms of polymers, are commonly processed in batch processes. Because of the need to save capital costs, sharing the existing units in the plant for the production of numerous different goods is a common practice. As a consequence, via a complex pipeline network, the process materials also need to be transferred from one device to another. In addition, in order to promote this utility purpose, any energy transfer duty needed in these production processes may often involve the transport of a heating or cooling) medium within another pipeline network. The tasks of finding possible material-transfer routes and then synthesizing the corresponding operating procedures are usually conducted on an ad hoc basis manually. The demand for time and effort can be daunting for a complex batch chemical process and furthermore the resulting recipe is always vulnerable to error. In order to relieve work load and also to improve operation safety, it is highly desirable to develop computer algorithms for automatic synthesis of the needed operation procedures.

For the batch processes, two basic types of synthesis problems have been addressed in the literature, i.e., (1) the generation of plant-wide operating procedures involving a full spectrum of unit operations, and (2) the sequencing of valve-switching steps for fluid movements in plants.

The former issue is essentially concerned with the difficulties involved in incorporating considerations of production planning/scheduling into recipe management. Such an analysis was carried out by Crooks and Macchietto (1992). The batch processes were modeled on the representation of the state-task network (STN) in their paper (Kondili et al. 1988). By conducting a sequence of sub-targeting steps, the overall operational objective was achieved. With a mixed integer linear programming (MILP) logic-based technique, these steps were obtained. Viswanathan et al. (1998a, b) established a hierarchical planning system for the synthesis of batch operating procedures on the basis of the ISA standard S88.01 (ISA, 1995). A discrete event modeling tool called Grafchart was used in their work to represent the declarative and also procedural information for additional knowledge to be inferred incrementally. The detailed operating procedures can then be synthesized accordingly. Kim and Moon (2000) adopted an automatic safety verification system, i.e., symbolic model verifier (SMV), to synthesize a feasible operation sequence and to verify its safety. More specifically, this method can be used to identify the embedded operation error (if any), to find a minimum makespan and to synthesize an error-free operating procedure at the same time. Several examples were presented to illustrate the effectiveness of their approaches. Finally, Hoshi et al. (2002) proposed a knowledge-based method on the basis of two separate graph models. One was used to represent the plant structure and the other the material-conversion procedures. A recursive search algorithm was developed accordingly to generate the operation recipe, and this approach was successfully tested in a case study.

As stated earlier, the transport of material in a pipeline network between source(s) and sink(s) is a fundamental task to be carried out in the batch processes and is already covered in the previous parts. It should be noted that for practical implementations, the above findings are still not mature enough. Generally speaking, for actual implementation, the plant-wide operating procedures developed in the first group of studies are often not presented in adequate detail, e.g., see Kim and Moon (2000) and Hoshi et al. (2002), and are not appropriate for simultaneous operations, e.g., see Crooks and Macchietto (1992). On the other hand, to achieve a multi-task schedule that is very normal in an industrial plant, it is difficult to apply the particular valve-sequencing steps obtained in the second group. In this case, it addresses the generation of operating procedures to accomplish multiple tasks of material transfer identified by a given Gantt chart. To this end, the Petri net is used as a modeling tool for the material-transfer operations in batch plants.

There have been a few related studies in the past to model and design the hierarchical supervisory control system in batch processes with Petri nets, e.g., see Tittus and Lennartson (1999) and Ferrarini and Piroddi (2003). However, none of them are suitable for the synthesis of material-transfer procedures in a complex pipeline network. In this case, each valve, pump, and piping fragment in the given process is first modeled with a component Petri net. These component models are then connected according to the P&ID. Next, in order to identify all reachable states from a chosen initial system state and their relationships, the reachability tree of the Petri net model is developed with a standard algorithm (Murata 1989). From this reachability tree, all possible operating procedures (to achieve the same operation goal) can be easily identified.

5.2.1 Component Models

As stated earlier, via different pipeline networks, the process materials and heating/cooling mediums can be transferred in a batch chemical process. The division of the system into separate elements should be the first critical issue in modeling any network. By treating process equipment, e.g., valves, pumps and storage tanks, as nodes and the pipeline branches between any two nodes as guided arcs, Uthgenannt (1996) constructed the corresponding digraph. This strategy also produces unrealistic recipes that may even require spurious acts in the process. Let us consider the network presented in Fig. 5.13 and its digraph model in Fig. 5.14 as an example. The material-transfer routes from tank T1 to tank T3 can be identified by inspecting Fig. 5.14, i.e. (1) T1 → V1 → V4 → T3 and (2) T1 → V1 → V3 → V4 → T3. Notice that the opening of V3 in the second procedure is an unnecessary step. This is due to the fact that the material-transfer flow through valve V3 is bidirectional in this example and, thus, each of the three pipeline branches between V1, V3, and V4 is associated with two different arcs.

To avoid the above problem, the concept of the piping *fragments* (Foulkes et al. 1988) is adopted in this work for the development of Petri net models. In particular, a fragment is defined as a collection of pipeline branches and/or processing units isolated by the valves, pumps, and other means of flow blockage in the pipeline network. Let us again consider Fig. 5.13. Six fragments can be identified with this

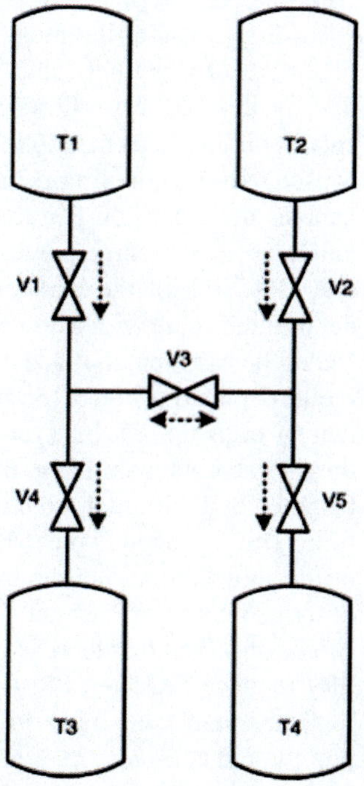

Fig. 5.13 An example plant (reprinted with permission from Wang et al. 2005. Copyright 2005 Elsevier)

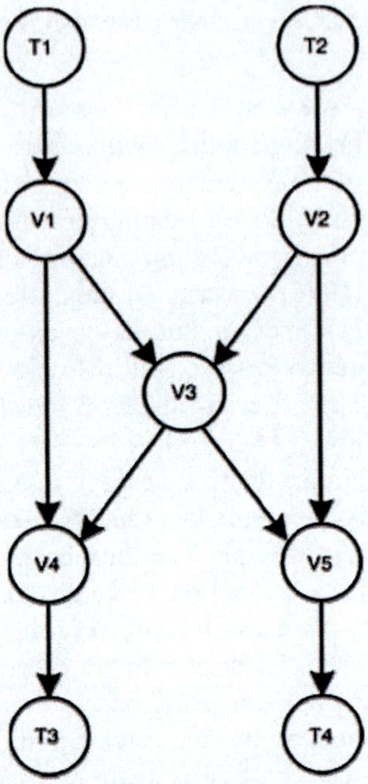

Fig. 5.14 Digraph model of example plant (reprinted with permission from Wang et al. 2005. Copyright 2005 Elsevier)

criterion, i.e., the shaded pipeline branches in Fig. 5.15. For illustration convenience, let us first consider the most basic structure of fragment, i.e., a section of pipeline FR isolated by an inlet valve VI and also an outlet valve VO (see Fig. 5.16a). Notice that, in this case, the flow in either valve is allowed only in one direction. The corresponding Petri net model is presented in Fig. 5.16b. The place FR in this model is used to reflect the fragment state. More specifically, a token entering such a place denotes the condition that the process material (or the heating/cooling medium) is delivered to the corresponding fragment from an upstream source fragment. The place PKI reflects the connection status of FR with its upstream fragment, and PKO denotes the same with a downstream fragment. On the other hand, if the outlet valve VO of the fragment in Fig. 5.16a permits bidirectional material transfer, the fragment model depicted in Fig. 5.16b should be changed to the one shown in Fig. 5.16c. Notice that the transitions TXO(1) and TXO(2) are used to denote, respectively, the material-transfer actions from FR to the downstream fragment and vice versa. It should also be noted that, although both transfer actions are allowed, only one can be taken at a time. Thus, a token is placed in PXO initially in every simulation run. On the basis of the above model-building convention, all mass-transfer paths in Fig. 5.15 can be described with the Petri net shown in Fig. 5.17. By inserting a token in place FR1, it can be observed that there is only one possible route for the token to flow from place FR1 to place FR5, i.e., FR $\rightarrow$ FR3 $\rightarrow$ FR5.

Other than the piping fragments, the creation of Petri net models for valves, pumps, and compressors is also required. In Fig. 5.18, the valve model is presented.

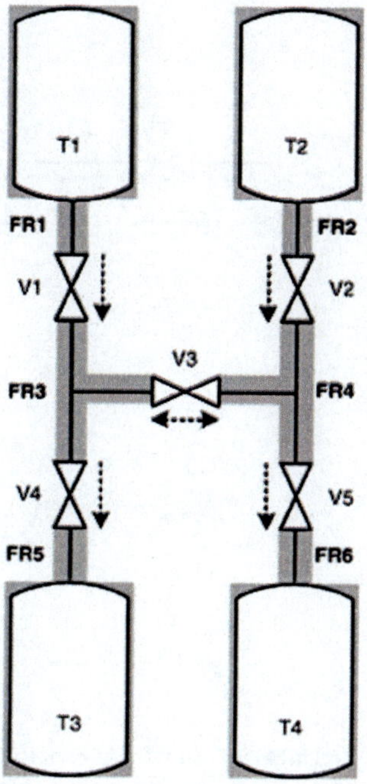

Fig. 5.15 Piping fragments of example plant (reprinted with permission from Wang et al. 2005. Copyright 2005 Elsevier)

The PV(O) and PV(C) locations here represent two opposite valve positions, i.e. open and close, respectively. The TV(O) and TV(C) transitions reflect the valve-switching behavior between PV(C) and PV(O) and vice versa. Note that the input position PC(O) of the transitions TV(O) can be interpreted as the valve-opening requirement once a particular material-transfer action is selected in a path. Similarly, the PC(C) position can be viewed as the prerequisite for valve-closing service. The PA(O) and PA(C) output positions of the two TV(O) and TV(C) transitions can be used to record the exact number of times the relevant valve-switching activities have been performed. A normal arc is adopted in the former case to avoid a token being permanently kept in place PC(O), while a test arc is used in the latter to prevent loss of the tokens in PV(O). Finally, note that the transition TR(C) is adopted for the same reason.

Since the operating procedures of pumps and compressors can be considered as well-established industrial practices, e.g., see Karassik and McGuire (1998), their detailed steps are not described in their component models for the sake of simplicity. Specifically, the valve model presented in Fig. 5.18 is also used for representing a power-generating device in this study. In this case, the places PV(O) and PV(C) represent two opposite states, i.e., on and off, of the device, respectively. The transitions TV(O) and TV(C) can be regarded as a series of standard operation actions to turn on and off the pump/compressor.

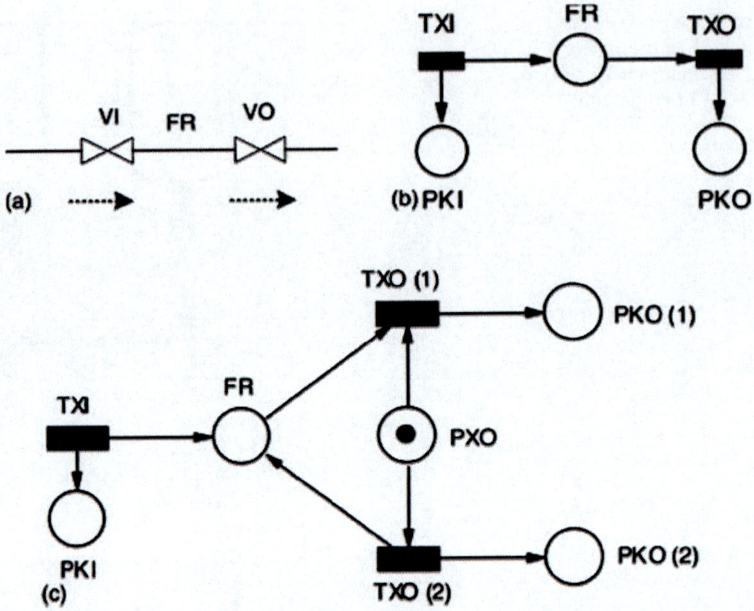

Fig. 5.16 **a** Basic structure of a piping fragment with two single-direction valves. **b** Petri net model of a basic fragment structure (two single-direction valves). **c** Petri net model of a basic fragment structure (one single-direction valve and one bidirection valve) (reprinted with permission from Wang et al. 2005. Copyright 2005 Elsevier)

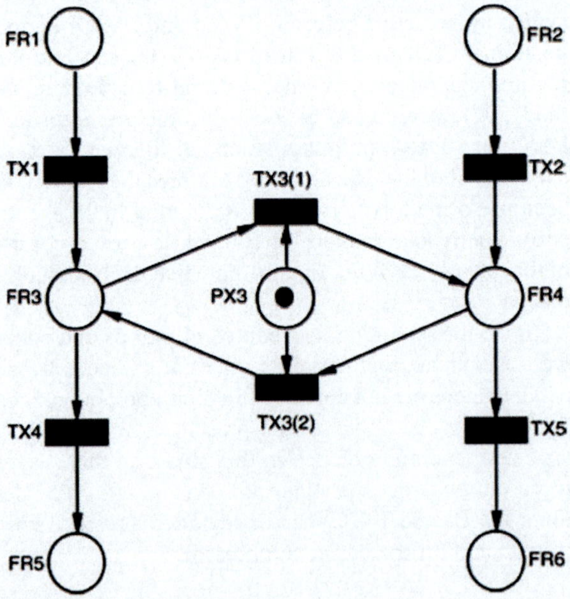

Fig. 5.17 Petri net model of material-transfer paths in the example plant (reprinted with permission from Wang et al. 2005. Copyright 2005 Elsevier)

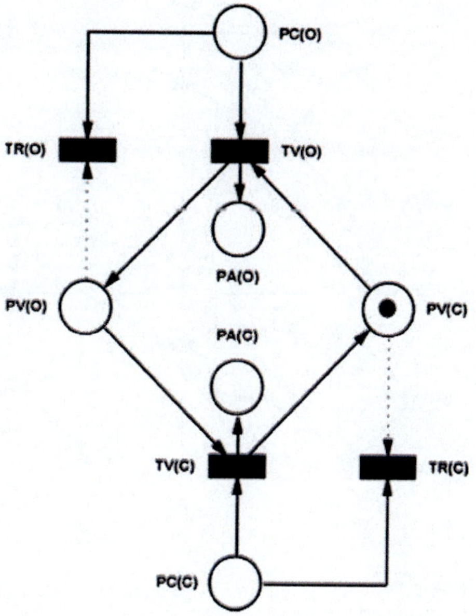

Fig. 5.18 Petri net model of a valve (reprinted with permission from Wang et al. 2005. Copyright 2005 Elsevier)

5.2.2 *Generation of Operation Steps*

Based on a system model, the operating procedure can be obtained. The part models should be assembled one-by-one according to the network configuration to create such a model. For example, using this method, the Petri net model of the plant mentioned in Fig. 5.13 can be generated easily (see Fig. 5.19). Note that the TX transitions are linked to PC(O)s places with standard arcs. This practice is intended to reflect the link between each operation of material transfer and the need to open the corresponding valve. A particular operating procedure can then be synthesized by performing simulation with the Petri net, given a route and a set of initial valve states. Let us consider the example in Fig. 5.13 again. Under the condition that all valves are closed initially, valves V1 and V4 should be switched to the open position for transferring material through the previously identified route FR1 → FR3 → FR5. Notice that the initial system condition of the Petri net model can be set by introducing a token in FR1 and also in the place representing the close position of each valve. The above operation steps can then be identified with the Petri net-based simulation. However, it should also be noted that, if a different set of initial valve states is adopted in the simulation run, the resulting operating procedure still remains the same. This is clearly undesirable. Thus, in order to guarantee the feasibility and safety of the material-transfer steps through a selected route, it is necessary to impose additional auxiliary control rules in operating the related valves, pumps and/or compressors. These *equipment operation rules* are summarized below.

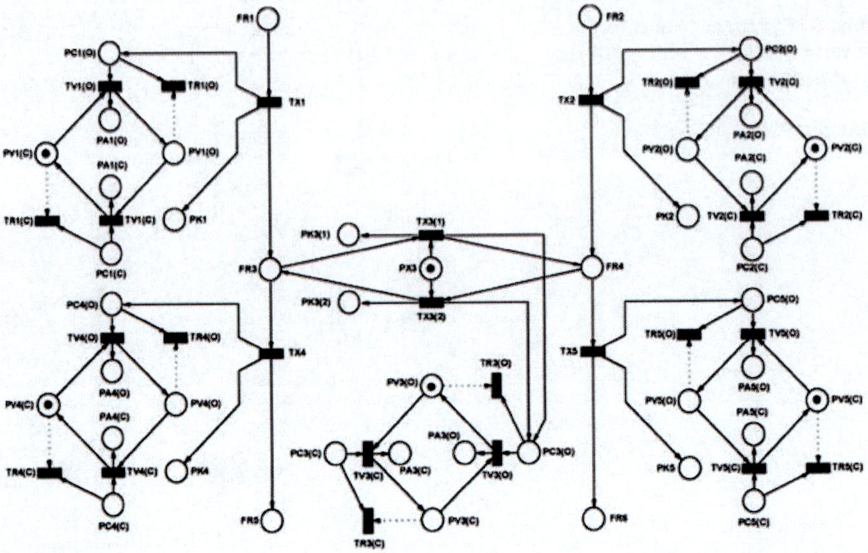

Fig. 5.19 The system model of example plant (reprinted with permission from Wang et al. 2005. Copyright 2005 Elsevier)

Equipment Operation Rules: *Given a specific material-transfer action, all valves and/or pumps surrounding its upstream fragment (except the one used for facilitating the present material transfer) should be closed/switched off.*

Additional normal output arcs should be added to connect the transition representing the given operation, i.e., TX, to the PC(C)s positions in the component models of the corresponding equipment, in order to realize this necessity of blocking all inputs and outputs not located on the selected path. In addition, an additional inhibitor arc should also be placed between place PK and the transition TV(C) in the component model of this equipment to preserve the equipment state required to execute the given operation. Additional normal and inhibitor arcs are inserted into the Petri net presented in Fig. 5.19 as a result of the implementation of these control laws. Figure 5.20 shows this updated version and for improved readability, the added arcs are drawn with bold solid lines. In Fig. 5.15, let us consider the material-transfer route FR1-FR3-FR5 once again. It is now possible to produce the right operating procedures for two separate sets of initial valve states with the updated net (see Table 5.8).

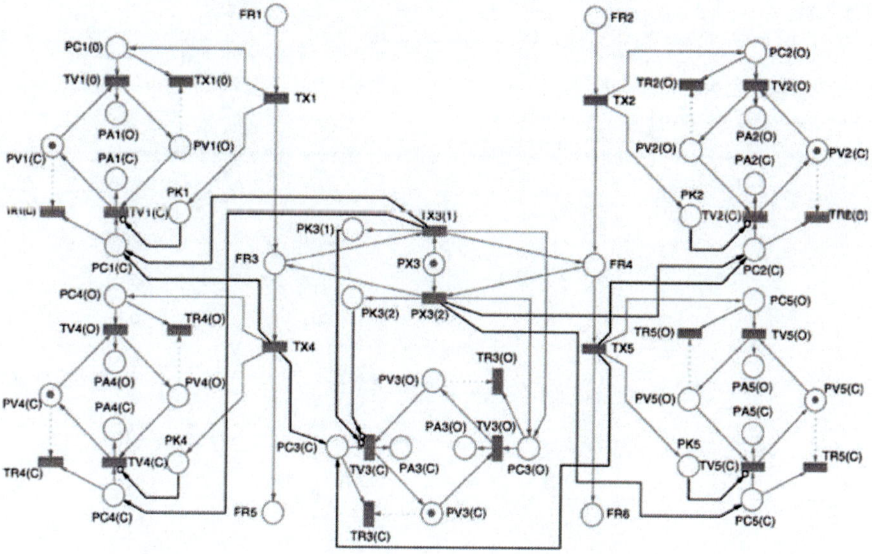

Fig. 5.20 The system model with equipment operation rules (reprinted with permission from Wang et al. 2005. Copyright 2005 Elsevier)

Table 5.8 The operating steps for two sets of initial valve states in the example plant

	Case 1					Case 2				
	V1	V2	V3	V4	V5	V1	V2	V3	V4	V5
Initial position[a]	C	C	C	C	C	C	O	O	O	C
Operation steps	open V1, open V4					close V3, open V1				

C denotes "close"; O denotes "open"

5.2.3 *Identification of Feasible Routes*

In the above discussions, before the necessary operation steps can be produced, the material-transfer route is assumed to be usable. Since there may be more than one path between a source fragment and a sink fragment, for the sake of operation versatility, it is always desirable to classify all possible routes. On the basis of the corresponding Petri net model, this task can be accomplished by creating a reachability tree. In particular, as many "new" markings as the number of allowed transitions according to the initial marking can be obtained first. One can then produce more markings from each new marking again. As elaborated in Sect. 5.1.2, repeating this process over and over results in an accessibility tree. In developing the reachability tree for our purpose, all enabled transitions associated with the valve-switching actions, i.e., those labeled as "TR" and "TV", must be fired *before* triggering any transition

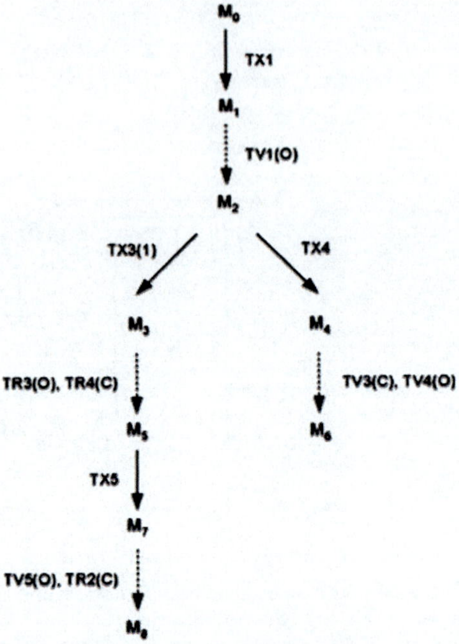

Fig. 5.21 A reachability tree of the Petri net in Fig. 5.20 (reprinted with permission from Wang et al. 2005. Copyright 2005 Elsevier)

representing material movement, i.e., the one labeled as "TX", to avoid generating duplicate branches.

Let us turn to the system described in Fig. 5.15. By assuming that the raw material is stored in tank T1 initially and all valves are closed except for valve V3, the corresponding reachability tree (shown in Fig. 5.21) can be generated according to the proposed algorithm. The markings associated with the nodes in this tree can be found in Table 5.9. In order to clearly identify the elements in a marking, the token numbers are grouped into six subsets and arranged sequentially in a vector, i.e.,

Table 5.9 The markings associated with the nodes in Fig. 5.21

Markings	Subsets
M0	{100,000}{00,000}{00,000}{00,100}{11,011}{000,000}
M1	{001,000}{00,000}{00,000}{00,100}{11,011}{100,000}
M2	{001,000}{10,000}{00,000}{10,100}{01,011}{100,000}
M3	{000,100}{10,000}{00,000}{10,100}{01,011}{101,000}
M4	{000,010}{10,000}{00,000}{10,100}{01,011}{100,010}
M5	{000,100}{10,000}{00,000}{10,100}{01,011}{101,000}
M6	{000,010}{10,010}{00,100}{10,010}{01,101}{100,010}
M7	{000,001}{10,000}{00,000}{10,100}{01,011}{101,001}
M8	{000,001}{10,001}{00,000}{10,101}{01,010}{101,001}

M_k = [$\mathbf{FR}_k$, $\mathbf{PA(O)}_k$, $\mathbf{PA(C)}_k$, $\mathbf{PV(O)}_k$, $\mathbf{PV(C)}_k$, $\mathbf{PK}_k$] and $k = 0,1,2, \ldots, 8$. Here, each subset label is identical to the place labels of its elements. For example, the elements stored in $\mathbf{FR}_k$ are the token numbers of places representing the fragment states. Based on this convention, useful information can be directly acquired from the marking of each terminal node in the reachability tree. Specifically, the sink fragment of a material-transfer route should be associated with one in the first subset $\mathbf{FR}_k$ of the terminal node. The actual valve-opening and valve-closing actions that make the transfer in each route possible can be found from the elements in the subsets $\mathbf{PA(O)}_k$ and $\mathbf{PA(C)}_k$, respectively. The final valve states can be obtained from the subsets $\mathbf{PV(O)}_k$ and $\mathbf{PV(C)}_k$. Finally, the connection status among various fragments in the network can be identified from $\mathbf{PK}_k$. Notice that the subsets $\mathbf{PC(O)}_k$, $\mathbf{PC(C)}_k$, and $\mathbf{PX}_k$ are not included in Table 5.9 since these data are not useful for the identification of the feasible operation steps. According to the reachability tree given in Fig. 5.21, it can be concluded that two material-transfer routes are originated from fragment FR1, i.e., FR1 → FR3 → FR5 (see M6 in Table 5.9) and FR1 → FR3 → FR4 → FR6 (see M8 in Table 5.9). From the subsets $\mathbf{PA(O)}_6$ and $\mathbf{PA(C)}_6$, it is clear that the former material-transfer task can be accomplished by closing V3 and then opening both V1 and V4. Similarly, the operation steps for the latter can be found in $\mathbf{PA(O)}_8$ and $\mathbf{PA(C)}_8$, i.e., opening V1 and V5. Notice that the valve-closing actions should be implemented at instances *earlier* than those of the valve-opening steps in the above procedures. This is due to the need to eliminate the possibility of transferring material to fragments that are not located on the selected route. Notice also that these operation steps can only be used to *facilitate* material transfer. The operating procedures to *terminate* the transfer tasks have not yet been discussed at this point.

Finally, it should be pointed out that the proposed route-searching algorithm may fail if the Petri net contains loops. For example, let us apply the above procedure to the network presented in Fig. 5.22. It can be found that there are five possible routes emanating from the fragment FR1. However, one of them forms an infinite loop, i.e., FR1 → FR3 → FR4 → FR6 → FR5 → FR3 → FR4 → FR6 → FR5 → ⋯. This loop is caused basically by the bidirectional valves V3, V4, V5, and V6. The effort to assemble these looping routes is obviously futile and, thus, they should be avoided in the search process. To this end, it is necessary to impose another set of control rules on the system model. These *connection enforcement rules* are summarized below.

Connection Enforcement Rules: *Given a specific material-transfer action, any additional transport into the upstream fragment of this action should be prohibited to avoid development of looping routes.*

Specifically, an inhibitor arc should be added between the place representing the connection status resulting from the transfer action under consideration and each transition representing an inlet connection of its upstream fragment.

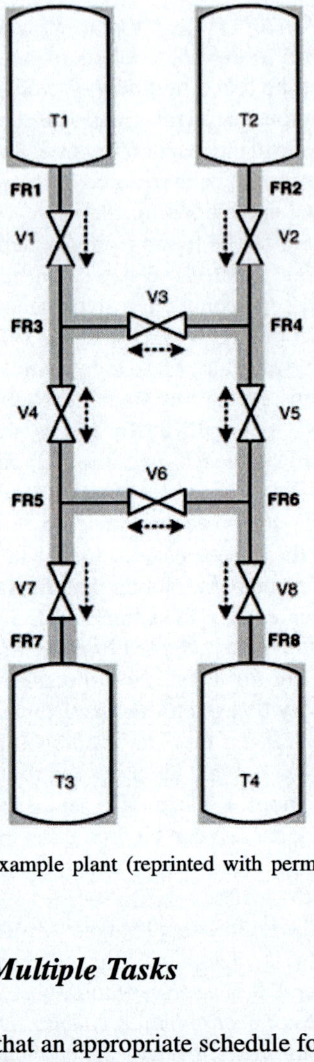

Fig. 5.22 Another fictitious example plant (reprinted with permission from Wang et al. 2005. Copyright 2005 Elsevier)

5.2.4 Execution of Multiple Tasks

In this case, it is assumed that an appropriate schedule for carrying out all material-transfer tasks is already available before the synthesis of specific operating procedures. For illustration convenience, let us consider two tasks scheduled according to the Gantt charts shown in Fig. 5.23a, b. They are referred to as *schedule A* and *schedule B*, respectively, in this paper. In essence, the key decision in generating the operating procedures to implement these two schedules is concerned with the issue of *fragment sharing*. In the former case, the two material-transfer routes adopted to accomplish the given tasks are allowed to be overlapping. To provide such opportunities, the token number in every place with initial label "PK" should be reset to zero before initiating the search process to create the operation steps of the second task. On the other hand, all place-resetting possibilities must be eliminated in schedule B.

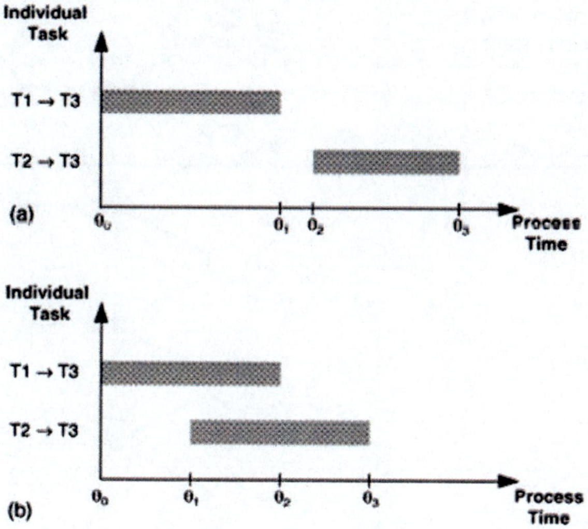

Fig. 5.23 **a** Schedule A. **b** Schedule B (reprinted with permission from Wang et al. 2005. Copyright 2005 Elsevier)

The above specifications can be interpreted in the Petri Net Model as the final set of control rules. Let us consider a mixing operation in the pipeline network given in Fig. 5.15 for illustration purposes. The raw materials for this process (say X and Y) are presumed to be stored in tank T1 and tank T2, respectively, while tank T3 is used as a mixer. The material-transfer tasks needed should therefore be (1) the transport of material X from tank T1 to tank T3, and (2) the transport of material Y from tank T2 to tank T3. By adding a fictitious schedule manager's part model to the system model, the corresponding control rules can be enforced. This component in present example consists of a *timer* model (see Fig. 5.24a) and the models of two *task coordinators* (see Fig. 5.24b).

The timer model can be constructed according to a Gantt chart. On the basis of the delay times assigned to the transitions TC(1)–TC(4) in Fig. 5.24a, it is clear that the places P(1), P(2), and P(3) can be used to reflect the time intervals (θ_0,θ_1), (θ_1,θ_2), and (θ_2,θ_3), respectively, for both schedules. The place PS(i) ($i = 1$, 2, 3, 4) marks the instance that a task begins or ends. If schedule A is adopted, PS(1) and PS(3) denote the initiation commands of the first and second tasks, respectively, and PS(2) and PS(4) the corresponding termination commands. On the other hand, PS(1) and PS(2) represent the initiation signals and PS(3) and PS(4) the termination signals in schedule B.

The Petri net model of a task coordinator can be divided into three smaller nets. They are referred to in this paper as the models of *route synthesizer*, *reset processor,* and *task terminator* according to their respective functionalities. Notice that, once a task initiation signal is generated by the timer, i.e., a token is introduced in place PS(i), a token should flow pass the place "reset" and then "start" in the route synthesizer.

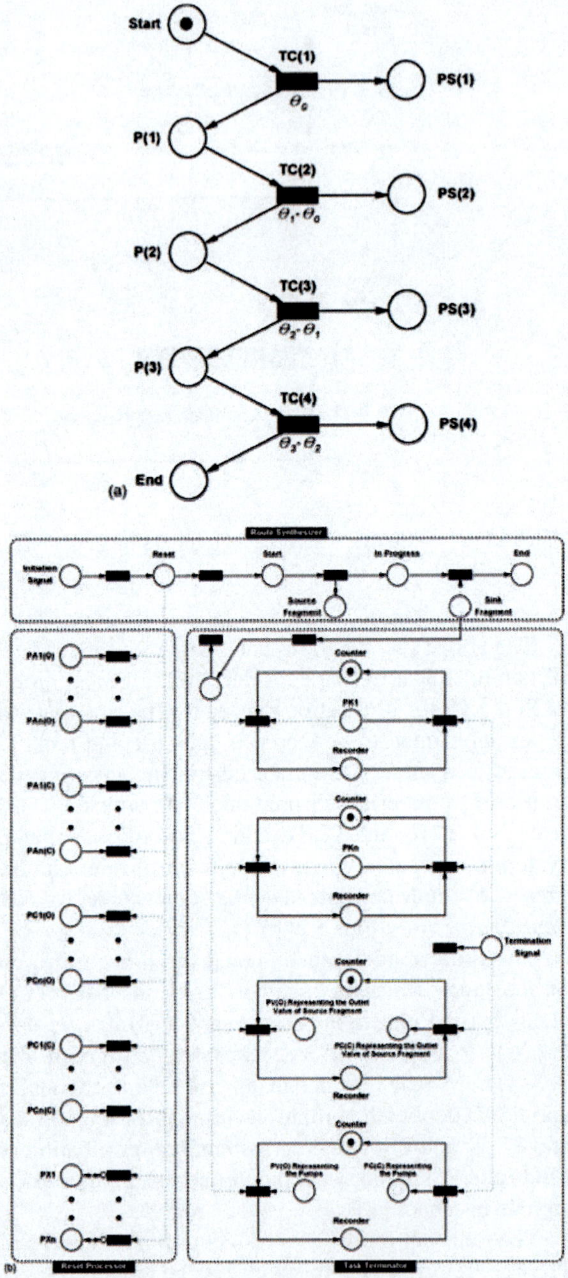

Fig. 5.24 **a** Petri net model of a timer. **b** Petri net model of a task coordinator (reprinted with permission from Wang et al. 2005. Copyright 2005 Elsevier)

The former is the entry point of reset processor, which performs basically two types of routine maintenance works on the system model:

1. clearing out all tokens in PA(O)s, PA(C)s, PC(O)s, and PC(C)s, and
2. feeding a token to each empty PX.

These practices are taken to avoid confusion in repeating the route synthesis process for multiple tasks.

The route synthesis mechanism mentioned in the previous section, on the other hand, is triggered after a token is given at the "start" location and, subsequently, at the source fragment of the designated material-transfer location. Note that if a token enters the position denoting the sink fragment status, the route synthesis process terminates. The task terminator is expected at this point to perform two types of book-keeping duties, i.e.,

1. storing the token number in every PK with the attached place representing a *recorder*, and
2. recording the token numbers in PV(O)s for the outlet valves of the source fragment and in PV(O)s for all pumps/compressors.

The operation steps to end a task can also be generated in the task terminator after a termination signal is issued by the timer. In particular, a valve (or pump/compressor) shall be closed (or switched off) if a token is residing in the corresponding *recorder* place. The tokens in PKs are also removed under the same condition. This mechanism is installed to allow overlapping routes in schedule A and also to produce non-overlapping ones in schedule B. Notice also that, since the second task begins at a time when the first is still in progress in schedule B, two material-transfer routes should have already been created before task 1 can be put to an end. It is thus necessary to provide a means to distinguish the two in determining the termination steps. This is achieved with the places denoted as the *counters*. Since only one set of counters are used in our model and these places are meant to be shared by the two task coordinators, the operation steps required to stop material transfer in the first route cannot be adopted in the second.

Let us assume again that all valves except V3 are initially closed in Fig. 5.15 and try to generate the operating procedures to achieve schedule A. With the proposed algorithm, the corresponding reachability tree can be constructed. Note that the first part of this tree, which is associated with the first task for material transfer, was previously generated in Fig. 5.21. In Fig. 5.25, the remainder is presented. It should be noted that this subtree is formed in Fig. 5.21 from the terminal node M6 as its labeling represents the device state reached after successfully synthesizing the material-transfer routes of the first mission. The node markings in Fig. 5.25 are given in Table 5.10. The operating procedures presented in Table 5.11 can be deduced from the knowledge embedded in the entire reachability tree.

In order to execute schedule B, let us consider the operating procedures next. Figure 5.26 and Table 5.12, respectively, contain the corresponding subtree and the markings of its nodes. Notice that it is possible to define only one material-transfer path, i.e., FR2 $\rightarrow$ FR4 $\rightarrow$ FR6, and its sink fragment does not contain tank T3. Thus,

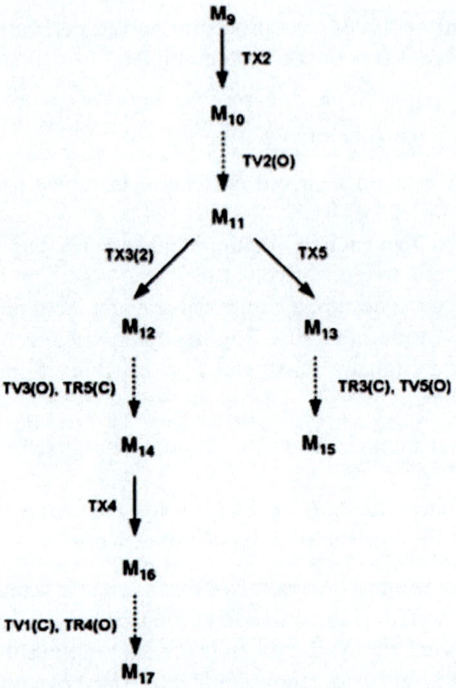

Fig. 5.25 The reachability tree associated with the second task in schedule A (reprinted with permission from Wang et al. 2005. Copyright 2005 Elsevier)

Table 5.10 The markings associated with the nodes in Fig. 5.25

Markings	Subsets
M9	{010,000}{00,000}{00,000}{00,010}{11,101}{000,000}
M10	{000,100}{00,000}{00,000}{00,010}{11,101}{010,000}
M11	{000,100}{01,000}{00,000}{01,010}{10,101}{010,000}
M12	{001,000}{01,000}{00,000}{01,010}{10,101}{010,100}
M13	{000,001}{01,000}{00,000}{01,010}{10,101}{010,001}
M14	{001,000}{01,100}{00,000}{01,110}{10,001}{010,100}
M15	{000,001}{01,001}{00,000}{01,011}{10,100}{010,001}
M16	{000,010}{01,100}{00,000}{01,110}{10,001}{010,110}
M17	{000,010}{01,100}{10,000}{00,110}{11,001}{010,110}

Table 5.11 The operating steps for executing schedule in the example plant

Task	Route	Operation steps
1	FR1 → FR3 → FR5	(1) Close V3 and open V1, V4 at θ0; (2) close V1 at θ1
2	FR2 → FR4 → FR3 → FR5	(1) Open V2, V3 at θ2; (2) close V2 at θ3

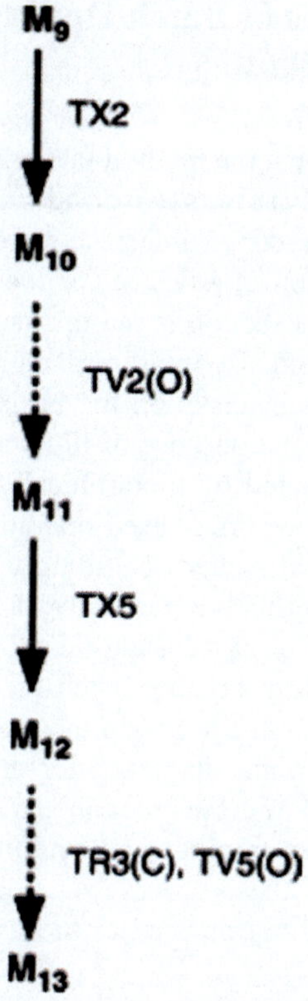

Fig. 5.26 The reachability tree associated with the second task in schedule B (reprinted with permission from Wang et al. 2005. Copyright 2005 Elsevier)

Table 5.12 The markings associated with the nodes in Fig. 5.26

Markings	Subsets
M9	{010,000}{00,000}{00,000}{10,010}{01,101}{100,010}
M10	{000,100}{00,000}{00,000}{10,010}{01,101}{110,010}
M11	{000,100}{01,000}{00,000}{11,010}{00,101}{110,010}
M12	{000,001}{01,000}{00,000}{11,010}{00,101}{110,011}
M13	{000,001}{01,001}{00,000}{11,011}{00,100}{110,011}

it can be seen that it is really not possible to execute the second task from the machine state achieved after the first one has been triggered.

5.3 Automatic Synthesis of Batch Operating Procedures in Beer Filtration Plant

As stated in the previous sections, the method taken in Chou and Chang (2005) and also in Wang et al. (2005) mainly suffers two disadvantages in practical applications, i.e., (1) the optimality of a selected feasible procedure cannot be guaranteed and (2) the time schedule must be given in advance for the performance of the material-transfer tasks. The focus of this section is put on issues relating to the synthesis of optimal operating procedures and also their implementation schedule to accomplish any set of given material-transfer tasks on the basis of Petri net models in order to resolve these problems. For the purpose of illustration, the same method for the example in Fig. 5.1 is constructed by hierarchically assembling the components. To produce either stage-based or time-based operating procedures, binary integer programs (BIPs) can then be formulated accordingly.

In particular, a system model with a hierarchy of four distinct component levels can be assembled. The first-level part (usually a programmable logic controller or a human operator) is used in any existing batch process to perform the operating steps specified in a recipe based on a defined time schedule or a collection of sensor measurements. In the second point, its acts alter the state of valves, pumps, and compressors. In turn, the states of these components decide the configuration of the process and therefore the operating mode and equipment condition of each process unit at the third stage. Finally, these process states are controlled at the last stage by sensors, which may or may not be used as the basis for further behavior of the controller.

For the purpose of recipe synthesis, it is obvious that the first-level components cannot be included in the Petri net model since the operating procedures are unavailable. Moreover, to simplify BIP formulation, the sensor models are ignored in this work by assuming that the process conditions and their measurements are always identical.

5.3.1 *Stage-Based Operating Procedures*

A simplified version of the recipe-synthesis problem is considered here for illustrative convenience. Let us briefly presume that in the model formulation, the elapsed times of all material-transfer activities are equal and can thus be ignored. Within a standard time frame, which is referred to as a stage in this document, the procedural steps taken to execute any task are needed to be completed and several tasks are allowed to be performed in a single stage. In other words, with the present approach, only the implementation *order* of the stages and the operational actions needed at each stage are produced.

5.3.1.1 Path Constraints

Logic constraints can be written to characterize the movements of tokens in the path model. In particular, two different types of binary variables, i.e., x_i and y_j^o, are adopted to represent the token numbers in places representing the fragment states (FR_i) and connection status (PV_j^o), respectively. To facilitate explanation of the constraint formulation, let us consider the generalized fragment model presented in Fig. 5.27. The causal relations between the fragment state of FR_i and those of its downstream and upstream fragments can be translated into the following two inequality constraints, respectively, according to the formulation techniques developed by Raman and Grossmann (1991), i.e.,

$$(1 - x_i) + \left(1 - y_{jd}^o\right) + x_{id} \geq 1 \quad jd \in \mathbf{JD}_i \quad id \in \mathbf{ID}_{jd} \tag{5.13}$$

$$(1 - x_i) + \left(1 - y_{ju}^o\right) + x_{iu} \geq 1 \quad ju \in \mathbf{JU}_i \quad iu \in \mathbf{IU}_{ju} \tag{5.14}$$

where $x_i, x_{id}, x_{iu} \in \{0, 1\}$; $y_{jd}^o, y_{ju}^o \in \{0, 1\}$; $\mathbf{JD}_i = \{jd_1, jd_2, \ldots\}$; $\mathbf{J}_i = \{ju_1, ju_2, \ldots\}$;

$\mathbf{ID}_{jd_k} = \{id_k\}$; $\mathbf{IU}_{jd_{k'}} = \{iu_{k'}\}$ In other words, the elements of set $\mathbf{JD}_i$ are used to distinguish the places representing the downstream connection states of fragment FR_i, i.e.,PV_{jd1}^O, PV_{jd2}^O, …, and the elements of set $\mathbf{JU}_i$ are for the places representing the upstream connection states of fragment FR_i, i.e.,PV_{ju1}^O, PV_{ju2}^O, $\cdots$; id_k and $iu_{k'}$ denote the indices of the places representing the k th downstream fragment and the k' th upstream fragment, respectively.

It is obvious that, other than the source and sink, there should be exactly one downstream connection and one upstream connection for any fragment FR_i on a material-transfer path. The corresponding logic constraints can be expressed as:

$$(1 - x_i) + \sum_{jd \in \mathbf{JD}_i} y_{jd}^o \geq 1 \tag{5.15}$$

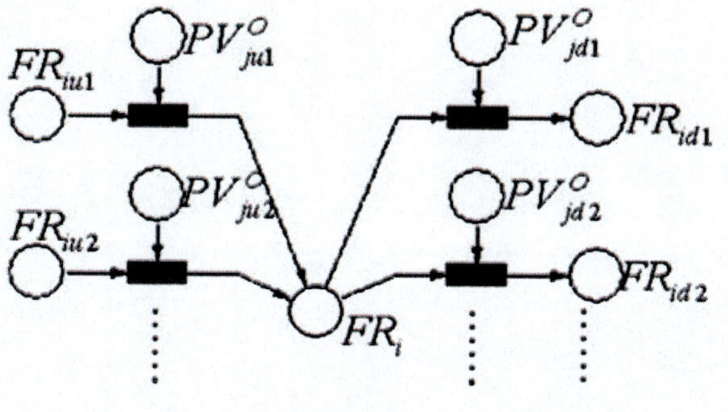

Fig. 5.27 Petri net representation of a fragment (FR_i) and its upstream and downstream connections (reprinted with permission from Lai et al. 2007. Copyright 2007 American Chemical Society)

$$(1 - x_i) + \sum_{j \in \mathbf{JU}_i} y^o_{ju} \geq 1 \tag{5.16}$$

$$\sum_{jd \in \mathbf{JD}_i} y^o_{jd} \leq 1 \tag{5.17}$$

$$\sum_{j \in \mathbf{JD}_i} y^o_{ju} \leq 1. \tag{5.18}$$

Notice that constraints (5.13)–(5.18) must be imposed upon all fragments except that (5.13), (5.15), and (5.17) cannot be used to describe the flow connections of sinks and (5.14), (5.16), and (5.18) are not applicable in the case of sources.

5.3.1.2 Operation Constraints

Only to describe the material-transfer paths can the above restrictions be used. It is also important to implement additional constraints that reflect the operational behavior of valves, pumps, and compressors for the purpose of producing actual operating steps. On the basis of the Petri net model given in Fig. 5.28, these constraints can be derived. Again with a different binary variable, the token number in each position is interpreted. Let us use the binary variables y^o_j and y^C_j to, respectively, denote the token numbers in the two places reflecting the states of equipment j, and use z^o_j and z^C_j to denote the token numbers in places representing the corresponding control commands. The token movements in equipment model can therefore be described as:

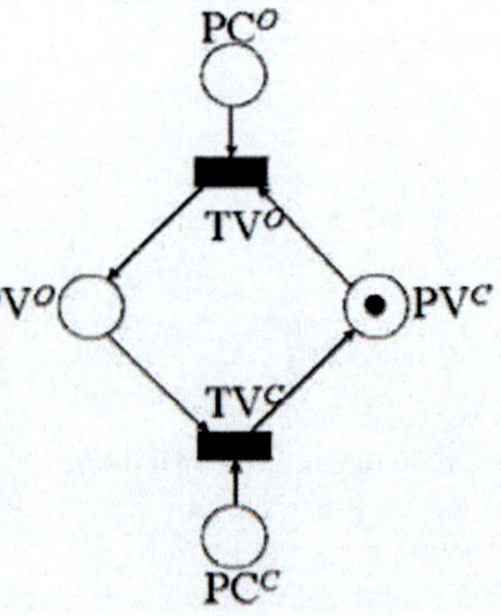

Fig. 5.28 Standard valve model (reprinted with permission from Lai et al. 2007. Copyright 2007 American Chemical Society)

$$\begin{aligned}&\left(1-y_j^{OI}\right)+\left(1-z_j^C\right)+y_j^C\geq 1\\&\left(1-y_j^{OI}\right)+z_j^C+y_j^o\geq 1\\&\left(1-y_j^{CI}\right)+\left(1-z_j^O\right)+y_j^o\geq 1\\&\left(1-y_j^{CI}\right)+z_j^O+y_j^C\geq 1.\end{aligned} \tag{5.19}$$

In the above constraints, y_j^{OI} and y_j^{CI} represent, respectively, the initial values of y_j^o and y_j^C. The first constraint is equivalent to the logic statement that, if the initial equipment state is "open" ($y_j^{OI} = 1$) and a "close" command is issued ($z_j^C = 1$), then the resulting state should be "close" ($y_j^C = 1$). On the other hand, if the "close" command is *not* given ($z_j^C = 0$) under the same initial condition ($y_j^{OI} = 1$), the equipment should remain at its "open" state ($y_j^O = 1$). Notice that the 3rd and 4th constraints can be interpreted in a similar way and thus are not repeated for the sake of brevity. It is assumed in the proposed model that, other than the source valves and the pumps, the initial states of all other valves should remain unchanged from those in the previous stage. On the other hand, it is also reasonable to institute the routine practices of closing all opened *source* valves and switching off all running pumps at the end of each operation stage. In other words, the corresponding binary variables y_j^{OI} s should be set to 0 and y_j^{CI} s should be 1 inall operation stages.

In addition to the constraints given above, it is necessary to impose the following auxiliary constraints to enhance search efficiency:

1. It is obvious that the state of any equipment must be unique, i.e.,

$$y_j^o + y_j^C = 1. \tag{5.20}$$

2. If an equipment is at its "open" (or "closed") initially, then it is meaningless to execute the operation step to open (or close) the same equipment. The following constraints are adopted to prevent such possibilities:

$$\begin{aligned}&y_j^{OI}+z_j^0\leq 1\\&y_j^{CI}+z_j^C\leq 1.\end{aligned} \tag{5.21}$$

3. In order to ensure practical applicability, it is assumed that every piece of level-2 equipment (except the source valves and power-generating devices) can be operated at most once. Thus, the corresponding constraints can be expressed as:

$$z_j^O + z_j^C \leq 1. \tag{5.22}$$

4. As mentioned before, the actions to close source valves and to switch off running pumps are assumed to be the routine steps performed at the end of each operation stage. The implied restrictions of this assumption can be written as:

$$\left(1-y_j^o\right)+z_j^C=1 \quad j\in \mathbf{JP}\cup\mathbf{JS} \tag{5.23}$$

where **JP** and **JS** denote, respectively, the sets of all power-generating units and source valves.

On the other hand, notice that the equipment model of a bidirectional valve is built with two standard equipment models. Extra constraints are thus needed to reconcile the conflicting control commands resulting from this modeling practice. In particular, the two corresponding fictitious valves cannot be both open, i.e.,

$$y_{j(1)}^o+y_{j(2)}^o\le 1. \tag{5.24}$$

Thus, the control commands to open or close these two fictitious valves should not be issued at the same time, i.e.,

$$\begin{aligned} z_{j(1)}^O+z_{j(2)}^O&\le 1\\ z_{j(1)}^C+z_{j(2)}^C&\le 1. \end{aligned} \tag{5.25}$$

Furthermore, notice that all possible states of a bidirectional valve can be classified according to the states of two fictitious valves, i.e.,

- State 1: $y_{j(1)}^0=0,\ y_{j(1)}^C=0,\ y_{j(2)}^0=0,\ y_{j(2)}^C=0;$
- State 2: $y_{j(1)}^o=1,\ y_{j(1)}^C=0,\ y_{j(2)}^o=0,\ y_{j(2)}^C=0;$
- State 3: $y_{j(1)}^0=0,\ y_{j(1)}^C=0,\ y_{j(2)}^0=1,\ y_{j(2)}^C=0.$

State 1 is associated with the "close" position of a bidirectional valve, while states 2 and 3 both correspond to the "open" position. The flows corresponding to the latter two states are opposite in direction and marked as (1) and (2), respectively. Due to the fact that a bidirectional valve can be considered to be closed as long as the flow in either direction is blocked, one can conclude that:

$$y_{j(1)}^C=y_{j(2)}^C=y_j^C. \tag{5.26}$$

Thus, Eq. (5.20) can be rewritten as:

$$y_{j(1)}^o+y_{j(2)}^o+y_j^C=1. \tag{5.27}$$

The transition from one state to another can be realized by manipulating the fictitious valves. It is clear that the maximum number of such transitions is six. To resolve the conflicting control commands required in these processes, let us use the binary variables u_j^o and u_j^C to represent, respectively, the *actual* control signals for opening and closing valve *j*. The correspondence between the fictitious and actual commands of a bidirectional valve is summarized in Table 5.1. Notice that, due to the need to satisfy the constraints in (5.25), not all combinations are included in this table. The fictitious commands listed in the first two rows of Table 5.1 are adopted to

Table 5.13 Correspondence between the Fictitious and Actual Commands of a Bidirectional Valve

$z^{O}_{j(1)}$	$z^{C}_{j(1)}$	$z^{O}_{j(2)}$	$z^{C}_{j(2)}$	u^{O}_{j}	u^{C}_{j}
1	0	0	0	1	0
0	1	0	0	0	1
0	0	1	0	1	0
0	0	0	1	0	1
1	0	0	1	0	0
0	1	1	0	0	0
0	0	0	0	0	0

change from state 1 to state 2 and vice versa, while those in rows 3 and 4 can be used to activate the forward and backward transitions between states 1 and 3 respectively. Since there is only one action required in each of the above four fictitious commands, the corresponding actual command should be the same. These logic relations can be expressed as:

$$\begin{aligned} \left(1 - z^{O}_{j(1)}\right) + z^{C}_{j(2)} + u^{O}_{j} &\geq 1 \\ \left(1 - z^{C}_{j(1)}\right) + z^{O}_{j(2)} + u^{C}_{j} &\geq 1 \\ \left(1 - z^{O}_{j(2)}\right) + z^{C}_{j(1)} + u^{O}_{j} &\geq 1 \\ \left(1 - z^{C}_{j(2)}\right) + z^{O}_{j(1)} + u^{C}_{j} &\geq 1. \end{aligned} \tag{5.28}$$

Notice that not all binary variables are included in these constraints. This is due to the fact that the values (0) of the missing variables can be directly inferred from constraints (5.22) and (5.25).

The fictitious commands in the 5th and 6th rows of Table 5.13 represent two separate sets of operation steps needed to change the two-valve system state from 3 to 2 and 2 to 3, respectively. However, if either set of operation steps are carried out in practice, the bidirectional valve is required to be first opened and then closed or vice versa. This implies that the actual valve position is unchanged and thus no real actions should be taken. Following are the inequality constraints representing the inference rules given in row 5:

$$\begin{aligned} \left(1 - z^{o}_{j(1)}\right) + \left(1 - z^{C}_{j(2)}\right) + \left(1 - u^{o}_{j}\right) &\geq 1 \\ \left(1 - z^{o}_{j(1)}\right) + \left(1 - z^{C}_{j(2)}\right) + \left(1 - u^{C}_{j}\right) &\geq 1. \end{aligned} \tag{5.29}$$

The constraints used to describe the logic in row 6 can be written as:

$$\begin{aligned} \left(1 - z^{C}_{j(1)}\right) + \left(1 - z^{O}_{j(2)}\right) + \left(1 - u^{C}_{j}\right) &\geq 1 \\ \left(1 - z^{C}_{j(1)}\right) + \left(1 - z^{O}_{j(2)}\right) + \left(1 - u^{o}_{j}\right) &\geq 1. \end{aligned} \tag{5.30}$$

Notice that the last row in Table 5.13 is associated with the possibility that the valve remains in its original state. In this situation, there should not be any actual action either. The corresponding constraints are:

$$\begin{aligned} z_{j(1)}^{O} + z_{j(2)}^{O} + z_{j(1)}^{C} + z_{j(2)}^{C} + \left(1 - u_j^o\right) \geq 1 \\ z_{j(1)}^{O} + z_{j(2)}^{O} + z_{j(1)}^{C} + z_{j(2)}^{C} + \left(1 - u_j^C\right) \geq 1. \end{aligned} \tag{5.31}$$

Finally, to facilitate consistent model formulation, the binary variables associated with the actual control commands of the single-directional valves and power-generating systems are also expressed with the same notations, i.e.,

$$\begin{aligned} u_j^o &= z_j^O \\ u_j^C &= z_j^C. \end{aligned} \tag{5.32}$$

5.3.1.3 Goal Constraints

In addition to the above-mentioned limitations, additional ones do need to be implemented in order to achieve the organizational objectives. The easiest objective is to perform a single task of material transfer between a pair of source and sink fragments. For example, let us consider the system presented in Fig. 5.1 and the task of transporting material from tank T_1 to tank T_4. In this case, the binary variables representing the fragment states of FR_1 and FR_8 should both be 1, while those associated with the other source and sink fragments should be set to 0, i.e.,

$$\begin{aligned} x_1 = x_8 = 1 \\ x_2 = x_7 = 0. \end{aligned} \tag{5.33}$$

Notice that, since the path constraints (5.15)–(5.18) have already been included in the integer program, one of the conditions given in the first part of Eq. (5.33) (i.e., $x_1 = 1$ or $x_8 = 1$) can in fact be neglected.

If there is a need to perform multiple tasks in a *single* stage, then a second subscript r can be added to the variables in path constraints given in equations (5.13)–(5.16) to distinguish the corresponding material-transfer routes. Specifically, x_i (the token number in FR_i) and y_j^O (the token number in PV_j^O) can be replaced by $x_{i,r}$ and $\zeta_{j,r}^O$. Since there is a one-to-one correspondence between a path and its source (or sink), the latter is used to identify the former in the present study. In other words, subscript r is both the label of a source fragment and that of the corresponding path. Let us again consider the system in Fig. 5.1 and two material-transfer tasks: (a) $T_1 \rightarrow T_7$, and (b) $T_2 \rightarrow T_8$. The corresponding goal constraints are:

$$
\begin{aligned}
x_{1,b} &= x_{2,a} = 0 \\
x_{7,a} &= x_{8,b} = 1.
\end{aligned} \tag{5.34}
$$

Notice that it is not necessary to stipulate the states of the source fragments here.

Since a fragment cannot be shared by more than one path and a level-2 component can cause the material flow along only one path, the following constraints must be valid:

$$
\sum_r x_{i,r} \leq 1. \tag{5.35}
$$

To avoid creating too many new variables and new operation constraints, the variables $\varsigma^O_{j,r}$ s can be related to the equipment states with the following equation:

$$
\sum_r \varsigma^O_{j,r} = y^0_j. \tag{5.36}
$$

As a result, there is no need to introduce the extra subscript r into the binary variables for characterizing the operation steps. More specifically, the variables y^o_j, y^C_j, z^o_j, z^C_j, u^o_j, and u^C_j in equations (5.17)–(5.32) should remain unchanged in multi-task applications.

In multi-task *multi-stage* applications, a third subscript t must be added to all variables in the path and operation constraints. Let us consider the problem of synthesizing operation steps for transporting material via separate routes in the pipeline network in Fig. 5.1 according to a *given* order, e.g., (1) $T_1 \rightarrow T_4$ and (2) $T_2 \rightarrow T_3$. The corresponding goal constraints can be written as:

$$
\begin{aligned}
x_{1,b,t} &= x_{2,a,t} = 0 \\
x_{8,a,1} &= x_{7,b,2} = 1.
\end{aligned} \tag{5.37}
$$

Notice that it is also possible to determine the execution order of operation stages and the detailed steps in each stage simultaneously with a BIP model. For example, let us consider four material-transfer tasks in the pipeline network in Fig. 5.1, i.e., $T_1 \rightarrow T_3$, $T_2 \rightarrow T_4$, $T_1 \rightarrow T_4$ and $T_2 \rightarrow T_3$. If the implementation order of these tasks is not specified a priori, one can formulate the goal constraints as follows:

$$
\begin{aligned}
& x_{1,b,t} = x_{2,a,t} = 0 \\
& \sum_{t=1}^{H} x_{7,a,t} = \sum_{t=1}^{H} x_{8,a,t} = \sum_{t=1}^{H} x_{7,b,t} = \sum_{t=1}^{H} x_{8,b,t} = 1
\end{aligned} \tag{5.38}
$$

where H is a sufficiently large positive integer. Finally, notice that it may not be necessary to specify the source and sink of every material-transfer path in certain multi-stage and multi-task operations. For example, if the operation goal is to clean the entire pipeline network by moving detergent through *every* fragment, then the following constraint should be imposed upon *all* fragments:

$$\sum_{t=1}^{\mathrm{H}} x_{i,t} \geq 1. \tag{5.39}$$

Notice that, in this constraint, the subscript r of the binary variable is dropped. This is due to our assumption that all source tanks are filled with detergent, and there is no need to stipulate a definite source fragment for every material-transfer route. Consequently, subscript r in the corresponding path and operation constraints must also be removed from the BIP model in the cleaning applications.

5.3.1.4 Objective Functions

An objective function is required in the formulation of any mathematical program. A reasonable choice may be:

$$\min_{\mathbf{U}^o,\mathbf{U}^c} \left[\sum_t \sum_r \sum_i x_{i,r,t} \right] \tag{5.40}$$

where,

$$\mathbf{U}^O = \begin{bmatrix} u^O_{1,1} & u^O_{1,2} & \cdots & u^O_{1,t} & \cdots & u^O_{1,\mathrm{H}} \\ u^O_{2,1} & u^O_{2,2} & \cdots & u^O_{2,t} & \cdots & u^O_{2,\mathrm{H}} \\ \vdots & \vdots & \cdots & \vdots & \cdots & \vdots \\ u^O_{j,1} & u^O_{j,2} & \cdots & u^O_{j,t} & \cdots & u^O_{j,\mathrm{H}} \\ \vdots & \vdots & \cdots & \vdots & \cdots & \vdots \end{bmatrix}$$

$$\mathbf{U}^C = \begin{bmatrix} u^C_{1,1} & u^C_{1,2} & \cdots & u^C_{1,t} & \cdots & u^C_{1,\mathrm{H}} \\ u^C_{2,1} & u^C_{2,2} & \cdots & u^C_{2,t} & \cdots & u^C_{2,\mathrm{H}} \\ \vdots & \vdots & \cdots & \vdots & \cdots & \vdots \\ u^C_{j,1} & u^C_{j,2} & \cdots & u^C_{j,t} & \cdots & u^C_{j,\mathrm{H}} \\ \vdots & \vdots & \cdots & \vdots & \cdots & \vdots \end{bmatrix}.$$

This operation objective is essential to minimize the total path length, i.e., the total number of fragments embedded in all material-transfer routes. Another candidate chosen for the present study is:

$$\min_{\mathbf{U}^o,\mathbf{U}^C} \left[\sum_t \sum_j u^O_{j,t} + \sum_t \sum_j u^C_{j,t} \right]. \tag{5.41}$$

This alternative objective is to minimize the total number of operation steps.

5.3.2 *Time-Based Operating Procedures*

Every task is expected to be performed within a specified stage in a stage-based operating procedure and each stage ends only when all its tasks are completed. Since the form, size, and length of the fragments in a practical pipeline network can be significantly different, the corresponding residence times of the material being transported should not be the same and thus the tasks at a point may end at different times. A typical stage-based schedule for three material-transfer tasks (labeled as *a, b,* and *c*) can be found in Fig. 5.29a. Due to the assumption that only paths *b* and *c* are partially overlapped in this system, these three tasks must be carried out in two consecutive stages. Since task *a* ends later than task *b* (i.e., $t_1 < t_2$), task c must start at time t_2 and end at t_4 in this schedule. It is obvious that the total operation time can be shortened to t_3 by starting task *c* at an earlier time t_1 according to the Gantt chart given in Fig. 5.29b. In order to be able to produce operating procedures to realize this type of schedules, the description of time must be incorporated into the BIP model. Following is the proposed approach to formulate the model constraints.

5.3.2.1 Time-Tracking Mechanisms

As a template for building the time-based BIP models, the Petri networks mentioned above can still be used. The only difference is that any transition in this net must be allocated a delay to reflect the residence time (or processing time) associated with its input position. In the stage-based binary integer program, the index *t* is now treated in terms of the number of time units in the time-based model as the *real time*. Consequently, the variable $x_{i,r,t}$ (which represents the state of fragment *i* on path *r*

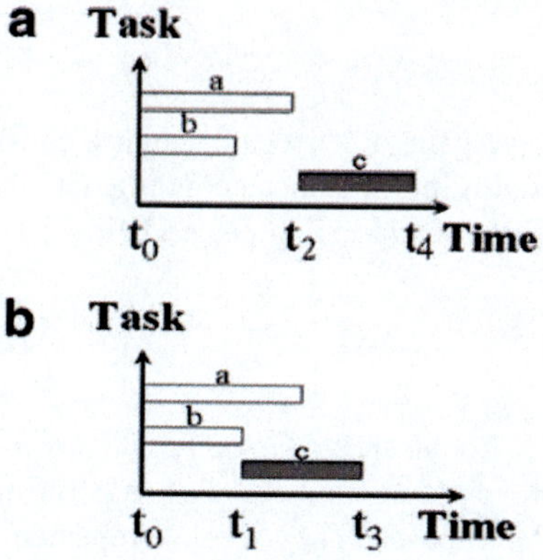

Fig. 5.29 **a** Typical schedule achieved with stage-based operating procedure. **b** Typical schedule achieved with time-based operating procedure (reprinted with permission from Lai et al. 2007. Copyright 2007 American Chemical Society)

in operation stage t) is replaced in the present case by a new variable $\sigma_{i,r,t}$ (which represents the state of fragment i on path r at actual time t). The former variable equals 1 inonly one single stage, while the latter may assume the value of 1 at several consecutive instances as long as the material-transfer task on path r is in progress. In the proposed model, the values of fragment states are controlled with two extra binary variables according to the following equation:

$$\sigma_{i,r,t+1} = \sigma_{i,r,t} + v_{i,r,t+1} - \omega_{i,r,t+1} \tag{5.42}$$

where $v_{i,r,t}$ is used to convert $\sigma_{i,r,t}$ from 0 to 1 and $\omega_{i,r,t}$ vice versa. The path constraints in Eqs. (5.13)–(5.16) are now replaced with two identical sets of inequalities to regulate $\sigma_{i,r,t}$ and $v_{i,r,t}$, i.e.,

$$\left(1 - \sigma_{i,r,t}\right) + \left(1 - \varsigma^{0}_{jd,r,t}\right) + \sigma_{id,r,t} \geq 1 \quad jd \in \mathbf{JD}_i \quad id \in \mathbf{ID}_{jd} \tag{5.43}$$

$$\left(1 - \sigma_{i,r,t}\right) + \left(1 - \varsigma^{O}_{ju,r,t}\right) + \sigma_{iu,r,t} \geq 1 \quad ju \in \mathbf{JU}_i \quad iu \in \mathbf{IU}_{ju} \tag{5.44}$$

$$\left(1 - \sigma_{i,r,t}\right) + \sum_{jd \in \mathbf{JD}_i} \varsigma^{O}_{jd,r,t} \geq 1 \tag{5.45}$$

$$\left(1 - \sigma_{i,r,t}\right) + \sum_{j \in \mathbf{J}_i} \varsigma^{O}_{ju,r,t} \geq 1 \tag{5.46}$$

$$\left(1 - v_{i,r,t}\right) + \left(1 - \varsigma^{O}_{jd,r,t}\right) + v_{id,r,t} \geq 1 \quad jd \in \mathbf{JD}_i \quad id \in \mathbf{ID}_{jd} \tag{5.47}$$

$$\left(1 - v_{i,r,t}\right) + \left(1 - \varsigma^{O}_{ju,r,t}\right) + v_{iu,r,t} \geq 1 \quad ju \in \mathbf{JU}_i \quad iu \in \mathbf{IU}_{ju} \tag{5.48}$$

$$\left(1 - v_{i,r,t}\right) + \sum_{jd \in \mathbf{D}_i} \varsigma^{O}_{jd,r,t} \geq 1 \tag{5.49}$$

$$\left(1 - v_{i,r,t}\right) + \sum_{j \in J_i} \varsigma^{O}_{ju,r,t} \geq 1. \tag{5.50}$$

Again, it is necessary to use constraint (5.36) to combine all fictitious connection states into the corresponding actual equipment state. For the sake of completeness, this constraint is repeated below:

$$\sum_{r} \varsigma^{o}_{j,r,t} = y^{o}_{j,t}. \tag{5.36$'$}$$

To ensure that exactly one downstream connection and one upstream connection for any fragment FR_i on a material-transfer path, constraints (5.17) and (5.18) should be imposed. They are also repeated as follows:

$$\sum_{jd \in \mathbf{JD}_i} y^{o}_{jd,t} \leq 1 \tag{5.17$'$}$$

$$\sum_{j \in \mathbf{JU}_i} y^{o}_{ju,t} \leq 1. \tag{5.18$'$}$$

Notice that a material-transfer task begins at the time when a feasible route can be chosen to satisfy all the path constraints mentioned above, i.e., equations (5.17′), (5.18′), (5.36′), and (5.43)–(5.50). At this initial time (say t_0), all state variables σ_{i,r,t_0} and v_{i,r,t_0} corresponding to the fragments on the selected path should be converted from 0 to 1 according to Eq. (5.42). If the task is still not completed at time t_0+1 (i.e., $\omega_{i,r,t_0+1} = 0$), it can be further deduced from Eq. (5.42) that the values of v_{i,r,t_0+1} s should all be switched back to 0 while those of σ_{i,r,t_0+1} s must remain unchanged at 1. Based on the same rationale, it can be concluded that the values of $\sigma_{i,r,t}$ and $v_{i,r,t}$ at the later time instances ($t = t_0 + 2, t_0 + 3, \ldots$) should be kept at 1 and 0, respectively, as long as the task is not terminated.

The control of $\omega_{i,r,t}$ s is achieved with still another set of binary variables $\tau_{i,r,t}$ s, which can be used to determine the total processing time of a material-transfer task via the selected path. In particular, the initial time of the task is recorded with the following equation:

$$\tau_{q,r,t} = v_{q,r,t} \quad q \in \mathbf{SC}_r \tag{5.51}$$

where $\mathbf{SC}_r$ is the set of source fragments of route r. In this study, it is assumed that the mean residence time needed for fluid particles to be transported through every fragment in the pipeline network can be determined in advance and the total processing time of a particular task can be estimated by summing these characteristic times associated with all fragments on the material-transfer path. Thus, the time-based token movements along the selected path can be described as:

$$\left(1 - \tau_{i,r,t}\right) + \left(1 - \varsigma^{O}_{jd,r,t}\right) + \tau_{id,rt+n_i} \geq 1 \quad jd \in \mathbf{JD}_i \quad id \in \mathbf{ID}_{jd} \tag{5.52}$$

where n_i denotes the mean residence time of fragment FR_i. Notice that the values of $\omega_{i,r,t}$ s on route r should all be zero except when the task terminates at time t. For the possible sink fragments,

$$\omega_{q,r,t+n_q} = \tau_{q,r,t} 4q \in \mathbf{SK}_r \tag{5.53}$$

where n_q denotes the mean residence time of sink fragment FR_q; $\mathbf{SK}_r$ is the set of all possible sink fragments of route r. For the other fragments on the material-transfer route,

$$\begin{aligned}&\left(1-\sigma_{i,r,t}\right)+\left(1-\sum_{q\in\mathbf{SK}_r}\omega_{q,r,t+1}\right)+\omega_{i,r,t+1}\geq 1\\&\left(1-\omega_{i,r,t}\right)+\sum_{q\in\mathbf{SK}_r}\omega_{q,r,t}\geq 1\\&\left(1-\omega_{i,r,t+1}\right)+\sigma_{i,r,t}\geq 1.\end{aligned}\tag{5.54}$$

These constraints are equivalent to the logic statement that if and only if fragment FR_i is used on path r at time t and the task is terminated at time $t+1$, then the corresponding $\omega_{i,r,t+1}$ must also be set to 1.

Finally, to ensure that none of the fragments can be shared by more than one route at the same time, Eq. (5.35) should be rewritten as

$$\sum_r \sigma_{i,r,t}\leq 1.\tag{5.55}$$

5.3.2.2 Time-Based BIP Model

In a sense, the topic provided in the subsection above is a summary of the limitations of the time-based direction. They are very different from their stage-based counterparts, it can be noted. On the other hand, with only slight modifications, the stage-based operation constraints can be implemented in time-based integer programs. First of all, the subscript t should of course be added to Eqs. (5.19)–(5.32). In the time-based version of constraint (5.7), $y_{j,t}^{OI}$ and $y_{j,t}^{CI}$ represent, respectively, the initial values of $y_{j,t}^{O}$ and $y_{j,t}^{C}$ associated with *every* equipment unit in the system. They can be determined from the equipment states at the previous time instance, i.e.,

$$\begin{aligned}y_{j,t}^{OI}&=y_{j,t-1}^{O}\\y_{j,t}^{CI}&=y_{j,t-1}^{C}.\end{aligned}\tag{5.56}$$

Since each task may last for a period of time, the routine operation steps described in constraint (5.23) must be performed at the termination time, i.e.,

$$\left(1-\sum_r\omega_{i,r,t}\right)+\left(1-y_{j,t}^{o}\right)+z_{j,t}^{C}=1 \quad j\in\mathbf{JP}\cup\mathbf{JS}\quad i\in\mathbf{IP}_j\cup\mathbf{IS}_j\tag{5.57}$$

where the element of set $\mathbf{IP}_j$ is the upstream fragment of power-generating unit j while $\mathbf{IS}_j$ represents the set of upstream fragment of source valve j. Also, for the purpose of improving the solution efficiency, let us assume that all operation actions can be performed only at the initial times or termination times of the material-transfer tasks. The corresponding requirements can be written as:

$$\begin{aligned}&\sum_r \left(v_{i,r,t} + \omega_{i,r,t}\right) + \left(1 - u^o_{j,t}\right) \geq 1 \quad j \in \mathbf{JP} \cup \mathbf{JV} \quad i \in \mathbf{IP}_j \cup \mathbf{IV}_j \\ &\sum_r \left(v_{i,r,t} + \omega_{i,r,t}\right) + \left(1 - u^c_{j,t}\right) \geq 1 \quad j \in \mathbf{JP} \cup \mathbf{JV} \quad i \in \mathbf{IP}_j \cup \mathbf{IV}_j\end{aligned} \tag{5.58}$$

where **JV** denotes the set of all valves in the pipeline network; $\mathbf{IV}_j$ represents the set of upstream fragment of source valve j.

Since the methods for formulating target constraints are the same as those used to create the stage-based BIP models, for the sake of brevity, they are not replicated here. In the time-based integer program for synthesizing an operating procedure with the fewest steps, the objective function presented in Eq. (5.41) can be used. It is possible to express the objective function for achieving the shortest path length as:

$$\min_{\mathbf{U}^o, \mathbf{U}^c} \sum_t \sum_r \sum_i v_{i,r,t}. \tag{5.59}$$

Finally, it should be noted that a new objective function can also be formulated to generate a procedure for completing the given tasks within the shortest time period. For this purpose, let us introduce the final set of binary variables f_t s to reflect if one or more material-transfer task is taking place at time t, i.e.,

$$\left(1 - \sigma_{i,r,t}\right) + f_t \geq 1. \tag{5.60}$$

To avoid creating a schedule with one or more interruption period, the following constraints must also be imposed:

$$(1 - f_{t+1}) + f_t \geq 1. \tag{5.61}$$

Thus, the objective function for yielding the minimum operation time can be written as:

$$\min_{\mathbf{U}^o, \mathbf{U}^C} \sum_{t=1}^{\mathrm{H}} f_t. \tag{5.62}$$

5.3.2.3 Beer Filtration Plant

Let us consider the beer filtration plant presented in Fig. 5.30a. This system consists of two filters (MMS1 and MMS2), two buffer tanks (T_1 and T_2), a supply and collection system for the cleanser or purger, and the interconnecting pipeline network. Notice that the filtration process is operated with 16 double-disk piston valves (DV1–DV16). Each valve can be switched to two alternative positions, i.e., on and off, according to Fig. 5.31a–c to manipulate the connections between its inlets and outlets.

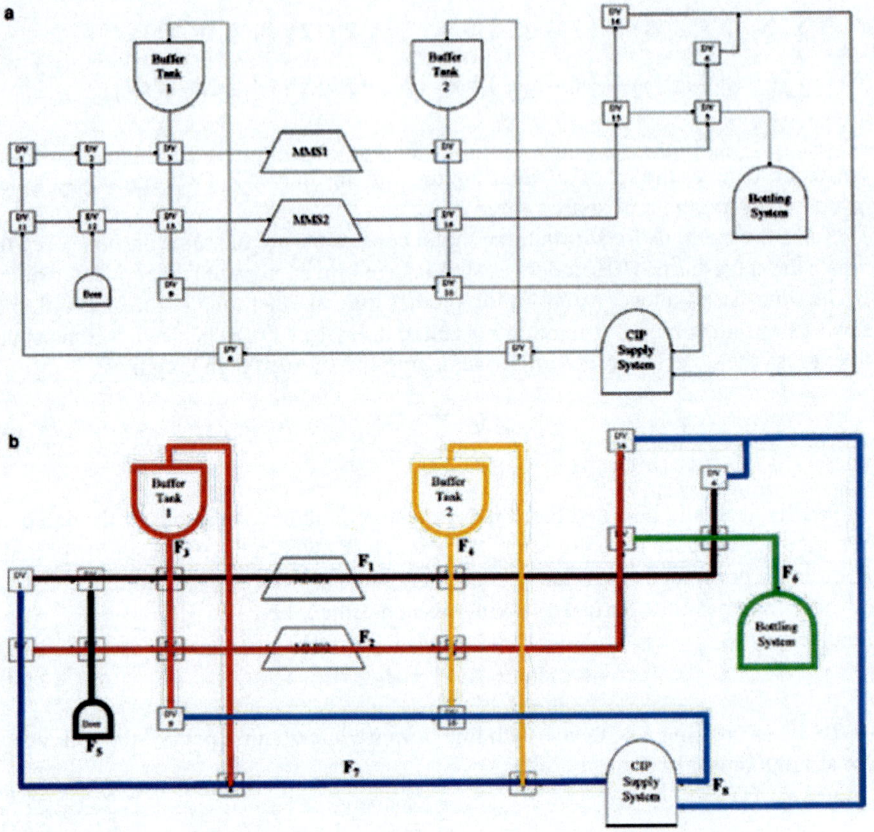

Fig. 5.30 **a** Beer filtration plant. **b** Piping fragments in beer filtration plant (Reprinted with permission from Lai et al. 2007. Copyright 2007 American Chemical Society)

Consequently, the process flow diagram in Fig. 5.30a can be divided into the 8 color-coded fragments presented in Fig. 5.30b. Notice that the path model of this system can be constructed accordingly in a straightforward fashion. This Petri net is not presented here for the sake of brevity.

There are four different operations in this beer filtration process, i.e., filling, filtering, bottling, and cleaning. The purpose of filling operation is to transport fresh beer from a source tank to the buffer tank T_1. In the filtering operation, the beer is transferred from tank T_1 to T_2 via filter MMS1 or MMS2. Clearly, the filtered beer in T_2 should then be moved to the bottling station in another material-transfer operation. The last plant procedure deals with the cleaning of the fragments with which beer has previously been manufactured. In this case, it is presumed that each fragment must be cleaned with a purger after a specified number of times has been used.

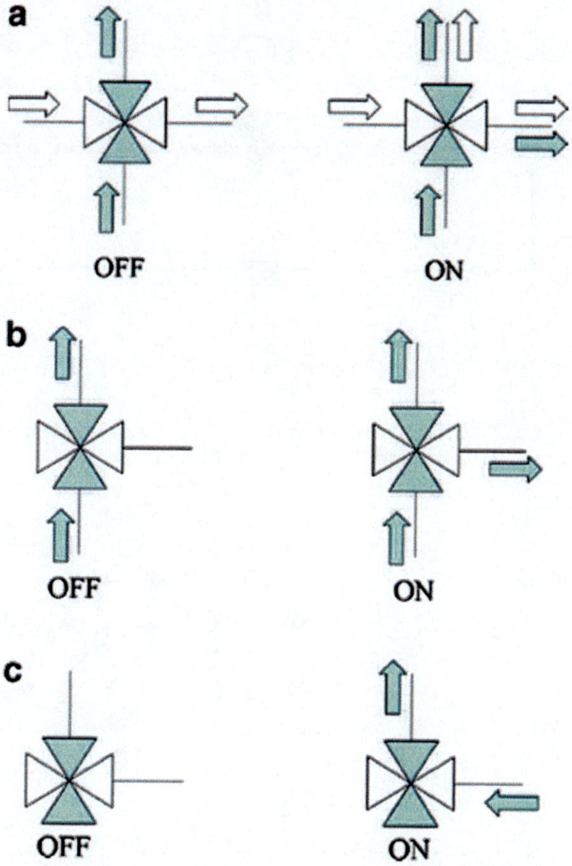

Fig. 5.31 Double-disk piston valve. **a** Two inputs and two outputs. **b** One input and two outputs. **c** One input and one output (reprinted with permission from Lai et al. 2007. Copyright 2007 American Chemical Society)

5.3.2.3.1 Petri Net Representations of Buffer Tanks and Filters

Specific Petri nets have been built on the basis of the process model given in Fig. 5.32 to represent the state transfer processes of tanks and filters in the beer filtration plant. As a general model of these two units, the Petri net in Fig. 5.33a can be considered. Note that eight places are on this net. Four of them are used to represent *equipment states* and the rest are *work states*. The former four states are described as "clean", "full with beer", "foul" and "full with purger", while the latter states can be interpreted as transferring " beer into", "beer out of", "purger into" and "purger out of" the given unit. If the purger does not stay in the process unit for a significant period of time during cleaning operation, then the three places on the right can be combined to form the Petri net given in Fig. 5.33b. This is the process model for the buffer tanks T_1 and T_2. Notice that the combined place can be regarded as the work state "cleaning" in this case. If it can be assumed that beer goes through the filters almost immediately, the Petri net model in Fig. 5.33b can be further simplified to the net in Fig. 5.33c.

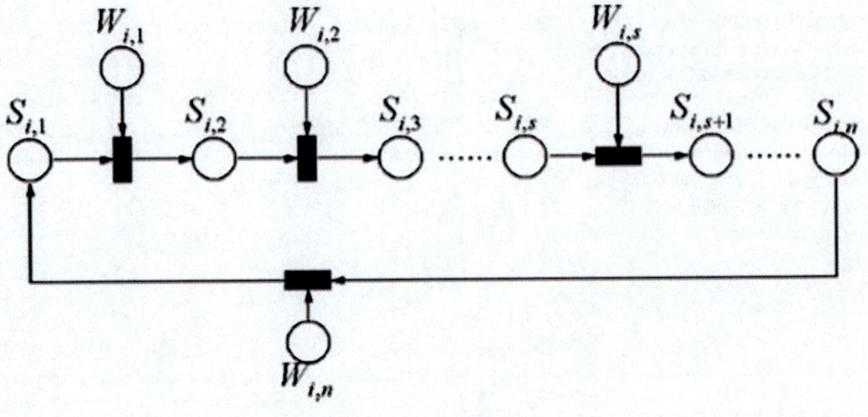

Fig. 5.32 General petri net representation of a process unit (reprinted with permission from Lai et al. 2007. Copyright 2007 American Chemical Society)

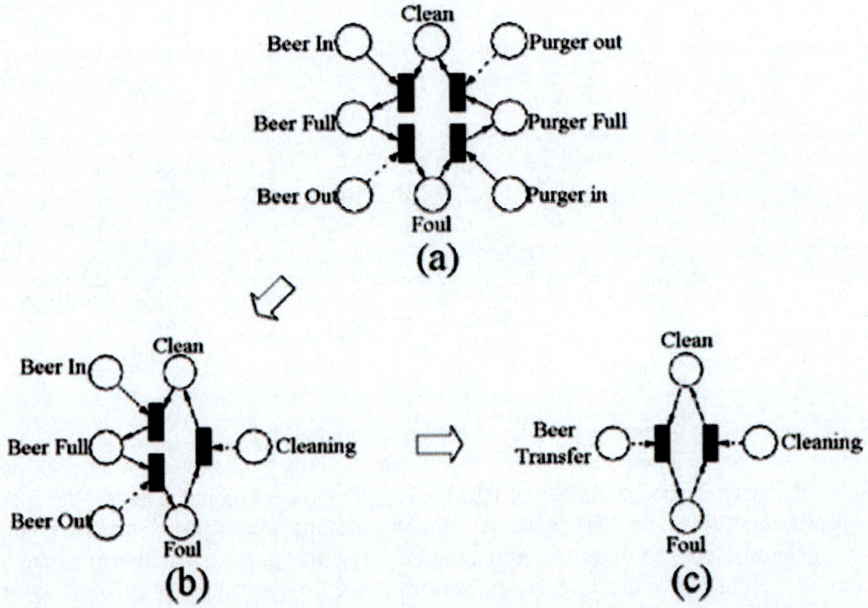

Fig. 5.33 Petri net models of process units. **a** Generalized model. **b** Tank model. **c** Filter model (reprinted with permission from Lai et al. 2007. Copyright 2007 American Chemical Society)

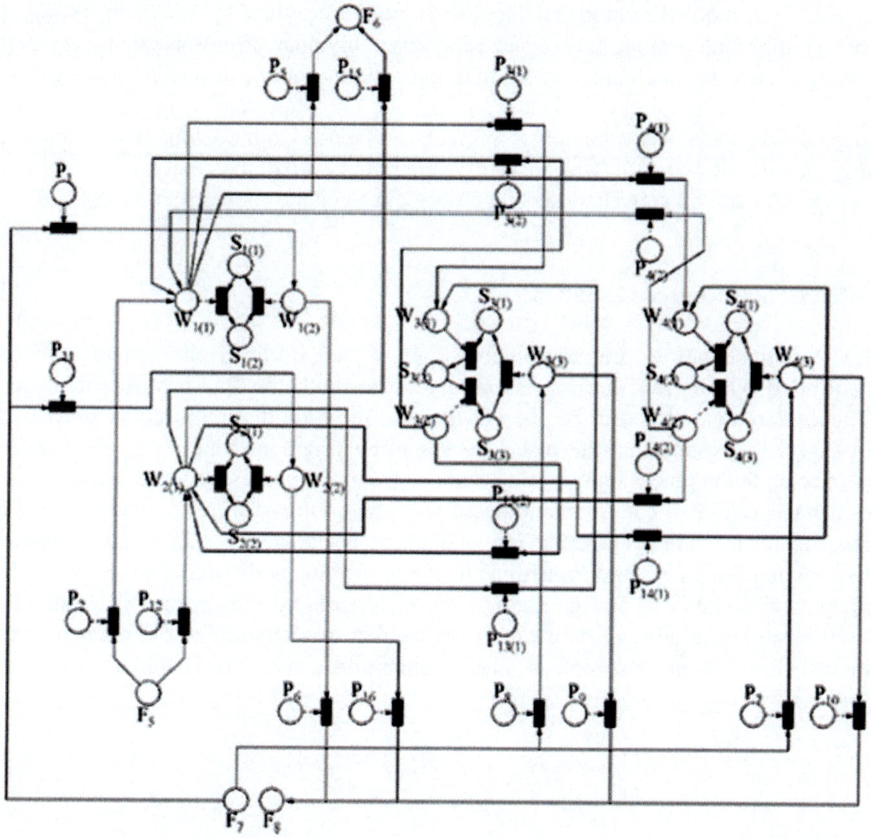

Fig. 5.34 Modified path model of the beer filtration plant (reprinted with permission from Lai et al. 2007. Copyright 2007 American Chemical Society)

The process models in Fig. 5.33b, c can be used to replace the places representing the corresponding fragments, i.e., $F_3(T_1)$, $F_4(T_2)$,F_1 (MMS1), and F_2 (MMS2), in the original path model. This modified version is presented in Fig. 5.34. In the 4 embedded process models, the places with label *S* are used to reflect the equipment states and those with label *W* represent the work states. The state of each of the remaining four fragments is modeled with a single place. In particular, F_5 and F_6 are the source and sink fragments of beer, while F_7 and F_8 are associated with the corresponding fragments of purger. From Fig. 5.34, it can be observed that the sources F_5 and F_7 are connected to the places representing the work states of filters and tanks, and they are then linked to the sinks F_6 and F_8. Thus, it is clear that the material-transfer paths should contain places representing the work states. It should also be noted that the complete system model can be obtained by attaching the valve models to the Petri net in Fig. 5.34.

Finally, it should be noted that the process models provided in Fig. 5.33b, c suggest that a buffer tank or filter may be used for reuse only once prior to cleaning. If a unit can be used without cleaning more than once, additional positions in these models need to be incorporated to reflect the additional unit states and working states. For instance, the Petri networks in Fig. 5.33a, b can be used to model the tank and filter, respectively, allowing for two consecutive operations. All of these networks and those in Fig. 5.33 can be viewed as special cases of the general model seen in Fig. 5.32.

5.3.2.3.2 Stage-Based Operation

As mentioned before, the equipment states of the 3rd-level components can be controlled by altering the process configuration with the 2nd-level components. This hierarchical structure can be observed in the modified path model presented in Fig. 5.34. Notice that the places representing fragment states (i.e., F_5–F_8) are connected to the places representing work states in the process models. The connection states (P_1–P_{16}) can be manipulated with the double-disk piston valves. Notice also that the modified path model in Fig. 5.34 can be converted to the original version by lumping the places and transitions in every process model into a single place. In other words, the path and operation constraints presented previously can still be used here, but additional constraints are needed to characterize the token movements within the process models. These constraints can be formulated according to Fig. 5.32, i.e.,

$$\begin{aligned} &\left(1-\lambda_{i,s,t}\right)+\left(1-\tilde{x}_{i,s,t}\right)+\lambda_{i,s',t+1}\geq 1 \\ &\left(1-\lambda_{i,s,t}\right)+\tilde{x}_{i,s,t}+\left(1-\lambda_{i,s',t+1}\right)\geq 1 \end{aligned} \tag{5.63}$$

where,

$$s'=\begin{cases} s+1 \text{ if } s=1,2,\cdots,n-1 \\ 1 \qquad\quad \text{if } s=n. \end{cases}$$

In the above constraints, $\lambda_{i,s,t}$ is the token number in place $S_{i,s}$ representing the sth equipment state of unit i at stage t, and $\tilde{x}_{i,s,t}$ is the token number in place $W_{i,s}$ representing the sth work state of unit i at stage t. Notice that the constraints in Eq. (5.63) are in a sense very similar to those given in equation (5.19). The first constraint above is equivalent to the logic statement that, given the corresponding work state, the equipment state must be switched from one to another in sequence. This statement is essentially the same as that implied by the first and third constraints in equation (5.19). On the other hand, if the aforementioned work state is absent, then the corresponding state-transition event should not occur. The 2nd constraint in equation (5.63) is imposed in the proposed process model to enforce this logic relation and its counterparts are the 2nd and 4th constraints in Eq. (5.19).

Additional constraints have also been added to enhance the solution efficiency. First of all, it is obvious that the equipment state of every unit at any given stage is unique. This condition can simply be described with the following constraint:

$$\sum_{s=1}^{n} \lambda_{i,s,t} = 1. \tag{5.64}$$

Notice that the rationale for imposing this constraint is the same as that for equation (8). Next, it is convenient to ensure that, if a particular unit state is not present, then the corresponding work state should be prohibited, i.e.,

$$\lambda_{i,s,t} + \left(1 - x_{i,s,t}\right) \geq 1. \tag{5.65}$$

Notice that the constraints in Eq. (5.17) can be interpreted from a similar standpoint.

It is conceivable that there are only two possibilities for a unit to reach a particular state at a given stage, i.e., (1) the unit was in a preceding state at the previous stage and switched to the present state afterwards and (2) the unit was in the present state at the previous stage and remained unchanged. The corresponding constraint can be written as:

$$\left(1 - \lambda_{i,s',t}\right) + \lambda_{i,s,t-1} + \lambda_{i,s',t-1} \geq 1. \tag{5.66}$$

Since this inequality is always valid in the case of two-state components, it is therefore not included in the operation constraints.

As stated earlier, it is possible to interpret the locations representing work states as members of material-transfer routes. Thus, on the basis of both fragment states and work states of different units, the objective constraints in this example are formulated. It should also be pointed out that there is no need to add the subscript r to separate the source fragments of the material-transfer paths in order to produce stage-based operating procedures in the present application. This is due to the fact that restriction (5.64) prohibits the same units from performing multiple operations. Consequently, the target constraints of the four separate operations described above can be summarized in the sequel:

- In filling operation, it is required to transfer fresh beer from source tank to buffer tank T_1. This requirement can be formulated as:

$$x_{5,t} = \sum_{n_1=1}^{N_1} \tilde{x}_{3,bi(n_1),t} \tag{5.67}$$

where N_1 is the number of times allowed for repeating the filling operation in T_1 without cleaning; $bi(n_1)$ denotes the work state of transferring beer into the buffer tank for the n_1 th time.

- In the filtering operation, the beer in tank T_1 must be sent to tank T_2 by way of a filter. The corresponding constraint is:

$$\sum_{n_1=1}^{N_1} \tilde{x}_{3,bo(n_1),t} = \sum_{n_2=1}^{N_2} \tilde{x}_{4,bi(n_2),t} \tag{5.68}$$

  where N_2 is the number of repeat operations allowed in T_2; $bi(n_2)$ represents the work state of transferring beer into the buffer tank T_2 for the n_2 th time; $bo(n_1)$ is corresponding to the work state of withdrawing beer from tank T_1 for the n_1 th time. It can be observed from Fig. 5.16 that the resulting path must contain the filter in fragment F_1 or F_2.

- The third operation is concerned with the transportation of filtered beer from tank T_2 to the bottling station in fragment F_6. The goal constraint in this case can be developed with essentially the same approach as before:

$$\sum_{n_2=1}^{N_2} \tilde{x}_{4,bo(n_2),t} = x_{6,t}. \tag{5.69}$$

- The final operation is cleaning. This operation can be characterized as the task of moving the cleanser from supply fragment (F_7) to collection fragment (F_8) via a path containing the fouled units. Thus,

$$x_{7,t} = x_{8,t}. \tag{5.70}$$

The operation objective in this application is to minimize the number of operation steps to produce a fixed amount of bottled beer in a given stage horizon H. Thus, the objective function of the corresponding BIP model can be expressed by equation (5.41) and an extra constraint must be included in this model to stipulate the given product quantity B, i.e.,

$$\sum_{t=1}^{\mathrm{H}} x_{6,t} = B. \tag{5.71}$$

The appropriate product quantity may be determined by maximizing B while satisfying all constraints [except (5.71)] in the aforementioned integer program.

It is presumed in the present application that 2 repeat operations are permitted in filters and 3 are permitted in tanks. In the BIP model, the total stage number () adopted is 15. The actual number of bottling operations (B) has been found to be 5

Table 5.14 Stage-based operation sequence of the beer filtration plant

Stage	Task
1	Replenish T_1 with beer via MMS2
2	Transfer beer from T_1 to T_2 via MMS1 (beer filtration operation)
3	Replenish T_1 with beer via MMS1
	Transfer filtered beer from T_2 to bottling system via MMS2
4	Clean MMS1
5	Transfer filtered beer from T_2 to bottling system via MMS1
6	Replenish T_1 with beer via MMS1
	Clean MMS2
7	Clean MMS1
8	Transfer beer from T_1 to T_2 via MMS2 (beer filtration operation)
9	Transfer filtered beer from T_2 to bottling system via MMS1
	Clean T_1
10	Replenish T_1 with beer via MMS2
	Clean T_2
11	Transfer beer from T_1 to T_2 via MMS2 (beer filtration operation)
12	Replenish T_1 with beer via MMS1
	Clean MMS2
13	Transfer filtered beer from T_2 to bottling system via MMS2
	Clean MMS1
14	Transfer beer from T_1 to T_2 via MMS1 (beer filtration operation)
15	Transfer filtered beer from T_2 to bottling system via MMS2

and the minimum number of operational steps is 74. It is possible to define a total of 15 stages and find the operations conducted at each stage in Table 5.14. Table 5.15 presents the resulting stage-based operating procedure.

5.3.2.3.3 Time-Based Operation

An additional index r is added to differentiate the variables representing the fragment states and also the work states of filters and tanks in order to evaluate the processing times of various operations. For characterizing the complex behaviors of fragment states in the present case, the time-tracking mechanisms described in equations (5.42)–(5.55) are still valid. On the other hand, an equation similar to the equation (5.42) should define the working states of process units:

$$\tilde{\sigma}_{i,s,r,t+1} = \tilde{\sigma}_{i,s,r,t} + \tilde{v}_{i,sy,t+1} - \tilde{\omega}_{i,s,r,t+1}. \tag{5.72}$$

Table 5.15 Stage-based operating procedure of the beer filtration plant

Stage	Operation actions
1	Open valves V_{12} and V_{13}
2	Close valve V_{12}
	Open valves V_3, V_4, and V_{14}
3	Close valves V_3 and V_4
	Open valves V_2 and V_5
4	Close valves V_2 and V_5
	Open valves V_1 and V_6
5	Close valves V_1, V_6, V_{13}, and V_{14}
	Open valves V_4 and V_5
6	Close valves V_4 and V_5
	Open valves V_2, V_3, V_{11}, and V_{16}
7	Close valves V_2, V_3, V_{11}, and V_{16}
	Open valves V_1, V_6, V_{13}, and V_{14}
8	Close valves V_1 and V_6
9	Close valve V_{13}
	Open valves V_4, V_5, V_8, and V_9
10	Close valves V_4, V_8, V_9, and V_{14}
	Open valves V_7, V_{10}, V_{12}, and V_{13}
11	Close valves V_7, V_{10}, and V_{12}
	Open valve V_{14}
12	Close valves V_5, V_{13}, and V_{14}
	Open valves V_2, V_3, V_{11}, and V_{16}
13	Close valves V_2, V_3, V_{11}, and V_{16}
	Open valves V_1, V_6, V_{14}, and V_{15}
14	Close valves V_1, V_6, and V_{15}
	Open valves V_3 and V_4
15	Open valve V_{15}

The time-tracking constraints of $\tilde{\sigma}_{i,s,r,t}$, $\tilde{V}_{i,s,r,t}$, and $\tilde{\omega}_{i,s,r,t+1}$ can be derived in a straightforward fashion on the basis of Eqs. (5.43)–(5.55). They are omitted in this paper for the sake of brevity.

Due to the aforementioned changes in expressing the work states, it is also necessary to modify the constraints used in the stage-based case for describing the process models. Specifically, Eq. (5.64) should be changed to:

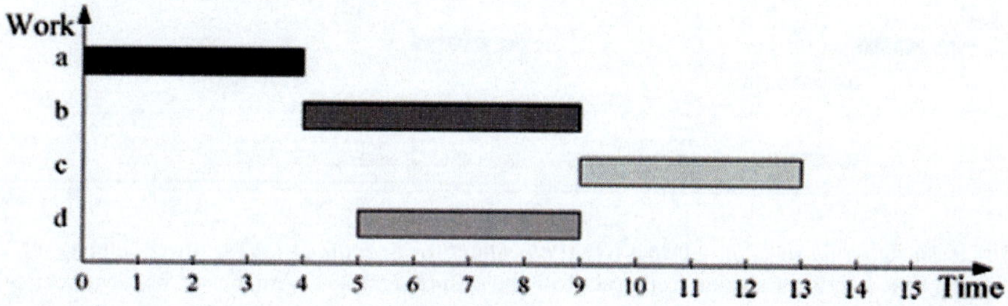

Fig. 5.35 Optimal schedule of time-based procedure for operating the beer filtration plant ($H = 15$) (reprinted with permission from Lai et al. 2007. Copyright 2007 American Chemical Society)

Table 5.16 Time-based operating procedure of the beer filtration plant

Time	Operation actions
0	Open P_2, P_3, and P_{14}
4	Close P_2 and P_3; open P_{13}
5	Open P_1 and P_6
9	Close P_1, P_6, P_{13}, and P_{14}; open P_4 and P_5
13	Close P_5

$$
\begin{aligned}
&(1-\lambda_{i,s,t}) + \left(1 - \sum_r \tilde{\omega}_{i,s,t}\right) + \lambda_{i,s't+1} \geq 1 \\
&(1-\lambda_{i,s,t}) + \sum_r \tilde{\omega}_{i,s,r,t} + (1-\lambda_{i,s',t+1}) \geq 1 \\
&\sum_r \tilde{\omega}_{i,s,r,t} \leq 1.
\end{aligned}
\tag{5.73}
$$

Equation (5.65) must also be rewritten as:

$$
\lambda_{i,s,t} + (1 - \tilde{\sigma}_{i,sy,t}) \geq 1. \tag{5.74}
$$

Equations (5.64) and (5.66) should remain the same.

In the time-based procedure case studies, it is presumed that all units must be cleaned after each operation. Thus, in Fig. 5.34, the modified path model of this method can be represented by the Petri net. 2 time units are selected to be the residence times of beer and cleanser in filters. It is all presumed that the residence times associated with all other fragments and units are 1.

In the BIP model, a time horizon (H) of 15 units of time was first used. It has been found that it is possible to conduct only one bottling process and the minimum number of operation steps is 15. Figure 5.35 and Table 5.16, respectively, show the related service schedule and operating procedure. The time horizon was then expanded to 35 units of time. As a consequence, with 47 operation measures, two batches of bottled beer can be made. Figure 5.36 displays the related schedule.

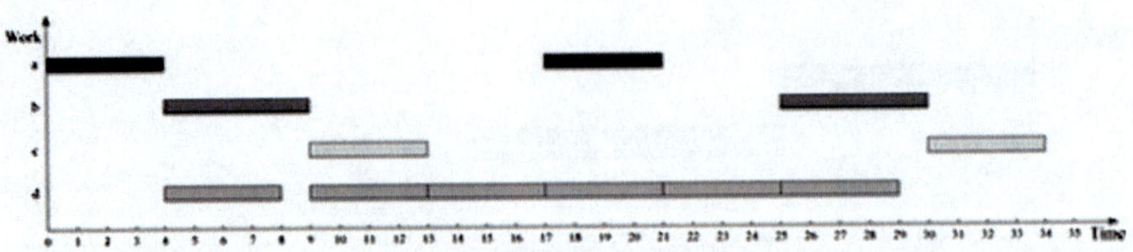

Fig. 5.36 Optimal schedule of time-based procedure for operating the beer filtration plant ($H = 35$) (reprinted with permission from Lai et al. 2007. Copyright 2007 American Chemical Society)

5.4 Semiconductor Manufacturing Process Scheduling Strategy

In recent years, the highly automated and capital-intensive nature of modern semiconductor manufacturing facilities has made the design and management of their production processes a significant area of research activity. In a standard campaign, multiple types of goods are usually processed simultaneously. Each product must go through more than one phase, and one out of several available instruments will perform the operations in each phase. Traditionally, based on previous experiences, the production schedules in such a setting are manually stipulated in an ad hoc fashion. This scheduling activity is also time-consuming, laborious, and error-prone, owing to the high degree of process complexity. There are strong incentives to establish a computer-aided approach to promote the systematic generation of optimal schedules for semiconductor production operations in order to circumvent these disadvantages.

There have been many scheduling-related studies reported in the literature. A few examples are given below. Uzsoy et al. (1991) tried to use disjunctive graph to model the final test procedure; Johri (1993) and Duenyas et al. (1994) outlined the major challenges in planning and scheduling semiconductor processes; Chen et al. (1995) constructed an integer programming model to describe the test process in IC production and then solved this model with the Lagrangian relaxation technique; Lu and Kumar (1991) and Narahari and Khan (1996) applied first-buffer-first-serve (FBFS) and last-buffer-first-serve (LBFS) strategies to the re-entrant processes; Kumar et al. (2004) analyzed the re-entrant wafer production schemes on the basis of queuing theory.

Due to the extreme complexity of multi-product, multi-stage, and multi-tool processes, it is clear that the wafer production schedule of any given system cannot be successfully synthesized without an appropriate model of its manufacturing activities. Since the Petri net has long been demonstrated to be suitable for characterizing and analyzing the discrete-event systems (Peterson 1981; David and Alla 1994), it was adopted in a number of subsequent works for modeling and simulating the semiconductor production processes. Cavalieri et al. (1997) used the colored Petri nets to describe the semiconductor production sequences and, then, identified the minimum-cost schedules according to the corresponding reachability trees with a trial-and-error approach; Lin and Huang (1998) modeled the furnace in an IC fab also with colored-timed Petri net; Allam and Alla (1998) simulated and analyzed the assembling and testing processes on the basis of hybrid Petri nets; Jeng et al. (1998) modeled the etching area in an IC lot fabrication system with Petri nets, and also evaluated its performance accordingly; Xiong and Zhou (1998) built the Petri net model for semiconductor test facility and applied heuristic search algorithm to identify test schedules; Jeng et al. (2000) developed Markovian timed Petri nets for performance analysis of semiconductor manufacturing systems; Kuo and Huang (2003) constructed colored-timed Petri net models to design flexible processing routes for multiple products in IC fabs; Chiang et al. (2006) and Chien and Chen (2007) proposed Petri net-based scheduling models and solved with genetic algorithm (GA).

From the aforementioned studies, it can be observed that although the Petri networks are very useful for modeling and simulating semiconductor manufacturing activities, the numerical algorithms necessary to identify a correct schedule (e.g., GA) may not be sufficiently efficient. On the other hand, it should be remembered that in the last two decades, the scheduling approaches used for batch chemical processes have progressed considerably. A large number of generalized models have successfully been developed and applied. For example, Pagageorgaki and Reklaitis (1990) proposed a MINLP model to generate the optimal schedules for multi-product batch processes; Kondili et al. (1993) solved this problem with a MILP program; Kim and Moon (2000) synthesized the multi-purpose schedules with symbolic model verifier (SMV). It is worth noting that the mathematical program reported in Kondili et al. (1993) was formulated with discrete-time representation according to a graphic model of the batch process, i.e., the so-called state-task network (STN). Ierapetritou and Floudas (1998a, b)later proposed a STN-based MILP model for producing an optimal batch schedule. An equivalent resource task network (RTN) representation was proposed by Pantelides (1994). Zhang and Sargent (1996) provided a unified mathematical formulation to determine the optimal operating conditions of RTN in continuous-time representation. The STN-based model later became a popular choice for many scheduling applications due to its capability to capture the equipment-sharing possibilities. Various other mathematical programs have also been constructed accordingly, e.g., Shah et al. (1993) and Maravelias and Grossmann (2003). An overview of the continuous-time versus discrete-time approaches for scheduling multi-product and/or multi-purpose batch processes was presented by Floudas and Lin (2004). Finally, a STN-based cyclic scheduling strategy has also been proposed by Wu and Ierapetritou (2004). They essentially modified the aforementioned short-term model (Ierapetritou and Floudas 1998a, b) with additional constraints to accommodate the unique requirements in periodic operations.

From the above discussions, it is obvious that before constructing the schedule-generating models, a particular state-task network must be built. However, since this representation is too easy to capture all the complex features embedded in the manufacturing processes of semiconductors, it is often difficult to create an effective STN model and systematically and efficiently verify its correctness. Remember, on the other hand, that the Petri net is not only an acceptable model for wafer development operations, but also a convenient method for simulation. In validating the model and also checking the viability of any given schedule, the latter capability can be very useful. It is therefore the purpose of this work to establish a scheduling protocol based on Petri net. You may summarize the key steps of this method as follows:

- Construction of a Petri net model for characterizing all manufacturing activities in the given semiconductor process,
- Formulation of a mathematical programming model for describing the token movements in Petri net, and
- Generation of the optimal production schedules from the solution of the above-mentioned mathematical program.

5.4.1 *Petri Net Models*

A generalized method of model building is defined here. It should be noted first that the Petri networks used in this research are colorless but timed. It is possible to classify the places in a net into two general forms, i.e., tool states and buffer states, while the transitions exclusively represent activity phases. Only positive integer values may assume the token number assigned to each position and the arc weights are all one. The transitions may be allocated real-time delays to indicate processing times required in the corresponding phases of service. For the following example, a Petri net model can be constructed easily with the aforementioned conventions according to the process data provided.

After completing the IC packaging operations, a series of final tests (FTs) have to be performed on the finished products. Several separate operation stages can consist of a standard FT phase. In each process, the hardware facilities necessary include tester, handler and other auxiliary equipment, and these facilities can be viewed as work center components. The Petri net models for final test (FT) jobs were developed by Xiong and Zhou (1998). Four distinct jobs were considered in one of their examples. Three separate stages are required to complete each job and each must be carried out in a dedicated work center. Table 5.1 lists the work centers and the corresponding processing times needed for the implementation of the activity stages in each job. In addition, let us presume that these work centers do not share similar components and can thus be independently run.

On the basis of Table 5.17, the corresponding Petri net model can be constructed easily according to the proposed conventions (see Fig. 5.37). In this model, transition

Table 5.17 Work centers and their processing times in FT process

Stage	Job 1	Job 2	Job 3	Job 4
1	(M1, 2)	(M3, 4)	(M1, 3)	(M2, 3)
2	(M2, 3)	(M1, 2)	(M3, 5)	(M3, 4)
3	(M3, 4)	(M2, 2)	(M2, 3)	(M1, 3)

(1 time unit = 10 min)

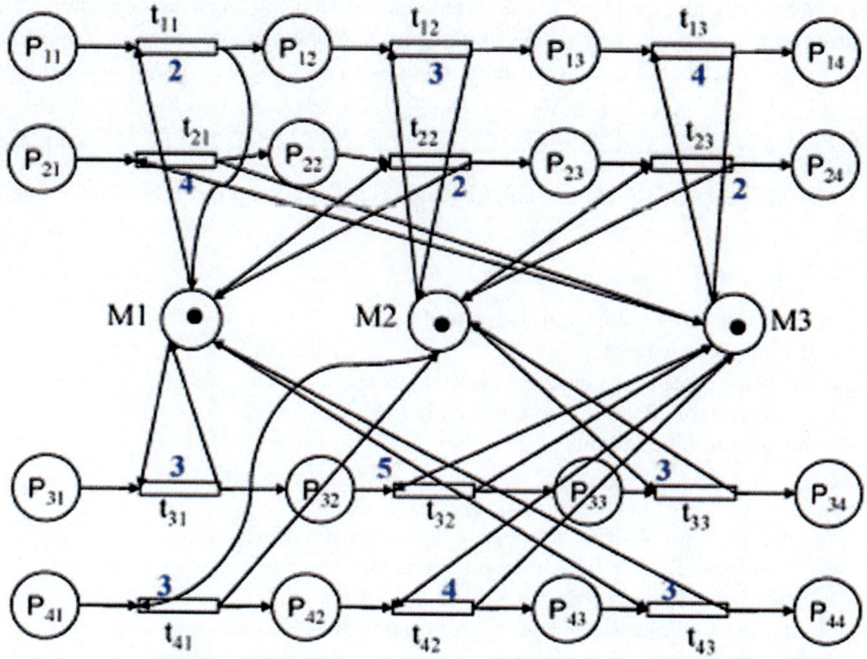

Fig. 5.37 Petri net model of a FT process with 4 jobs, 3 stages and 3 work centers (reprinted with permission from Lee et al. 2011. Copyright 2011 Elsevier)

t_{ij} represents the operation in stage j of job $i (i = 1, 2, 3, 4;\; j = 1, 2, 3)$; place P_{ik} denotes the buffer state k of job i $(i = 1, 2, 3, 4;\; k = 1, 2, 3, 4)$; place M_l denotes the state of l th machine or work center (i.e., tool state). Finally, notice that the processing times are used as the delay times of the transitions.

5.4.2 *Optimal Scheduling Strategy*

The Petri net model can be used as the basis for simulating development activities in a semiconductor manufacturing plant in the light of a feasible schedule. Since the best development schedule from such a graphic model cannot always be easily defined, this study instead describes the token movements in a Petri net with the limitations of the mathematical programming model. By resolving this model according to any given objective function, an optimal schedule can then be created.

Ierapetritou and Floudas (1998a, b) introduced a conceptual term, *event point*, to denote the instance when a task (and also the use of corresponding unit) begins or ends. On the basis of this idea, a binary variable $am(t, m, n) \in \{0, 1\}$ is adopted in this work to reflect the conditions of transition t and the place associated with tool *m*. Specifically, $am(t, m, n') = 1$ denotes that, starting from the event point n', transition t is *enabled* by the state of place m, i.e., one or more token is present in this place. These conditions are usually maintained for a finite period of time, which is denoted as $delay(t, m)$, until the enabled transition t can be fired. On the other hand, $am(t, m, n'') = 0$ simply means that transition t cannot be enabled at event point n'' by the state of place m.

The indices, sets, parameters, and variables used in the proposed model are defined below.

- Indices

  m: the place label associated with a tool;
  n: the label of an event point;
  p: the place label representing an actual buffer;
  s: the place label denoting an artificial buffer;
  t: the label of a transition.

- Sets

  $\mathcal{M}$: the set of all places representing the states of available tools;
  $\mathcal{M}_t$: a subset of $\mathcal{M}$ in which the places are all connected to transition$t$;
  $\mathcal{N}$: the set of all event points, i.e.,$\{1, 2, \ldots, N\}$;
  $\mathcal{P}$: the set of all places representing the states of actual buffers;
  $\mathcal{S}$: the set of all places representing the states of artificial buffers;
  $\mathcal{S}_p$: a subset of $\mathcal{S}$ in which all places are associated with actual buffer $p \in \mathcal{P}$;
  $\mathcal{T}$: the set of all transitions;
  $\mathcal{T}_m$: a subset of $\mathcal{T}$ in which all transitions are connected to place $m \in \mathcal{M}$;
  $\mathcal{T}_s^{\text{in}}$: a subset of $\mathcal{T}$ which contains all input transitions of place $s \in \mathcal{S}$;
  $\mathcal{T}_s^{\text{out}}$: a subset of $\mathcal{T}$ which contains all output transitions of place $s \in \mathcal{S}$.

- Parameters

  delay(t, m): the delay time needed to fired transition $t \in \mathcal{T}_m$ after it is enabled;
  $st^I(s)$: the initial token number in place $s \in \mathcal{S}$;
  $st^U(p)$: the maximum token number allowed in place $p \in \mathcal{P}$.

- Variables

  $am(t, m, n)$: the binary variable used to reflect if transition $t \in \mathcal{T}$ is enabled by the state of place $m \in \mathcal{M}_t$ at event point $n \in \mathcal{N}$;
  H: the time horizon of the production campaign;
  $st(s, n)$: the token number in place $s \in \mathcal{S}$ at event point $n \in \mathcal{N}$;
  $T^E(t, m, n)$: the time at event point $n \in \mathcal{N}$ when transition $t \in \mathcal{T}$ can be enabled by the state of place $m \in \mathcal{M}_t$;
  $T^F(t, m, n)$: the time at event point $n \in \mathcal{N}$ when transition $t \in \mathcal{T}$ can be fired by the state of place $m \in \mathcal{M}_t$.

5.4.3 Token Movements

The token movements in a Petri net are created by *firing* transitions. Since an event point is regarded in this work as the instance when a transition starts to be enabled, the difference in token numbers at every place $s \in \mathcal{S}$ between two consecutive event points can therefore be expressed as:

$$st(s,n) - st(s,n-1) = \sum_{t \in \mathcal{T}_s^{\text{in}}} \sum_{m \in \mathcal{M}_t} am(t,m,n-1) - \sum_{t' \in \mathcal{T}_s^{\text{out}}} \sum_{m' \in \mathcal{M}_t} am(t',m',n) \quad \forall s \in \mathcal{S} \;\; \forall n \in \mathcal{N} \tag{5.75}$$

where $st(s,0) = st^I(s)$. To improve solution efficiency of the proposed optimization problem, it is necessary to set all unused binary variables to be zeros, i.e.,

$$am(t,m,n) = 0 \quad \forall t \in \mathcal{T} \;\; \forall m \in \mathcal{M}_t/\mathcal{M} \;\; \forall n \in \mathcal{N}. \tag{5.76}$$

Since a transition can only be fired after it has been enabled for a specified period of delay time, the firing time $T^F(t,m,n)$ should be calculated according to the enabling time $T^E(t,m,n)$ on the same event point, i.e.,

$$T^F(t,m,n) = T^E(t,m,n) + delay(t,m) \cdot am(t,m,n) \quad \forall t \in \mathcal{T} \;\; \forall m \in \mathcal{M}_t \;\; \forall n \in \mathcal{N}. \tag{5.77}$$

For the same transition$t$, the enabling time and firing time at the current event point should be earlier than the corresponding times at the subsequent event point, i.e.,

$$T^E(t,m,n+1) \geq T^E(t,m,n) \tag{5.78}$$

$$T^F(t,m,n+1) \geq T^F(t,m,n) \quad \forall t \in \mathcal{T} \;\; \forall m \in \mathcal{M}_t \;\; \forall n \in \mathcal{N} \;\; n \neq N. \tag{5.79}$$

At a particular time instance, if a transition has been enabled but not fired for an event point, the same transition is not allowed to be enabled again for the subsequent event point. In other words, the following constraint should be imposed:

$$T^E(t,m,n+1) \geq T^F(t,m,n) - H \cdot [1 - am(t,m,n)] \quad \forall t \in \mathcal{T} \;\; \forall m \in \mathcal{M}_t \;\; \forall n \in \mathcal{N} \;\; n \neq N. \tag{5.80}$$

If two or more transitions can be enabled by the same place $m \in \mathcal{M}$, then one of them (i.e., $t \in \mathcal{T}_m$) can be enabled at an event point only after another one is enabled and then fired at the prior event point, i.e.,

$$\begin{aligned} &T^E(t, m, n+1) \geq T^F(t', m, n) - H \cdot [1 - am(t', m, n)] \\ &\forall m \in \mathcal{M} \quad \forall t, t' \in \mathcal{T}_m \quad t \neq t_t' \quad \forall n \in \mathcal{N} \quad n \neq N. \end{aligned} \tag{5.81}$$

Since the precedence order of enabling and then firing any pair of neighboring transitions (say $t \in \mathcal{T}_m, t' \in \mathcal{T}_m$ and $t \neq t'$) is unambiguously specified in the given Petri net, this inherent feature must be stipulated in the model as:

$$\begin{aligned} &T^E(t, m, n+1) \geq T^F(t', m', n) - H \cdot [1 - am(t', m', n)] \\ &\forall m \in \mathcal{M}_t \forall m' \in \mathcal{M}_{t'} m \neq m' \quad \forall n \in \mathcal{N} \quad n \neq N. \end{aligned} \tag{5.82}$$

At event point $n+1$, the enabling action of transition $t \in \mathcal{T}_m$ should occur after every transition in the same set, i.e., $\forall t' \in \mathcal{T}_m$, is fired at all prior event points, i.e.,

$$\begin{aligned} &T^E(t, m, n+1) \geq \sum_{\substack{n' \in \mathcal{N} \\ n' \leq n}} \sum_{t' \in \mathcal{T}_m} [T^F(t', m, n') - T^E(t', m, n')] \\ &\forall m \in \mathcal{M} \quad \forall t \in \mathcal{T}_m \quad n \in \mathcal{N} \quad n \neq N. \end{aligned} \tag{5.83}$$

Obviously, enabling and firing of all transitions must be completed within the production horizon, i.e.,

$$T^E(t, m, n) \leq H \tag{5.84}$$

$$\begin{aligned} &T^F(t, m, n) \leq H \\ &\forall t \in \mathcal{T} \quad \forall m \in \mathcal{M}_t \quad \forall n \in \mathcal{N}. \end{aligned} \tag{5.85}$$

5.4.4 Shared Resources

Since more than one transition may be present in set $\mathcal{T}_m$, it is necessary to make sure that at most one transition can be enabled by place m at every event point:

$$\begin{aligned} &\sum_{t \in \mathcal{T}_m} am(t, m, n) \leq 1 \\ &\forall m \in \mathcal{M} \quad \forall n \in \mathcal{N}. \end{aligned} \tag{5.86}$$

Since more than one place may be present in set $\mathcal{M}_t$, it is also necessary to make sure that only the token in one of them is removed after firing transition t at any event point:

$$\sum_{m \in \mathcal{M}_t} am(t, m, n) \leq 1 \\ \forall t \in \mathcal{T} \quad \forall n \in \mathcal{N}. \tag{5.87}$$

Finally, the capacity limit of an actual buffer can also be imposed

$$\sum_{s \in \mathcal{S}_p} st(s, n) \leq \text{stmax}(p) \\ p \in \mathcal{P} \quad n \in \mathcal{N}. \tag{5.88}$$

5.4.5 *Linearization of Minimum-Horizon Model*

Several different objective functions can be adopted to produce the optimal production schedule. One of them is for achieving the minimum horizon, i.e., min H. Due to equations (5.80)–(5.82), the resulting optimization problem becomes a mixed-integer nonlinear program (MINLP). These nonlinear constraints can be converted into linear forms by introducing extra real variables and additional logic constraints. For illustration convenience, let us consider Eq. (5.80) as an example. In this case, the new variable can be defined as:

$$h_{cm}(t, m, n) = H \cdot am(t, m, n). \tag{5.89}$$

Equation (5.80) can then be transformed to:

$$T^s(t, m, n+1) \geq T^f(t, m, n) - H + h_{am}(t, m, n). \tag{5.90}$$

The value of the new variable $h_{am}(t, m, n)$ can be set with the following two logic constraints:

$$0 \leq h_{am}(t, m, n) \leq M_1 \cdot am(t, m, n) \tag{5.91}$$

$$0 \leq H - h_{am}(t, m, n) \leq M_2 \cdot [1 - am(t, m, n)] \tag{5.92}$$

where M_1 and M_2 are large enough positive numbers.

5.4.6 *Final Test Process*

For particular workloads, a minimum-horizon schedule can be created. Figures 5.38 and 5.39 display the optimization results obtained for a simple example, i.e., the

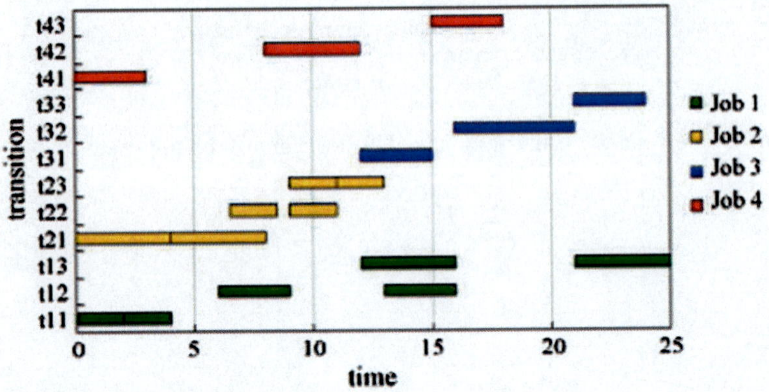

Fig. 5.38 Gantt chart for transitions in FT process–Scenario 1 (reprinted with permission from Lee et al. 2011. Copyright 2011 Elsevier)

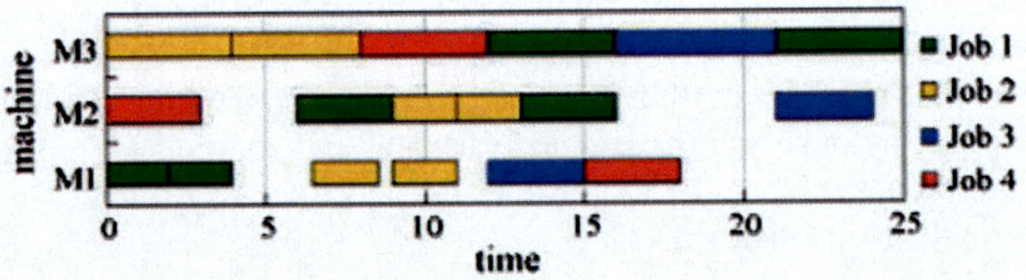

Fig. 5.39 Gantt chart for machines in FT process–Scenario 1 (reprinted with permission from Lee et al. 2011. Copyright 2011 Elsevier)

testing of 2, 2, 1, and 1 batches of items in tasks 1 to 4, respectively. Figure 5.38 shows the Gantt chart for the allowed transitions in the Petri net (or operating phases in various FT process jobs). In this case, it is assumed that the time unit is 10 min and so the minimum horizon is 250 min. Note that the service phases in the same job cannot be performed consecutively because of the need to share machines. According to Fig. 5.37 and Fig. 5.38, the work schedule of machines can be defined (see Fig. 5.39). As the processing time of M3 is the longest of the three machines, it can be observed that the key emphasis of scheduling arrangements should therefore be put on the minimization of its total running time. For a larger scheduling assignment, the same optimization run was also replicated. In particular, in the second case, the workloads of the four test workers are increased to 12, 12, 8, and 8 batches, respectively. In this case, the minimum horizon is 168 time units (1680 min), and Figs. 5.40 and 5.41, respectively, contain the corresponding Gantt charts for the transitions and machines.

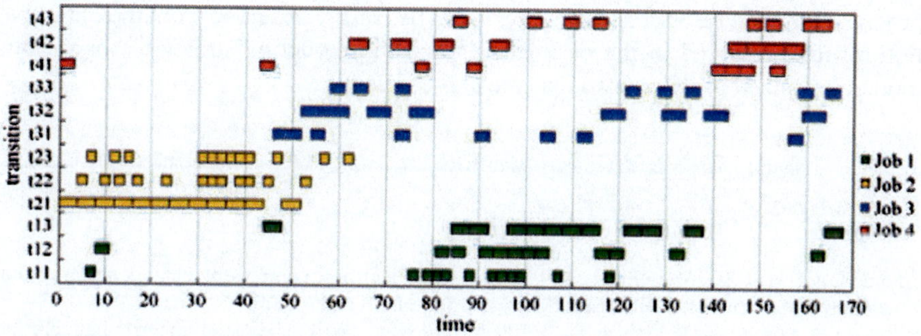

Fig. 5.40 Gantt chart for transitions in FT process–Scenario 2 (reprinted with permission from Lee et al. 2011. Copyright 2011 Elsevier)

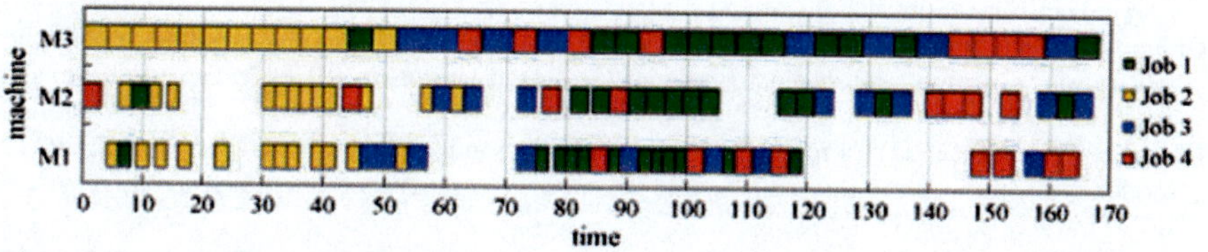

Fig. 5.41 Gantt chart for machines in FT process–Scenario 2 (reprinted with permission from Lee et al. 2011. Copyright 2011 Elsevier)

5.5 Concluding Remarks

In order to produce comprehensive operating procedures for many applications, systematic methods have been presented on the basis of Petri net representation. The cleaning routes are chosen for pipeline network cleaning on the basis of the Petri net representation of all material-transfer paths. From the simulation results obtained with the Petri net model of the whole system, the operation steps for transporting detergent via a specified route can be described. The multi-route cleaning recipes can also be created with the proposed simulation techniques by connecting this net with another one representing the schedule manager. Both feasible material-transfer routes and the corresponding activity steps can be established from the accessibility trees of the resulting Petri net for multiple material-transfer tasks in a batch process by implementing additional control rules. By solving the mathematical programming model according to various objective functions, a binary integer program can be formulated for optimal recipes containing the detailed procedure steps. It is possible to distinguish two distinct classes of operation modes: (1) stage-based operating procedures and (2) time-based operating procedures. With the example of the beer filtration facility, the viability of the Petri net-based method is clearly checked. In addition, in the latter segment, the use of Petri net in traditional semiconductor manufacturing activities was seen. To create the most productive work schedules, a mathematical program can then be built accordingly. It can be clearly observed from the optimization results presented in the case studies that the modeling strategies for various moderately sized problems are feasible and efficient.

References

Allam M, Alla H (1998) Modeling and simulation of an electronic component manufacturing system using hybrid petri nets. IEEE Trans Semicond Manuf 11(3):374–383

Cavalieri S, Mirabella O, Marroccia S (1997) Improving flexible semiconductor manufacturing system performance by a colored Petri net-based scheduling algorithm. IEEE sixth international conference on emerging technologies and factory automation proceedings. Los Angeles, CA, pp 369–374

Chen TR, Chang TS, Chen CW (1995) Scheduling for IC sort and test with preemptiveness via Lagrangian-relaxation. IEEE Trans Syst Man Cybern 25:1249–1250

Chiang TC, Huang AC, Fu LC (2006) Modeling, scheduling, and performance evaluation for wafer fabrication: a queuing colored petri-net and GA-based approach. IEEE Trans Autom Sci Eng 3(3):330–337

Chien CF, Chen CH (2007) Using genetic algorithms (GA) and a coloured timed Petri net (CTPN) for modeling the optimization-based schedule generator of a generic production scheduling system. Int J Prod Res 45(8):1763–1789

Chou HH, Chang CT (2005) Petri-net based strategy to synthesize the operating procedures for cleaning pipeline networks. Ind Eng Chem Res 44:114

Crooks CA, Macchietto S (1992) A combined MILP and logic-based approach to the synthesis of operating procedures for batch plants. Chem Ical Eng Commun 114:117

David R, Alla H (1994) Petri Nets for modeling of dynamic-systems—a survey. Automatica 30:175–202

Duenyas I, Fowler JW, Schruben LW (1994) Planning and scheduling in Japanese semiconductor manufacturing. J Manuf Syst 13:323–332

Ferrarini L, Piroddi L (2003) Modular Design and Implementation of a Logic Control System for a Batch Process. Comput. Chem. Eng. 27:983

Floudas CA, Lin X (2004) Continuous-time versus discrete-time approaches for scheduling of chemical processes: a review. Comput Chem Eng 28:2109

Foulkes NR, Walton MJ, Andow PK, Galluzzo M (1988) Computer-aided synthesis of complex pump and valve operations. Comput Chem Eng 12:1035

Hoshi K, Nagasawa K, Yamashita Y, Suzuki M (2002) Automatic generation of operating procedures for batch production plants by using graph representations. J Chem Eng Jpn 35(4):377

Ierapetritou MG, Floudas CA (1998a) Effective continuous-time formulation for short-term scheduling. 1. Multipurpose batch processes. Ind Eng Chem Res 37:4341–4359

Ierapetritou MG, Floudas CA (1998b) Effective continuous-time formulation for short-term scheduling. 2. Multipurpose/multiproduct continuous processes. Ind Eng Chem Res 37: 4360

ISA (1995) S88.01 batch control, part 1: models and terminology. ISA, Research Triangle Park, NC

Jeng MD, Xie XL, Chou SW (1998) Modeling, qualitative analysis, and performance evaluation of the etching area in an IC lot fabrication system using Petri nets. IEEE Trans Semicond Manuf 11:358–373

Jeng MD, Xie XL, Hung WY (2000) Markovian timed Petri nets for performance analysis of semiconductor manufacturing systems. IEEE Trans Syst, Man, Cybern-Part b: Cybern 30(5):757–771

Johri PK (1993) Practical issues in scheduling and dispatching in semiconductor wafer fabrication. J Manuf Syst 12:474–485

Karassik IJ, McGuire JT (1998) Centrifugal pumps, 2nd edn. New York, USA: Chapman & Hall, pp 885–887

Kim J, Moon I (2000) Synthesis of safe operating procedure for multi-purpose batch processes using SMV. Comput Chem Eng 24:385

Kondili E, Pantelides CC, Sargent RWH (1988) A general algorithm for scheduling batch operations. In: Proceedings of the process systems engineering '88, Sydney, Australia

Kondili E, Pantelides CC, Sargent RWH (1993) A general algorithm for short-term scheduling of batch operations. I. MILP formulation. Comput Chem Eng 17: 211

Kumar R, Tiwari MK, Allada V (2004) Modeling and rescheduling of a re-entrant wafer fabrication line involving machine unreliability. Int J Prod Res 42(21): 4431–4455

Kuo CH, Huang HP (2003) Distributed performance evaluation of a controlled IC fab. IEEE Trans Robot Autom 19(6):1027–1033

Lai JW, Chang CT, Hwang SH (2007) Petri-net based binary integer programs for automatic synthesis of batch operating procedures. Ind Eng Chem Res 46(9):2797–2813

Lee YH, Chang CT, Wong DSH, Jang SS (2011) Petri-net based scheduling strategy for semiconductor manufacturing processes. Chem Eng Res Des 89(3):291–300

Lin SY, Huang HP (1998) Modeling and emulation of a furnace in IC fab based on colored-timed Petri net. IEEE Trans Semicond Manuf 11(3):410–420

Lu SH, Kumar PR (1991) Distributed scheduling based on due dates and buffer priorities. IEEE Trans Autom Control 36:1406–1416

Maravelias CT, Grossman IE (2003) A new general continuous-time state task network formulation for the short term scheduling of multi-purpose batch plants. Ind Eng Chem Res 42:3056

Murata T (1989) Petri nets: properties, analysis and applications. Proc. IEEE 77(4):541

Narahari Y, Khan LM (1996) Performance analysis of scheduling policies in re-entrant manufacturing systems. Comput Oper Res 23(1):37–51

O'Shima E (1978) Safety supervision of valve operation. J Chem Eng Jpn 11:390

Pagageorgaki S, Reklaitis GV (1990) Optimal design of multipurpose batch plants. 1. Problem formulation. Ind Eng Chem Res 29:2054–2062

Pantelides CC (1994) Unified frameworks for the optimal process planning and scheduling. In: Rippin DWT, Hale J (eds) Proceedings of the 2nd conference on foundations of computer-aided process operations, p 253

Peterson JL (1981) Petri net theory and modeling of systems. Prentice-Hall, Englewood Cliffs, New Jersey

Raman R, Grossmann IE (1991) Relation between MILP modeling and logical inference for chemical process synthesis. Comput Chem Eng 15:73

Rivas JR, Rudd DF (1974) Synthesis of failure-safe operation. AIChE J. 20:320

Shah N, Pantelides CC, Sargent R (1993) A general algorithm for short-term scheduling of batch operations. II. Computational issues. Comput Chem Eng 17(2): 229–244

Tittus M, Lennartson B (1999) Hierarchical supervisory control for batch processes. IEEE Trans Control Syst Technol 7(5):542

Uthgenannt JA (1996) Path and equipment allocation for multiple, concurrent processes on networked process plant units. Comput Chem Eng 20:1081

Uzsoy R, Martinvega LA, Lee CY (1991) Production scheduling algorithms for a semiconductor test facility. IEEE Trans Semicond Manuf 4:270–280

Viswanathan S, Johnsson C, Venkatasubramanian V, Ärzen KE (1998a) Automating operating procedure synthesis for batch processes: Part I. Knowledge representation and planning framework. Comput Chem Eng 22(11): 1673

Viswanathan S, Johnsson C, Venkatasubramanian V, Ärzen KE (1998b) Automating operating procedure synthesis for batch processes: Part II. Implementation and application. Comput Chem Eng 22(11): 1687

Wang YF, Chang CT (2003) A hierarchical approach to construct petri nets for modeling fault propagation mechanisms in sequential operations. Comput Chem Eng 27:259

Wang YF, Wu JY, Chang CT (2002) Automatic hazard analysis of batch operations with Petri nets. Reliab Eng Syst Saf 76(1):91

Wang YF, Chang CT (2004) Petri-net based deductive reasoning strategy for fault identification in batch processes. Ind Eng Chem Res 43:2704

Wang YF, Chou HH, Chang CT (2005) Generation of batch operating procedures for multiple material-transfer tasks with Petri net. Comput Chem Eng 29: 1822

Wu D, Ierapetritou M (2004) Cyclic short-term scheduling of multiproduct batch plants using continuous-time representation. Comput Chem Eng 28:2271

Xiong HH, Zhou MC (1998) Scheduling of semiconductor test facility via petri nets and hybrid heuristic search. IEEE Trans Semicond Manuf 11(3):384–393

Zhang X, Sargent RWH (1996) The optimal operation of mixed production facilities—a general formulation and some approaches for the solution. Comput Chem Eng 20:897–904

Chapter 6
Normal Operating Procedures Obtained with Untimed Automata

As mentioned before, other than the process flow diagram (PFD) and the piping and instrumentation diagram (P&ID), the sequential function chart (SFC) of every standard operating procedure (SOP) should also be documented for process design. Despite the fact that the modern plants are becoming more and more complex, their operating procedures are still generated manually on the basis of the designer's experience. Manual synthesis of the operating procedure in a realistic system **can** be very difficult since it is both time-consuming and error-prone. It is thus desirable to develop a systematic approach to automatically conjecture viable steps so as to achieve given production goal(s) (Lu et al. 2017).

An operating procedure must be synthesized according to the initial system state and the final operational goal. To overcome the difficulties caused by combinatorial pathway explosion, many available studies focused on issues concerning systematic and efficient procedure synthesis. The original problem was defined by Rivas and Rudd (1974), and subsequent works were carried out extensively in the later years. O'Shima (1978) devised an algorithm to search for a series of valve operations that allow fluid flow between any two chosen locations in a chemical plant. Foulkes et al. (1988) constructed the so-called "condition lists" to describe all pipeline fragments and utilized AI-based search strategies to identify all possible routes between storage tanks for material transfer. Crooks and Macchietto (1992) formulated an MILP model with embedded logic constraints to synthesize operating procedures for the batch processes. Uthgenannt (1996) used digraphs to characterize the process networks and applied an existing search technique to uncover the material-transfer routes and the corresponding operating procedures. Yang et al. (2001) made use of the symbolic model verifier to synthesize safe operating procedures and, furthermore, configured the safety interlocks accordingly. Ferrarini and Piroddi (2003) suggested characterizing any given SFC with a Petri net to validate the corresponding operation schedule and to detect the presence of deadlock. Lai et al. (2007) built binary integer programs (BIPs) based on Petri net models for the automatic generation of batch operating procedures. Yeh and Chang (2012a, b) developed a systematic approach to generate procedures according to untimed automata under normal conditions and

C.-T. Chang et al., *Process Plant Operating Procedures*, Advances in Industrial Control,
https://doi.org/10.1007/978-3-030-70978-5_6

also for emergency responses. Li et al. (2014) developed an improved modeling strategy based on timed automata to create both the cyclic operation steps and the corresponding time schedules in terms of Gantt chart. Cochard et al. (2011) presented a timed-automata-based method to synthesize safe operation sequences.

Since all aforementioned studies emphasized only the synthesis aspects, the resulting SFCs may not be readily acceptable in actual applications. To be specific, these SOPs were not validated either in simulation studies with credible software or in the pilot plant experiments. Furthermore, if several candidates can be generated, it is clearly necessary to evaluate them with a collection of different criteria so as to identify the most suitable one. Generally speaking, the previous works not only lacked efforts in the verification and assessment of the synthesized procedures but also did not produce benchmark examples to establish their credibility for actual implementations. To fully address these practical concerns, a comprehensive design approach is presented in the present chapter for synthesis, validation, and evaluation of alternative SFCs.

The untimed automata are utilized in the present chapter for procedure synthesis. In particular, all components in a given system are first characterized with automata according to simple modeling principles (Kang and Chang 2014; Wang et al. 2017). On the other hand, the intended operation is further divided into several distinguishable stages and the unique intrinsic features of each stage, e.g., stable operation, condition adjustment, phase change, reaction, and material charging and/or unloading, are then identified. The so-called "control specifications" of every stage can be described accordingly with natural language and automata so as to set the target state, to create different operation paths via state splitting, to limit feasible operations to those that follow only the designated partial sequences, to avoid unsafe operations by stipulating illegal strings, etc. A system model and the corresponding observable event traces (OETs) can then be generated by synchronizing all aforementioned automata via the embedded functions of free software Supremica (Åkesson et al., 2006). For any practical application, one or more operating procedures can be easily extracted from these traces and formally summarized with SFCs. The popular commercial package, i.e., Aspen Plus Dynamics®, is used next to validate these SFCs in simulation studies. Since more than one candidate may be generated, they are evaluated on the basis of several economic performance indices, e.g., the completion time, the energy consumption level, and the total amount of off-spec products. Finally, it is also possible to further fine-tune the best procedure by adjusting the critical parameters in SFCs. Three realistic examples, i.e., the semi-batch reaction process, and the startup operations of flash drum and distillation column, are presented to demonstrate the merits of the proposed design approach.

6.1 Extended Finite Automata

It should be noted first that a brief review of the conventional automaton has already been given in Subsection 3.2.2. The extended finite automaton (EFA) is an improved

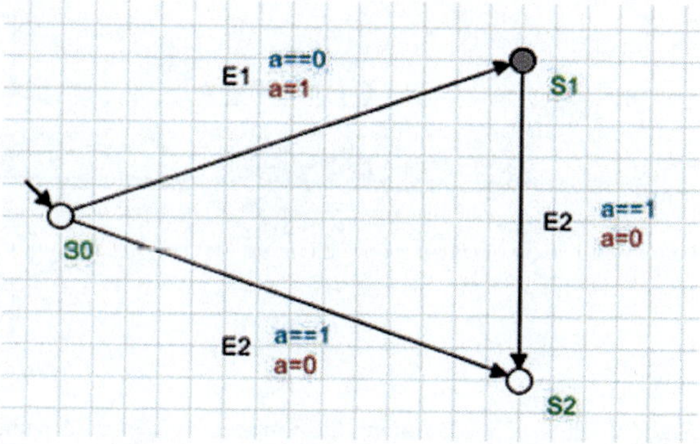

Fig. 6.1 Graphic representation of an extended finite automaton (Reprinted with permission from Chen and Chang 2019. Copyright 2019 American Chemical Society)

version of the traditional structure. Specifically, each event in EFA is equipped with two extra attributes, i.e., variable and guard, and their functions are further explained as follows:

- An integer variable (with user-specified upper and lower limits) can be used to update the equipment state after completing an event-driven transition. An example is shown in Fig. 6.1, in which variable a is updated to 1 according to the equation "$a = 1$" via event E1.
- A guard is the prerequisite of the corresponding state transition. Let us again consider Fig. 6.1 as an example and assume that the initial value of variable a is 0. Therefore, only event E1 is allowed at the initial state S0 due to the logic constraint "a == 0" and, when S1 is reached after state transition, this variable should be updated to 1.

6.2 An Illustrative Example

To facilitate clear illustration of the process structure, let us consider the startup operation of the continuous flash process in Fig. 6.2 as an example. It is assumed that at a steady state, the feed is a mixture of 30 wt% water and 70 wt% methanol and its flowrate, temperature, and pressure are kept at 26,000 kg/h, 20 °C and 1.1 bar, respectively. The steady-state temperature and pressure in the flash drum are set at 75 °C and 1.01 bar, respectively, while the corresponding liquid level is 2.5 m. The heating medium in the heater is assumed to be low-pressure steam. It is also required that the concentration of methanol in the top product should not be lower than 83 wt%. In this system, there are four PID controllers (FC01, TC01, PC01, and LC01) for controlling the feed rate, the temperature, the vapor pressure, and the liquid level in the flash drum, respectively. The corresponding actuators are control valves, i.e., Vin, Vlps, Vvap, and Vliq. It is also assumed that, initially, all valves are closed, all controllers are on MANUAL mode, and the flash drum is empty and at room temperature.

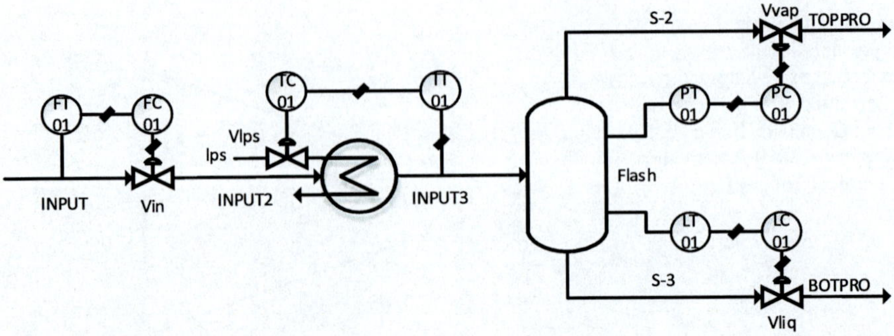

Fig. 6.2 PFD of a continuous flash process (Reprinted with permission from Chen and Chang 2019. Copyright 2019 American Chemical Society)

Basically, each of the identifiable items in the PFD can be treated as a component, and they are classified into a five-level hierarchy according to Fig. 3.1 in Chap. 3. The top-level component is usually a programmable logic controller (PLC) or a human operator; the PID controllers and their actuators (i.e., FC01/Vin, TC01/Vlps, PC01/Vvap, and LC01/Vliq) are classified as the second-level components; the material and energy flows surrounding each unit (i.e., INPUT, TOPPRO, BOTPRO, and the energy flow from the heater to the flash drum) in the subsequent level are viewed as the third-level components; every processing unit, such as the flash drum in the present example, is regarded as a fourth-level component; all online sensors (i.e., FT01, TT01, PT01, and LT01) are grouped into level 5.

6.3 Component Models

Every component in a given process is modeled with an automaton. To build this model for a component, all its normal conditions and failed states should be enumerated and represented with distinct places. The initial state should be indicated by pointing to the corresponding place with an arrow, but there is no need to assign the marked states in a component model. All events that facilitate state transitions should then be identified and each represented with a directed arc between its input and output places (states). Also, if necessary, the guard(s) of every event and the updated variable value(s) can be attached to the corresponding arc.

It is not necessary to construct an automaton to describe a level-1 component, i.e., PLC or human operator, since the operating procedure is not known in advance in the present applications. All component models in the other levels can be found in Appendix 6.1. For the sake of brevity, let us consider only the level-2 components, i.e., the actuators and the PID controllers, as examples.

6.3.1 Actuators

To be specific, let us construct an automaton to characterize control valve Vin according to the model-building principles mentioned above (see Fig. 6.3). The places Vin_full_close and Vin_full_open, respectively, denote two extreme states of Vin, i.e., the fully closed and fully open positions, while the other three places between them are used to represent the partial openings of 25, 50, and 75%. It is also assumed that Vin is fully closed before the startup operation. Any valve state can be driven to another via a series of adjustment steps, i.e., the valve opening actions (oVin_0to1,oVin_1to2,oVin_2to3, and oVin_3to4) and the valve closing actions (cVin_4to3, cVin_3to2, cVin_2to1, and cVin_1to0). The two additional attributes of events, i.e., variable and guard, are also utilized on the corresponding arcs. An integer variable is used to update the component state after completing an event-driven transition, while the guard(s) is used to stipulate the sufficient condition(s) of state transition. Let us consider event oVin_0to1 as an example. Its guards (prerequisites) are expressed as s_flow == 1&s_flow > A_Vin&A_Vin! = 4, and they can be interpreted as follows:

- s_flow == 1: The controller output signal is at the qualitative value of 1.
- s_flow > A_Vin: The output signal of flow controller FC01 is larger than the air pressure corresponding to the current position of valve Vin.
- A_Vin! = 4: The current air pressure at valve Vin does not reach maximum.

Note that the guards on other arcs between Vin_full_close and Vin_full_open can be interpreted in a similar fashion. Two additional places, Vin_SC and Vin_SO, are included in this model to characterize the failures when Vin is stuck at the closed and

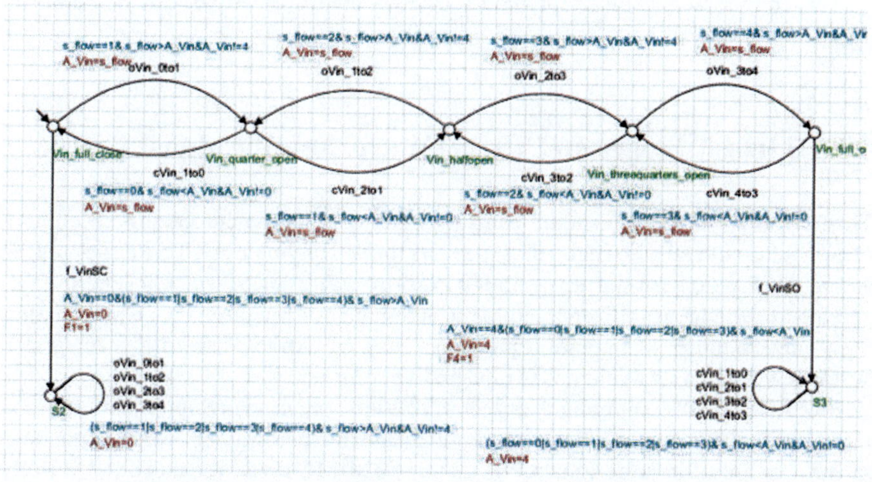

Fig. 6.3 Traditional model of control valve Vin in flash startup example (Reprinted with permission from Chen and Chang 2019. Copyright 2019 American Chemical Society)

open positions, respectively. These failed states can be reached via events f_VinSC and f_VinSO, and the resulting values of A_Vin should be fixed at 0 and 4, respectively. In the present chapter, all failed states are ignored because the objective here is to synthesize normal operating procedure. Since at least two places must be adopted to characterize a component condition, e.g., the positions of a control valve in the above automaton, it is clear that the above approach is only effective for modeling simple systems with relatively few state variables. To facilitate the easy construction and concise representation of automata, the component models have been "compressed", and the component Vin is modeled alternatively with an automaton using significantly fewer places and transitions (see Fig. 6.16 in Appendix 6.1).

6.3.2 PID Controller

Generally speaking, the role of a PID controller in executing the steps in an SOP depends on its actuator in two scenarios:

(i) If the actuator is a hand valve or a solenoid valve, then the PID controller is obviously not needed.

(ii) If the actuator is a control valve, then the PID controller can be utilized to vary its opening in two alternative modes:

  (a) If the controller is on AUTO mode, it is necessary for the operator or PLC to alter its set point so as to adjust the output signal indirectly.

  (b) If the controller is set at MANUAL mode, the operator or PLC should be able to directly adjust the output signal.

The compressed component model of PID controller FC01 can be found in Fig. 6.17 in Appendix 6.1.

6.4 Intrinsic Stages and Their Control Specifications

Theoretically, the system model can be constructed by synchronizing all component models with an automaton that specifies the operation target. This target-setting automaton for the flash startup operation is given in Fig. 6.4. Since only generic engineering knowledge is utilized to build the automata in Appendix 6.1, an overwhelmingly large number of operation pathways may be extracted from this integrated system even when its dynamics is moderately complex.

More specifically, applying the synchronization operation in Supremica (Åkesson et al. 2006) yielded a complicated and unmanageable pathway network for the flash startup example. Since the dynamic behavior of a MIMO system cannot be adequately

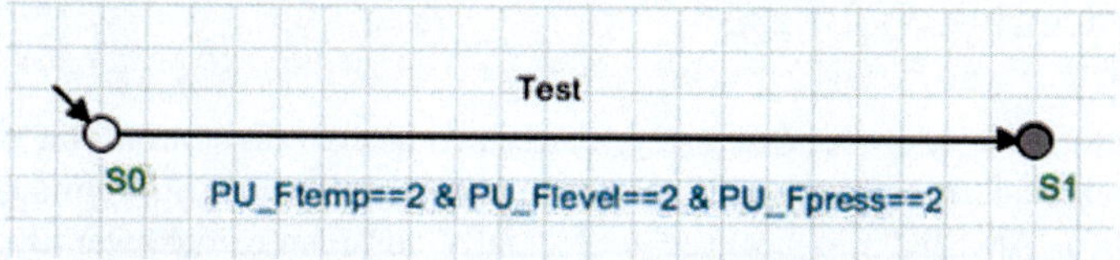

Fig. 6.4 Final target of flash startup operation (Reprinted with permission from Chen and Chang 2019. Copyright 2019 American Chemical Society)

described with the untimed automata developed on the basis of qualitative information only, this network consists of not only multiple feasible routes but also an extremely large number of unnecessary and impractical scenarios.

Although the final target of a specific operation can be unambiguously stipulated (e.g., Fig. 6.4), it may only be approached properly via a series of intermediate stages with interim goals which are often not explicitly stipulated a priori. It is thus important to uncover these embedded subtasks and identify their unique features in advance. These features may be broadly classified as (1) material charging, (2) material unloading, (3) reaction, (4) state adjustment, (5) phase change, and (6) stable operation. For illustration purposes, let us revisit the flash startup example. Based on engineering knowledge and operational experience, it is clearly necessary to place a small quantity of raw material in the flash drum in the initial stage and allow the liquid level reaching a height which is safe for heating. In the next stage, the temperature and pressure in the drum should be adjusted to the set points and the input and output flow rates be raised to the steady-state levels. Notice also that whenever such adjustments in operating conditions are called for, it is always beneficial to assess the pros and cons of alternative pathways that facilitate the required state transitions. Finally, the stable operating conditions should be maintained for a long enough period of time with the PID controllers. Thus, the feature sets of the above three stages may be characterized as follows: (1) state transfer and material charging, (2) state transfer, phase change, and material charging and unloading, and (3) stable operation.

All the features in a stage should be expressed first as the "control specifications" in natural language and then translated to automata. Cassandras and Lafortune (2008) suggested that five different types of automata may be constructed for use to set the target state (type A), to perform state splitting (type B), to impose a partial sequence (type C), to suppress an illegal substring (type D), and to ensure alternation between two particular events (type E). Let us again use the flash startup process as an example for illustration and, for the sake of conciseness, the control specifications in all stages are detailed in Appendix 6.2.

6.5 Procedure Synthesis

The operating procedures of every stage can be produced by synchronizing all component models and the corresponding control specifications with Supremica (Åkesson et al. 2006). An SFC can be used to formally summarize the overall procedure obtained by piecing together the steps in all stages of the given operation. A total of four SFCs have been generated for the flash startup process. The control actions and their responses in the three stages of all four SFCs are listed below:

- **SFC-6.1**: (1) Raise the liquid level in the flash drum to 1.45m by opening the inlet valve (Vin) fully; (2) Raise the temperature and level directly to their set points (i.e., 75 °C and 2.5m) by adjusting the inlet valve (Vin), the outlet valves (Vvap and Vliq), and the steam valve (Vlps); (3) Switch all PID controllers from MANUAL to AUTO modes and maintain stable operation at the targeted set points.
- **SFC-6.2**: (1) Raise the liquid level in the flash drum to 1.45m by opening the inlet valve (Vin) fully; (2) Raise the temperature to 40 °C and then to 75 °C in two consecutive steps and simultaneously raise the level to 2.5m in one step by adjusting the inlet valve (Vin), the outlet valves (Vvap and Vliq), and the steam valve (Vlps); (3) Switch all PID controllers from MANUAL to AUTO modes and maintain stable operation at the targeted set points.
- **SFC-6.3**: (1) Raise the liquid level in the flash drum to 2.5m by opening the inlet valve (Vin) fully; (2) Raise the temperature directly to 75 °C by adjusting the inlet valve (Vin), the outlet valves (Vvap and Vliq), and the steam valve (Vlps); (3) Switch all PID controllers from MANUAL to AUTO modes and maintain stable operation at the targeted set points.
- **SFC-6.4**: (1) Raise the liquid level in the flash drum to 2.5m by opening the inlet valve (Vin) fully; (2) Raise the temperature to 40 °C and then to 75 °C in two consecutive steps by adjusting the inlet valve (Vin), the outlet valves (Vvap and Vliq), and the steam valve (Vlps); (3) Switch all PID controllers from MANUAL to AUTO modes and maintain stable operation at the targeted set points.

As an example for illustration, let's take a closer look at SFC-6.1 in Fig. 6.5. The initial settings of PID controllers and actuators in this procedure are implemented according to S_0. The first activation conditions in SFC-6.1 are basically the initial sensor readings specified in AC_1. After AC_1 is verified, valve Vin is supposed to be opened fully as required in S_1 to raise the liquid level in the flash drum as quickly as possible. Upon observing the level reading of 1.45m, i.e., AC_2, utility heating should be applied and, at the same time, the inlet and outlet flows also begin via the operation steps specified in S_2. The subsequent activation conditions in AC_3 are sensor readings at the set points of temperature (75 °C) and level (2.5m) at steady state. The final steps of the startup operation, i.e., S_3, are operator (or PLC) actions to switch all PID controllers from MANUAL to AUTO modes and adjust their set points to the intended steady-state values, respectively. Finally, it should be noted that the valve opening of Vlps (93%) in this SFC is only an approximated (interpolated) value. This is due to the fact that, in Aspen Plus Dynamics, only the heater duty can

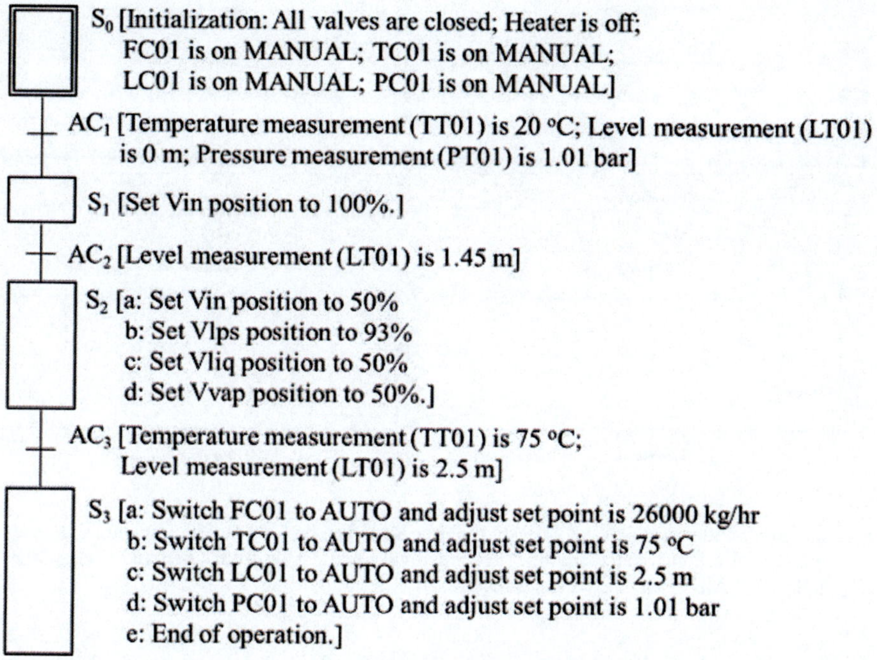

Fig. 6.5 SFC-6.1 obtained in flash startup example (Reprinted with permission from Chen and Chang 2019. Copyright 2019 American Chemical Society)

be adjusted for simulation. If SFC-6.1 is to be implemented in practical application, the exact opening of the steam valve must be determined experimentally.

6.6 Dynamic Simulation

The above candidate procedures can be validated in simulation studies with Aspen Plus Dynamics. The simulation results can be compared and the best one then selected accordingly.

First of all, it should be noted that the simulation studies reveal that the last two SFCs, i.e., SCF-6.3 and SFC-6.4, may be unsafe in practical applications due to delayed heating in startup operation. Specifically, the heater is turned on in these two cases only after the liquid level reaching 2.5m (instead of 1.45m in SFC-6.1 and SFC-6.2). Let us use SFC-6.4 as an example for illustration. The simulated time profiles of temperature and liquid level in the flash drum, the heater duty, and the concentration of methanol in overhead product are presented in Fig. 6.6. It can be observed that the heater duty is adjusted first at around 0.7h when the liquid level reaches 2.5m and then at around 0.92h when the temperature is 40 °C. Notice also that the level

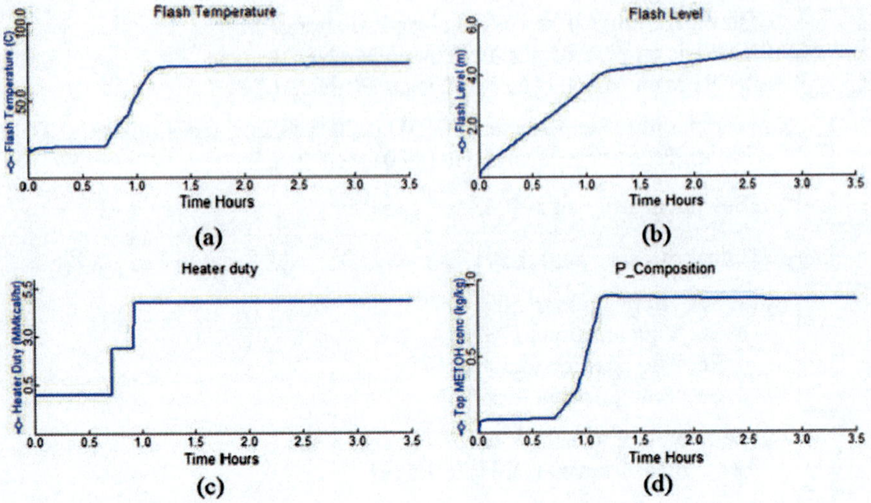

Fig. 6.6 Simulation results of the flash startup process driven by SFC-6.4: **a** temperature; **b** level; **c** heater duty; **d** product concentration (Reprinted with permission from Chen and Chang 2019. Copyright 2019 American Chemical Society)

rises continuously to 5m (which is the height of the flash drum) at around 2.49h and stays unchanged afterwards. Based on the observations concerning the dangerously high liquid level in the simulation results of SFC-6.3 and SFC-6.4, only SFC-6.1 and SFC-6.2 are compared in Table 6.1 according to three performance indices, i.e., the total amount of off-spec product, the total amount of energy consumed, and the total operation time. It can be found that SFC-6.1 outperforms SFC-6.2 essentially in every aspect.

The Aspen simulation results in the flash startup process driven by SFC-6.1 are presented in Fig. 6.7. It can be clearly verified that the concentration spec (83 wt% methanol) of the overhead product is reached at 0.51 h in Fig. 6.7d, and the corresponding temperature stabilizes around 75 °C at about the same time in Fig. 6.7a. It can also be confirmed from Fig. 6.7c that the liquid level eventually approaches 2.5m, which is well below the height of the flash drum (5 m).

Table 6.1 Performance indices of flash startup processes driven by SFC-6.1 and SFC-6.2

	Total amount of off-spec product (kg)	Total amount of energy consumed (MMkcal)	Total operation time (h)
SFC-6.1	150.20	0.6688	0.51
SFC-6.2	175.31	0.7882	0.59

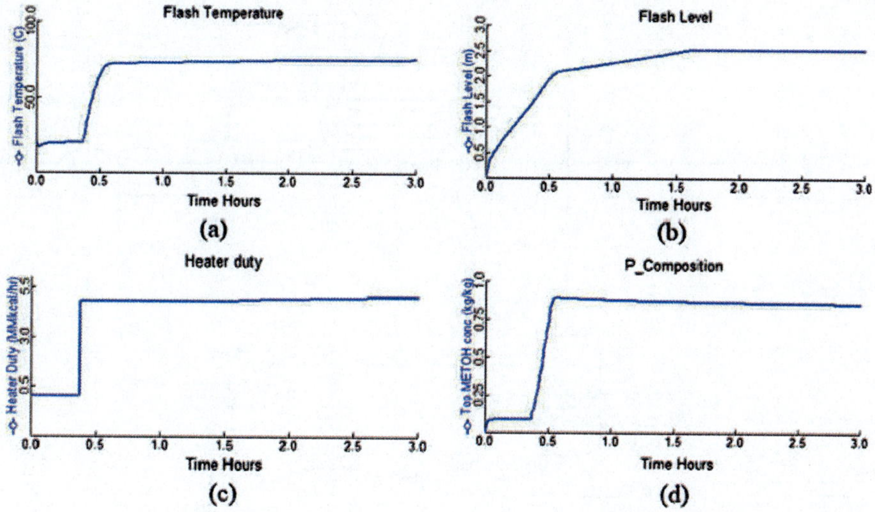

Fig. 6.7 Simulation results of the flash startup process driven by SFC-6.1: **a** temperature; **b** level; **c** heater duty; **d** product concentration (Reprinted with permission from Chen and Chang 2019. Copyright 2019 American Chemical Society)

6.7 Additional Examples

To demonstrate the feasibility of the proposed approach in more practical applications, two realistic examples are presented in this section.

6.7.1 Semi-batch Reaction

The chemical reaction considered in this example is $CHCl_3 + Cl_2 \rightarrow CCl_4 + HCl$, and the corresponding PFD is given in Fig. 6.8. To satisfy the product demand and ensure operational safety, the entire batch of 13,800 kg of $CHCl_3$ is first transported into the reactor and then the more toxic and corrosive chlorine is fed at a very low flowrate to facilitate its quick consumption. A period of stable operation can be maintained as the steady transferring process of Cl_2 begins. The reactor temperature and pressure are, respectively, kept steady at 80 °C and 4.89bar in this period, while the liquid level, temperature, and pressure in the separator at 0.75m, 13 °C, and 4.78bar, respectively. The concentrations of $CHCl_3$ and CCl_4 in the reactor and the separator, however, should vary with time during this "stable" period. Finally, it should be noted that the present operation is essentially the modified version of a built-in example of Aspen Plus Dynamics. Modifications have been introduced since the existing example did not incorporate the product discharge stage and, also, adopted only the simple flow-driven mode in dynamic simulation.

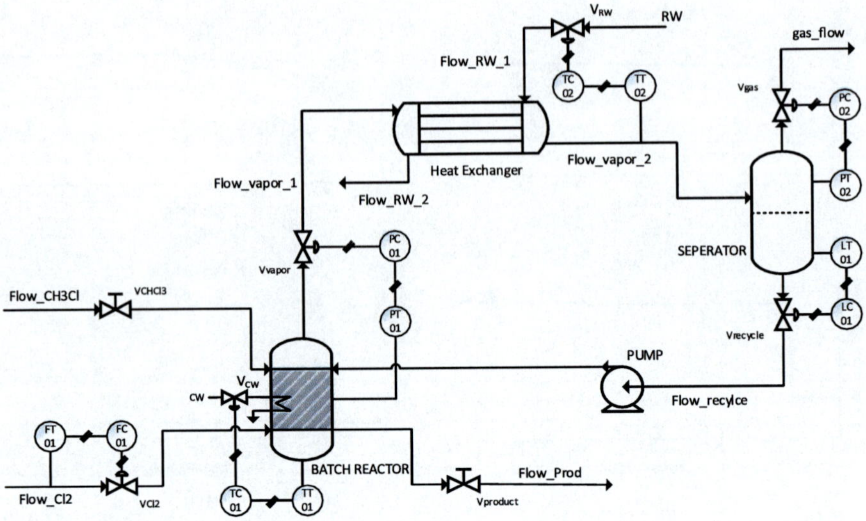

Fig. 6.8 PFD of a semi-batch reaction process (Reprinted with permission from Chen and Chang 2019. Copyright 2019 American Chemical Society)

6.7.1.1 Identification of Intrinsic Stages

As mentioned before, the two reactants in the present example are charged one at a time in sequence. The entire amount of the first reactant (chloroform) is placed into the reactor first, while the other (chlorine) is next fed continuously at a low flowrate. Since the reaction is exothermic, the reactor temperature and pressure can be raised by reaction heat to their anticipated set points and, at which instance, all controller modes in this system must be switched from MANUAL to AUTO modes. The resulting stable operating conditions (i.e., temperature and pressure) are supposed to last until some observable signs of the reaction end point are observed online. Subsequently, the product in the reactor should be cooled to room temperature and then discharged.

Based upon the above insights, the entire operation may be divided into five consecutive stages and their respective features are as follows: (1) charging chloroform into a reactor that results in a rise in the liquid level, (2) activating the exothermic reaction by charging the chlorine gas, and also removing the reaction heat with cooling water after the set points are reached, (3) stable operation, (4) lowering the temperature and pressure to the safe conditions, and (5) unloading the product.

6.7.1.2 Synthesis of Operating Procedures

Four SFCs, labeled, respectively, as SFC-6.5, SFC-6.6, SFC-6.7, and SFC-6.8 can be generated for the semi-batch reaction process. It should be noted that only the steps in the second stage of each SFC are different from those of the others. These steps are outlined as follows:

- **SFC-6.5**: Set flow controller FC01 on MANUAL mode and fix the chlorine flowrate to 1000 kg/h by adjusting valve VCl2; set temperature controller TC01 on MANUAL mode and allow the reactor temperature to rise first to 40 °C and then to 80 °C in two steps by adjusting valve Vcw manually.
- **SFC-6.6**: Set flow controller FC01 on MANUAL mode and fix the chlorine flowrate to 1500 kg/h by adjusting valve VCl2; set temperature controller TC01 on MANUAL mode and allow the reactor temperature to rise first to 40 °C and then to 80 °C in two steps by adjusting valve Vcw manually.
- **SFC-6.7**: Set flow controller FC01 on MANUAL mode and fix the chlorine flowrate to 1000 kg/h by adjusting valve VCl2; set temperature controller TC01 on MANUAL mode and allow the reactor temperature to rise first to 60 °C and then to 80 °C in two steps by adjusting valve Vcw manually.
- **SFC-6.8**: Set flow controller FC01 on AUTO mode and fix the chlorine flowrate to 1500 kg/h by adjusting valve VCl2; set temperature controller TC01 on MANUAL mode and allow the reactor temperature to rise first to 60 °C and then to 80 °C in two steps by adjusting valve Vcw manually.

For the sake of brevity, only SFC-6.5 is presented in Fig. 6.9 in detail.

6.7.1.3 Simulation, Validation, and Performance Assessment

All SFC-driven semi-batch reaction processes have been simulated with Aspen Plus Dynamics. To save space; only the simulation results generated according to SFC-6.5 and SFC-6.6 are presented in Figs. 6.10 and 6.11, respectively. Since in the latter case a higher chlorine flowrate results in a drastic rise in reactor pressure (see Fig. 6.11c) and more violent fluctuation in reactor temperature (see Fig. 6.11a), SFC-6.5 should be viewed as a safer and a more operable procedure.

It should be noted that the simulation results for SFC-6.7 and SFC-6.8 are quite similar to those for SFC-6.5 and SFC-6.6, respectively. Therefore, only the performance indices of the more feasible SFC-6.5 and SFC-6.7 are compared in Table 6.2. Notice that, since the generation rate of reaction heat is primarily governed by the chlorine feed rate, the outcomes of these two procedures are essentially the same and, thus, both are acceptable for applications.

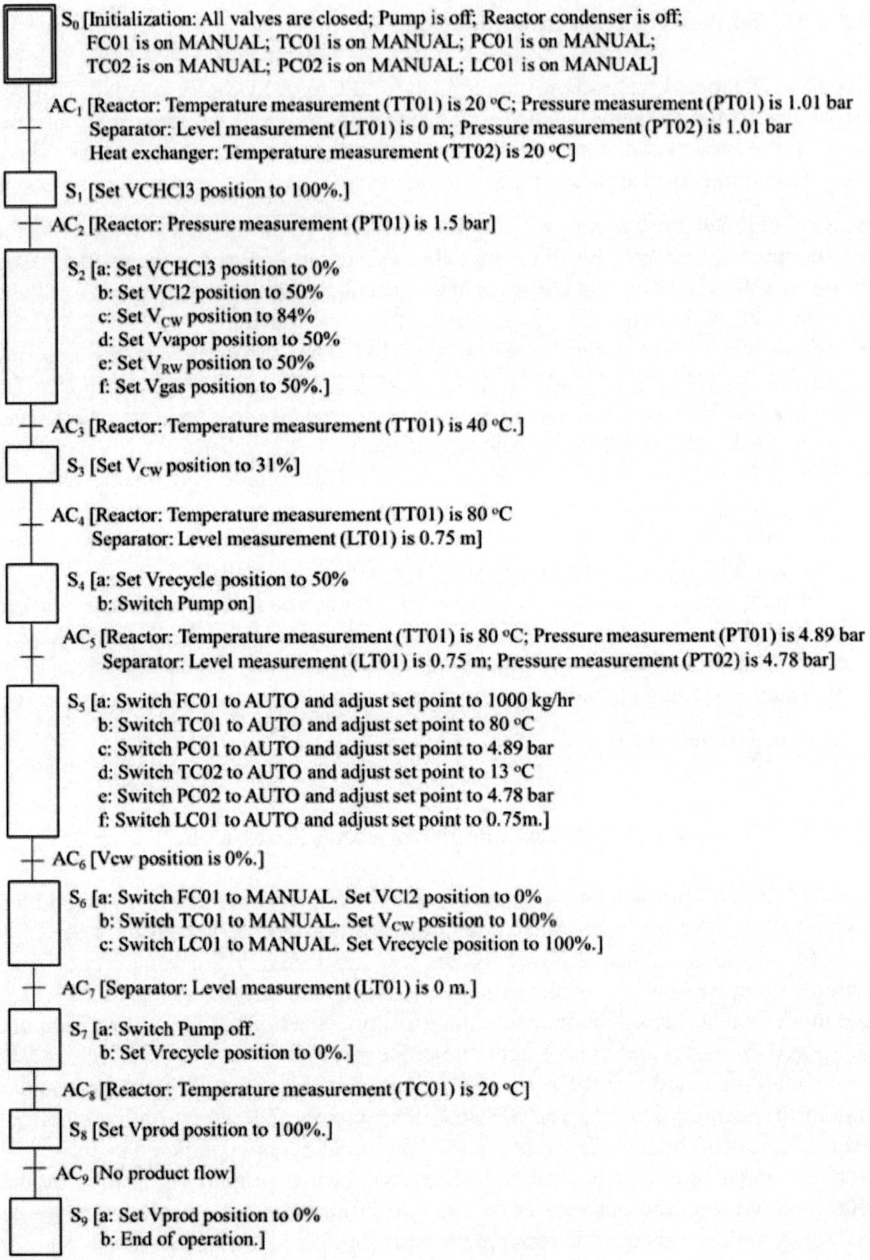

Fig. 6.9 SFC-6.5 obtained in semi-batch reaction example (Reprinted with permission from Chen and Chang 2019. Copyright 2019 American Chemical Society)

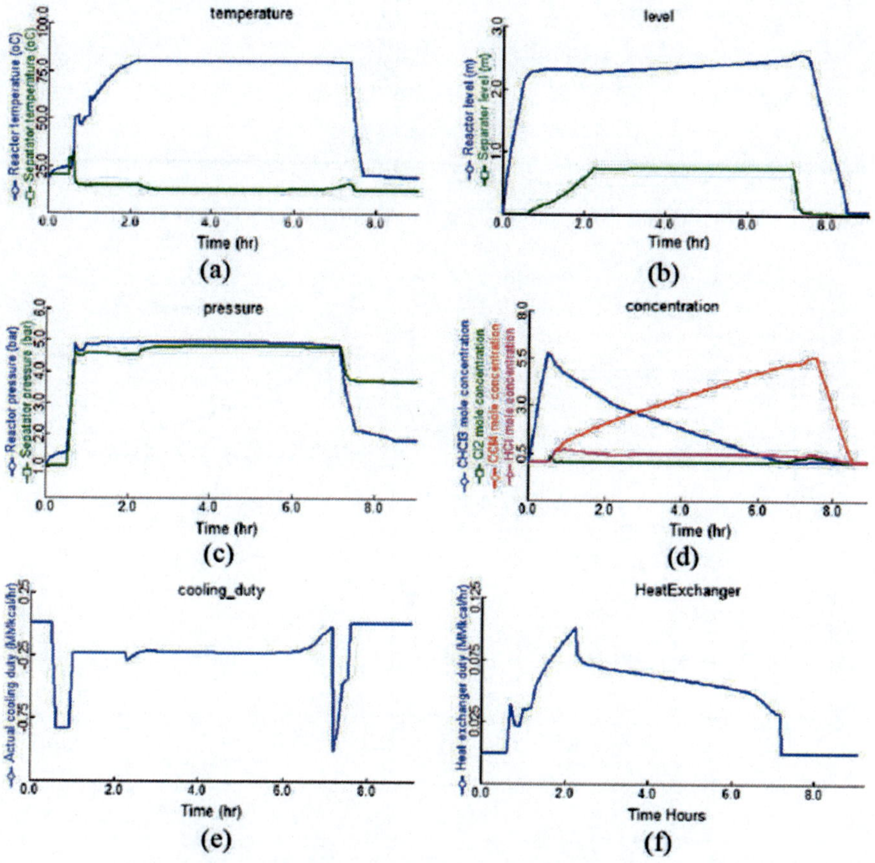

Fig. 6.10 Simulation results of the batch reaction process driven by SFC-6.5: **a** temperature; **b** level; **c** pressure; **d** concentration; **e** reactor cooling duty; **f** heat exchanger cooling duty (Reprinted with permission from Chen and Chang 2019. Copyright 2019 American Chemical Society)

6.7.2 *Distillation Startup*

In this example, let us consider the startup operation of the continuous distillation process described in Fig. 6.12. It is assumed that, at steady state, the feed is a mixture of 6 wt% CH_2Cl_2, 54 wt% $CHCl_3$, and 40 wt% CCl_4 and its flowrate, temperature, and pressure are kept at 10,000 kg/h, 20 °C and 6 bar, respectively. The total number of plates in the distillation column is 20, while the feed is directed to the 10th plate. The steady-state set-point temperatures at the 7th plate and 16th plate are set at 87.4 °C and 101.5 °C, respectively. The column is equipped with a condenser at the top and a reboiler at the bottom. The steady-state reflux ratio is 5 mol/mol. The steady-state pressure settings at plate 1/condenser and plate 2 are chosen to be 2.00 and 2.02bar, respectively, while the column pressure drop from the bottom

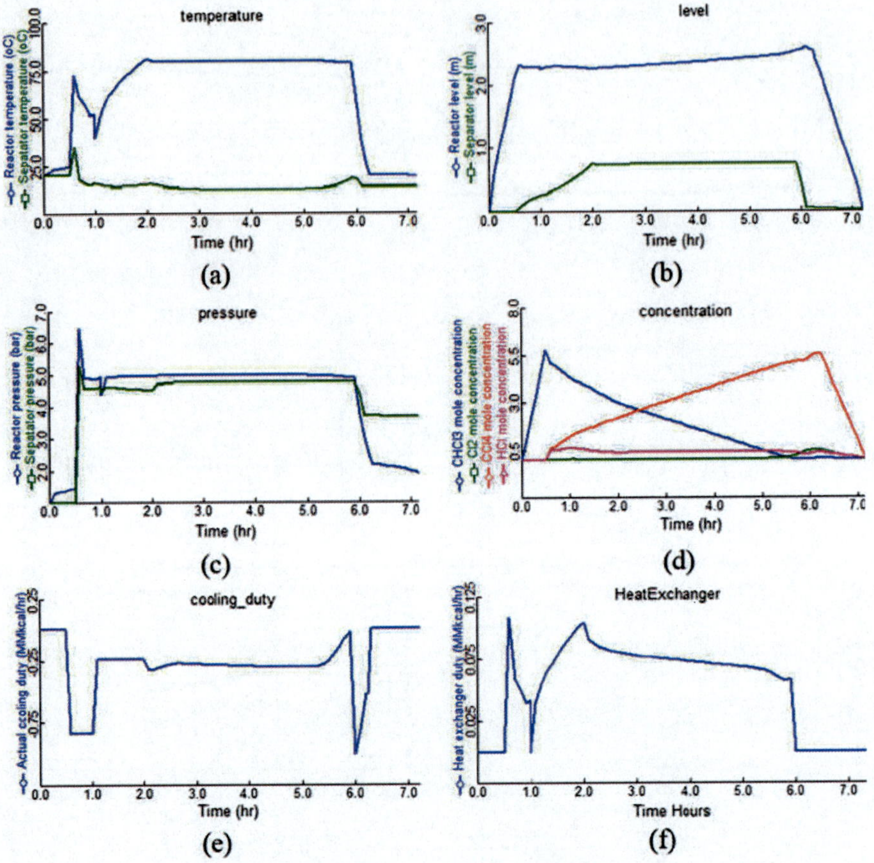

Fig. 6.11 Simulation results of the batch reaction process driven by SFC-6.6: **a** temperature; **b** level; **c** pressure; **d** concentration; **e** reactor cooling duty; **f** heat exchanger cooling duty (Reprinted with permission from Chen and Chang 2019. Copyright 2019 American Chemical Society)

Table 6.2 Performance indices of flash startup processes driven by SFC-6.5 and SFC-6.7

SFC No.	Total cooling duty (MMkcal)	Total yield of CCl_4 (kg)	Total operation time (h)
SCF-6.5	2.3604	17913	9.25
SFC-6.7	2.3528	17921	9.30

is 2.35bar. At steady state, the heights of liquid levels in the reflux drum and the column sump are controlled at 1.25 m and 1.0 m, respectively. It is also required that the concentration of light key ($CHCl_3$) in the top product should be greater than 81 mol% and that of heavy key (CCl_4) in the bottom product should not be lower than 97 mol%. In this system, there are six PID controllers (FC01, TC07, TC16, PC01, LC01, and LC02) for controlling the feed rate, the temperatures on the 7th and 16th

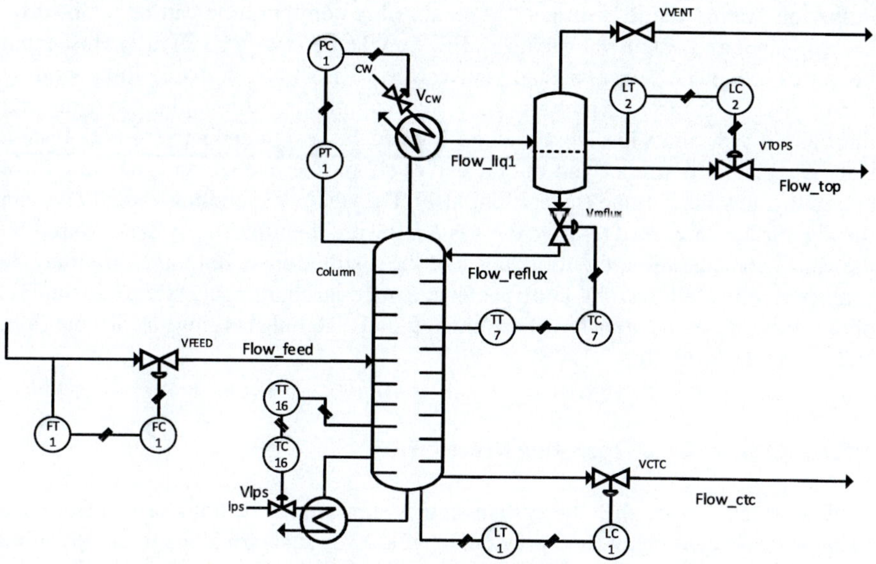

Fig. 6.12 PFD of a continuous distillation process (Reprinted with permission from Chen and Chang 2019. Copyright 2019 American Chemical Society)

plates, and the condenser pressure and the liquid levels in the column sump and the reflux drum, respectively. The corresponding control valves are as follows: VFEED, Vreflux, Vlps, Vcw, VCTC, and VTOPS. It is assumed that, before the startup, all valves are closed, all controllers are on MANUAL mode, and the reflux drum and the column sump are both empty and at room temperature. Finally, it should be noted that this startup operation is also included in Aspen Plus Dynamics as a built-in illustrative example. The objective of the present case study is to generate additional procedures with the proposed modeling approach and then compare them with the existing one.

6.7.2.1 Identification of Intrinsic Stages

Similar to the flash startup operation, it is also necessary to put enough feedstock into the column sump first so as to allow the liquid level reach a reasonable height which is safe for heating. Since the present example is concerned with startup, the desired steady-state conditions should eventually be maintained with the PID controllers. Therefore, the operating conditions of the distillation column must be transferred from those at the end of initial stage to their targeted set points in the final stage and the input and output flow rates be raised to the steady-state levels. This transition process is supposed to take place in two intermediate stages, i.e., stage 2 and stage 3. Firstly, in order to establish good contact between the liquid and the vapor in the entire distillation column, the adjustments of operating conditions are carried out in the total reflux mode in stage 2. A total reflux configuration can be realized by keeping input and output valves (i.e., VFEED, VTOPS, and VCTC) fully closed and the reflux valve (i.e., Vreflux) and vent valve (VVENT) open, while the operating conditions are adjusted simultaneously by manipulating the cooling and heating utilities via Vcw and Vlps. Next in stage 3, when the designated intermediate system state is reached, the input and output valves should be opened and the vent valve closed to allow the column run continuously. The steam and cooling water flows can then be further increased to drive the system toward the final steady state. Based on the above consideration, the four stages of the distillation startup operation may be characterized as follows: (1) state transfer and material charging; (2) state transfer and phase change; (3) state transfer, phase change, and material charging and discharging; and (4) stable operation.

6.7.2.2 Synthesis of Operating Procedures

In either stage 2 or stage 3, the system state is supposed to be transformed from one to another. As also mentioned previously, whenever such adjustments in operating conditions are called for, it is beneficial to follow pathways that facilitate smooth transitions. Two SFCs, labeled, respectively, as SFC-6.9 and SFC-6.10, have been generated according to this principle. Basically, the temperature at plate 16 is manipulated in three steps in the second stage of each SFC by adjusting the reboiler duty, i.e., from 20 to 50 °C, next to 75 °C, and finally to 85 °C. In stage 3, this temperature is further altered in SFC-6.9 and SFC-6.10, respectively, as follows: (1) from 85 to 95 °C and then to the target set point 101.5 °C in two steps; and (2) from 85 °C directly to 101.5 °C in one step. For the sake of brevity, again only the former case (i.e., SFC-6.9) is presented in detail in Fig. 6.13 for use as an illustration example.

6.7.2.3 Simulation, Validation, and Performance Assessment

The distillation startup process has been simulated with Aspen Plus Dynamics according to SFC-6.9 and SFC-6.10. Only the simulation results generated in the former case are presented in Fig. 6.14 for illustration brevity, while a comparison of various performance indices of the two procedures is given in Table 6.3.

First of all, it can be observed from Fig. 6.14d that, for SFC-6.9, the steady-state concentrations of overhead and bottom products are both on spec (which are stipulated at the beginning of Subsection 6.7.2). Notice also that the same trends in product concentrations also appear in the simulation results generated according to SFC-6.10. Although the ultimate goals of the startup operation can be achieved with the above two procedures, it can be observed from Table 6.3 that SFC-6.9 essentially outperforms SFC-6.10 in every aspect.

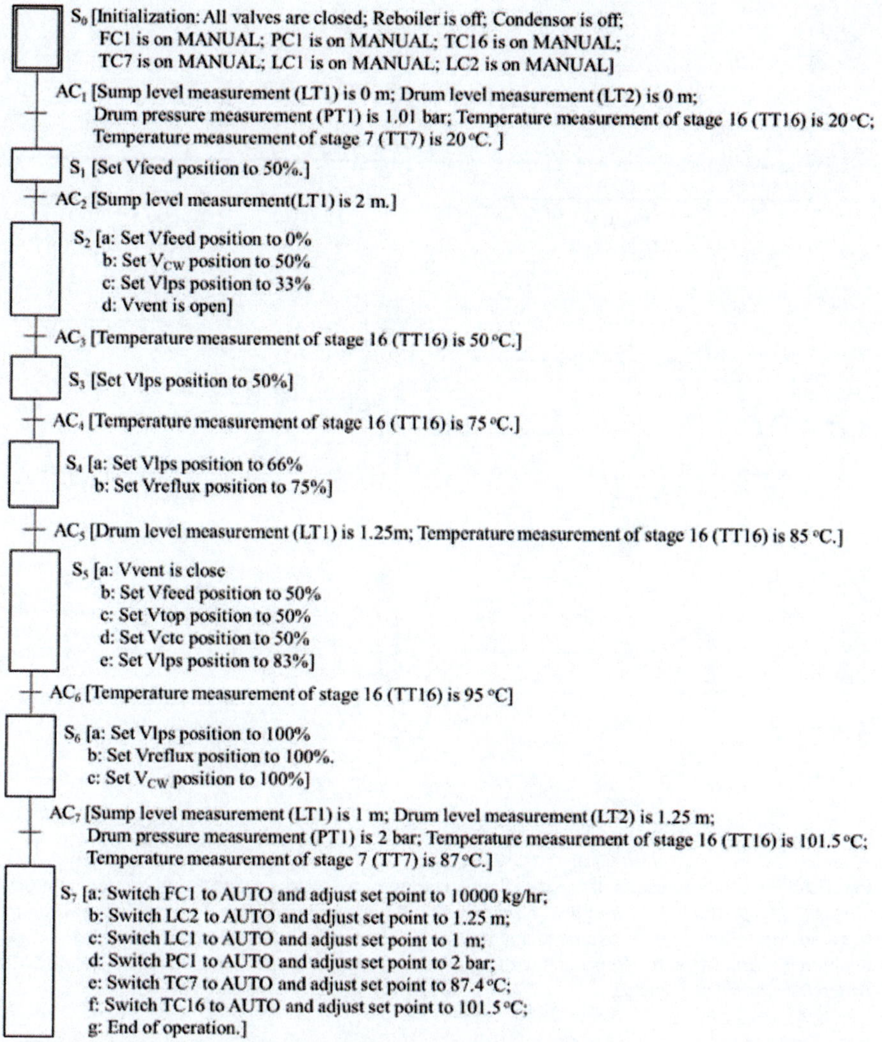

Fig. 6.13 SFC-6.9 obtained in distillation startup example (Reprinted with permission from Chen and Chang 2019. Copyright 2019 American Chemical Society)

6.7.2.4 Tuning of Total Reflux End Point

As mentioned previously in SFC-6.9 (see Fig. 6.13), the plate-16 temperature reached at the end of total reflux operation is 85 °C. In order to search for an improved end point of stage 2, two additional simulation studies have been carried out for temperatures in its neighborhood, e.g., 80 °C and 90 °C. A comparison of these two scenarios is given in Table 6.4, and they are referred to as SFC-6.11 and SFC-6.12,

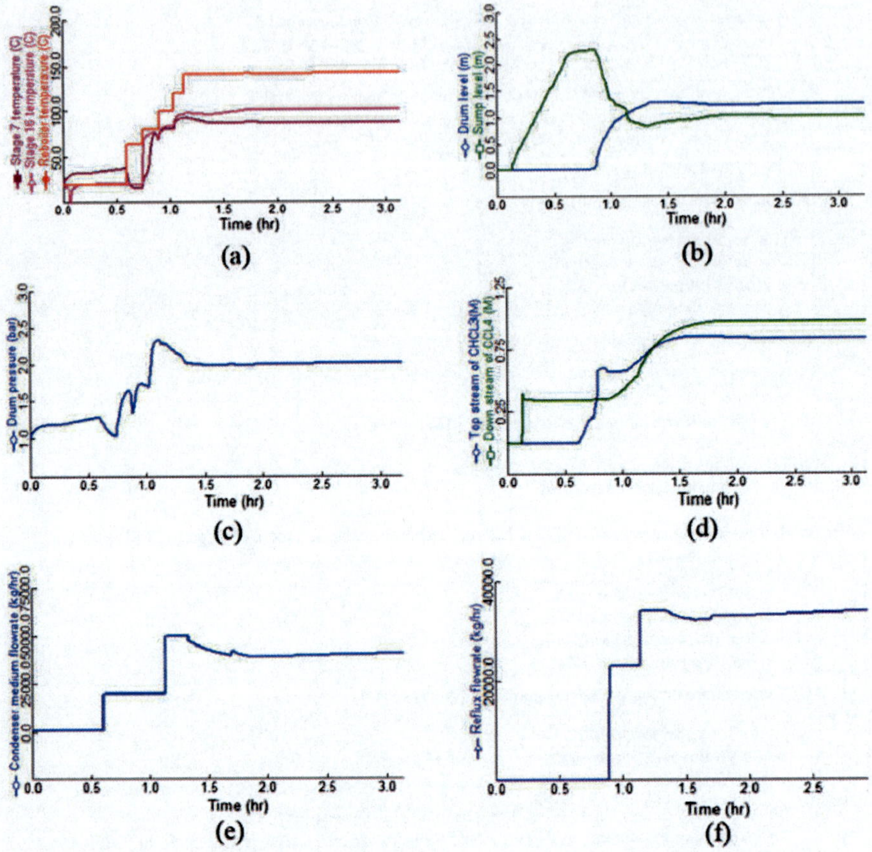

Fig. 6.14 Simulation results of the distillation startup process driven by SFC-6.9: **a** temperatures of stage 7, stage 16, and reboiler heating medium; **b** liquid levels in reflux drum and column sump; **c** pressure in reflux drum; **d** overhead and bottom concentrations; **e** flowrate of condenser cooling medium; **f** reflux flowrate (Reprinted with permission from Chen and Chang 2019. Copyright 2019 American Chemical Society)

Table 6.3 Performance indices of distillation startup processes driven by SFC-6.9 and SFC-6.10

SFC No.	Total amount of off-spec overhead product (kg)	Total amount of off-spec bottom product (kg)	Total amount of heating energy (MMkcal)	Total amount of cooling medium (kg)	Total operation time (h)
SFC-6.9	5630	1681	23.3437	107950	1.78
SFC-6.10	7044	3457	25.7482	119650	2.06

Table 6.4 Performance indices of distillation startup processes driven by SFC-6.11, SFC-6.12, and Aspen built-in procedure

SFC No.	Total amount of off-spec overhead product (kg)	Total amount of off-spec bottom product (kg)	Total amount of heating energy (MMkcal)	Total amount of cooling medium (kg)	Total operation time (h)
SFC-6.11	7617	3867	28.5956	132710	2.20
SFC-6.12	4632	203	21.7141	99864	1.54
Aspen	6105	897	28.6867	121640	2.16

respectively. From Tables 6.3 and 6.4, one can see that it is beneficial to slightly raise the end-point temperature.

6.7.2.5 Comparison with Existing Procedure

Notice that the performance indices of SFC-6.11 and SFC-6.12 in Table 6.4 are also compared with those resulting from the Aspen built-in procedure (see Fig. 6.15). It can be observed from Table 6.4 that the operation time of SFC-6.12 is significantly shorter than that of the Aspen built-in procedure. As a result, the total amounts of off-spec products and the total heating and cooling duties of the former operation are all smaller than those of the latter. From the SFC presented in Fig. 6.15, it may be deduced that the longer operation time of the Aspen built-in procedure is probably due to the more conservative startup practices. First of all, the liquid levels of the reflux drum and column sump are both brought back to 2m during the total reflux operation (stage 2) to ensure that there are enough inventories for the subsequent continuous operation. Secondly, in each heating step it is required to wait for 0.1 or 0.15h, while the corresponding activation conditions in SFC-6.12 are simply the designated online measurement values of temperatures at plate 16. In addition, the operating policy of SFC-6.12 concerning the switching actions between MANUAL and AUTO modes of the PID controllers is more straightforward than that of the Aspen built-in procedure. In the former case, MANUAL actions are always adopted to manipulate the actuators and the AUTO modes can only be activated when the designated set-point conditions are reached. On the other hand, the controller settings are switched from MANUAL to AUTO modes and vice versa throughout the startup process in the latter case, and this practice inevitably prolongs the operation time.

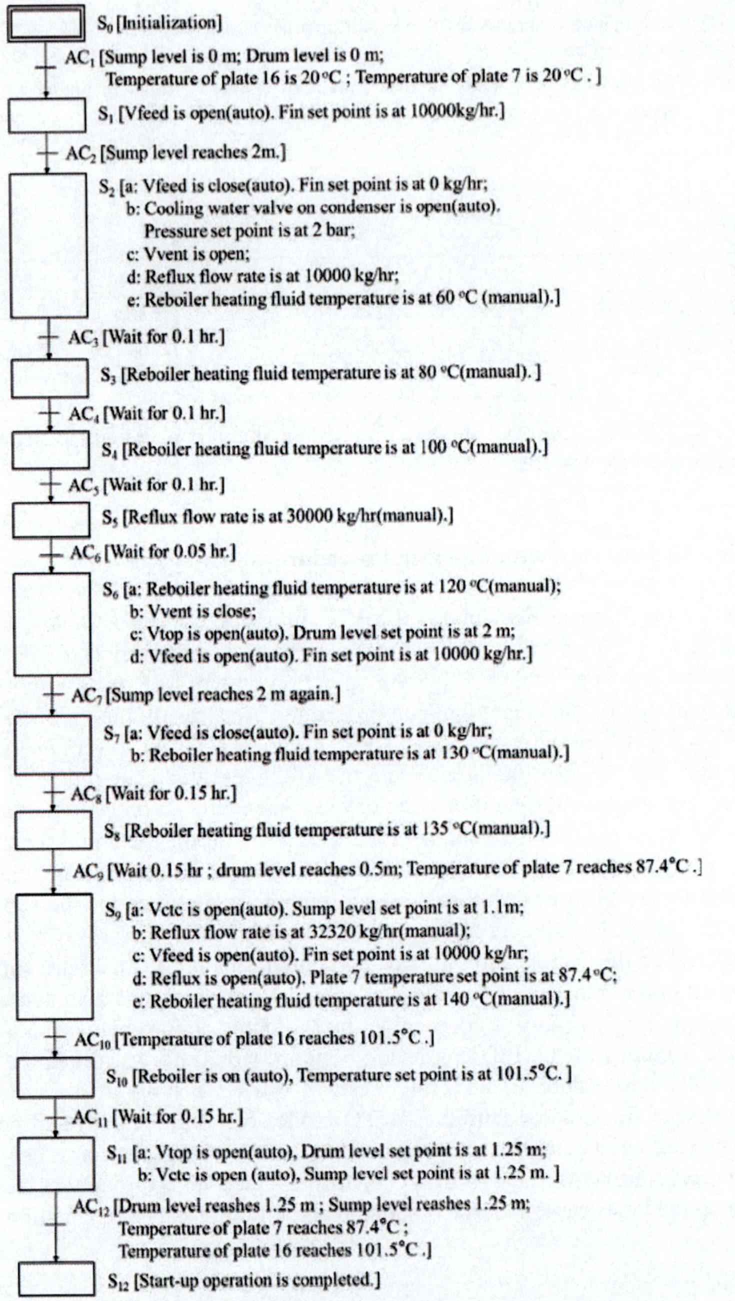

Fig. 6.15 Aspen Built-in Procedure in the Distillation Startup Example (Reprinted with permission from Chen and Chang 2019. Copyright 2019 American Chemical Society)

6.8 Concluding Remarks

A generic approach has been presented in this chapter for systematically creating operating procedures based on untimed automata. The proposed procedure synthesis steps include the following: (1) constructing an automaton model for each component in a given PFD; (2) dividing the operation into stages and developing automata to represent the control specifications of every stage, and (3) assembling the system model of each stage and consolidating them into a single SFC. The commercial software Aspen Plus Dynamics was used to validate and evaluate the candidate SFCs. Finally, this approach has been tested extensively and successfully on realistic cases.

Appendix 6.1 Component Models Used in Flash Startup Example

- **Level 2 (actuators and PID controller):**

The compressed models of control valve Vin and the corresponding PID controller are given in Figs. 6.16 and 6.17, respectively. The place Vin_normal in the former model in Fig. 6.16 can be obtained by merging all five places that represent normal valve positions in the traditional model presented in Fig. 6.3. The valve opening and closing actions are also combined and represented with two corresponding events,

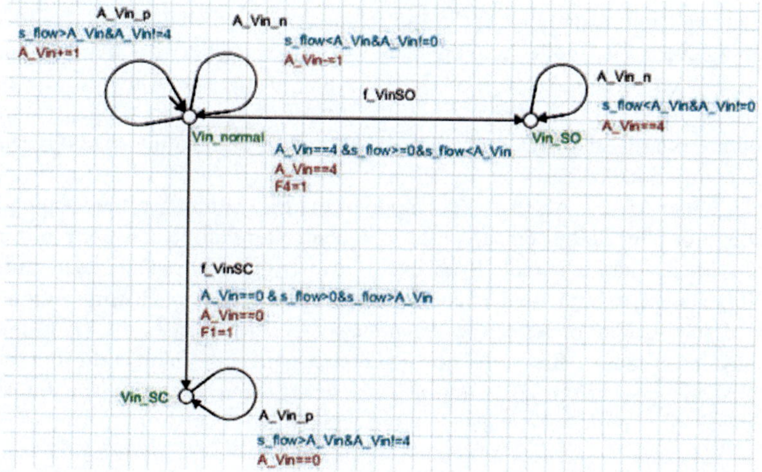

Fig. 6.16 Compressed model of control valve Vin in flash startup example (Reprinted with permission from Chen and Chang 2019. Copyright 2019 American Chemical Society)

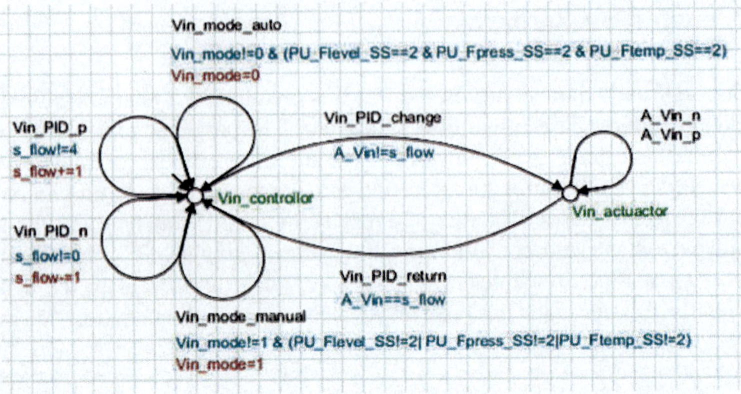

Fig. 6.17 Compressed model of PID controller FC01 in flash startup example (Reprinted with permission from Chen and Chang 2019. Copyright 2019 American Chemical Society)

i.e., A_Vin_p and A_Vin_n, in Fig. 6.16 on the looping arcs of Vin_normal, respectively. The variable A_Vin on either loop is updated according to a C-like code, i.e., A_Vin+ = 1 (or A_Vin = A_Vin + 1) and A_Vin− = 1 (or A_Vin = A_Vin − 1), while the guards can be interpreted in the same way as those in the traditional model. Finally, it should be noted that in fact the failed states in this model, i.e., Vin_SC and Vin_SO, and also the directed arcs attached to them can be omitted since the present goal of procedure synthesis is to automatically generate SOPs under the normal operating conditions.

It should be noted that every actuator is assumed to be manipulated via the MANUAL mode of the corresponding PID controller in the beginning, i.e., Vin_mode = 1 initially in Fig. 6.17. As soon as the anticipated set point is reached, the looping event Vin_mode_auto on the place Vin_controller is activated to signify the switch from MANUAL to AUTO modes and, at the same time, to update the value of Vin_mode to 0. Furthermore, if the guards of event Vin_mode_manual are satisfied afterwards, the PID controller FC01 could return to the MANUAL mode and reset Vin_mode to 1.

Under either MANUAL or AUTO mode of the PID controller, the looping event Vin_PID_p or Vin_PID_n on the place Vin_controller can be activated to increase or decrease the output signal. Both events can trigger the subsequent transition Vin_PID_change, which then enables the valve opening or closing actions,.i.e., A_Vin_p and A_Vin_n. After the valve state (A_Vin) reaches the same level of the controller output (s_flow), the original component state (Vin_controller) can be reestablished via event Vin_PID_return. The looping event Vin_PID_p or Vin_PID_n on the place Vin_controller can be again activated to alter the controller output signal.

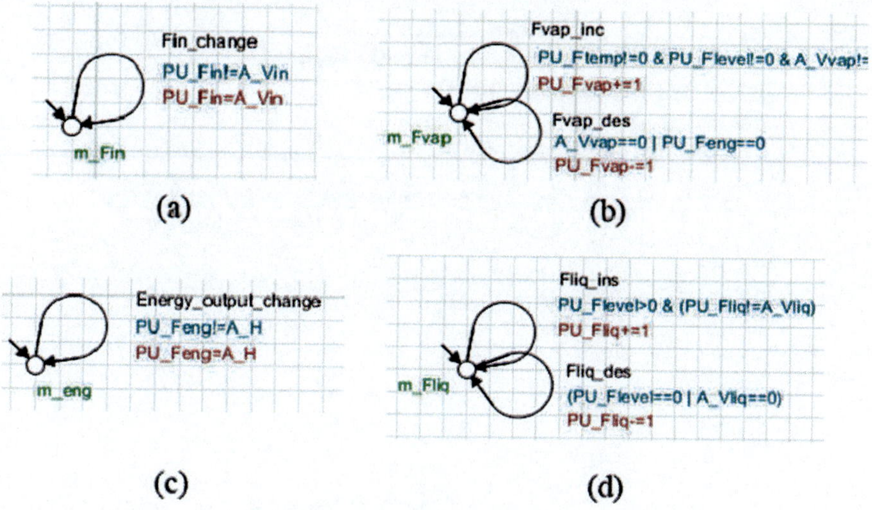

Fig. 6.18 Component models in the third level: **a** feed flowrate; **b** vapor product flowrate; **c** energy flowrate from heater; **d** liquid product flowrate (Reprinted with permission from Chen and Chang 2019. Copyright 2019 American Chemical Society)

- **Level 3 (process configuration)**:

As mentioned in Sect. 6.2, the material and energy flows surrounding each unit are viewed as the third-level components. Since in the flash startup example there is only one processing unit, i.e., flash drum, it is only necessary to consider its inputs and outputs, i.e., flowrates of feedstock (Fin), vapor product flowrate (Fvap), energy flowrate from the heater (Feng), and liquid product flowrate (Fliq). Their models are presented in Figs. 6.18a–d, respectively.

- **Level 4 (processing unit)**:

Since only liquid level, temperature, and pressure of the flash drum are measured online, the variations of these variables are characterized separately with the automata according to the basic principles of dynamic material and energy balances. These automata are presented below in Fig. 6.19.

- **Level 5 (online sensors)**:

Figure 6.20 shows the sensor models used to represent online measurements of level, temperature, and pressure in the flash drum.

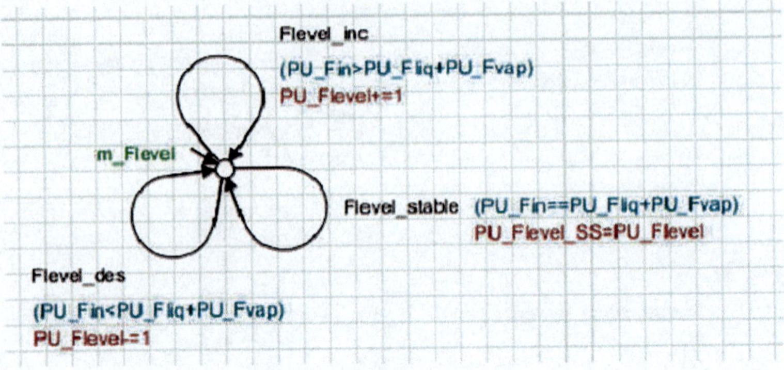

(a)

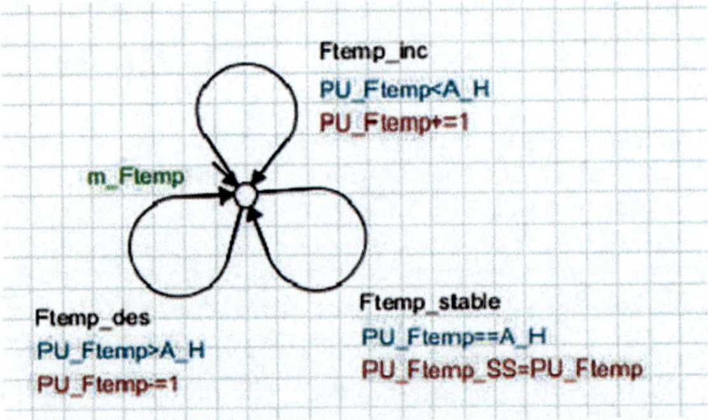

(b)

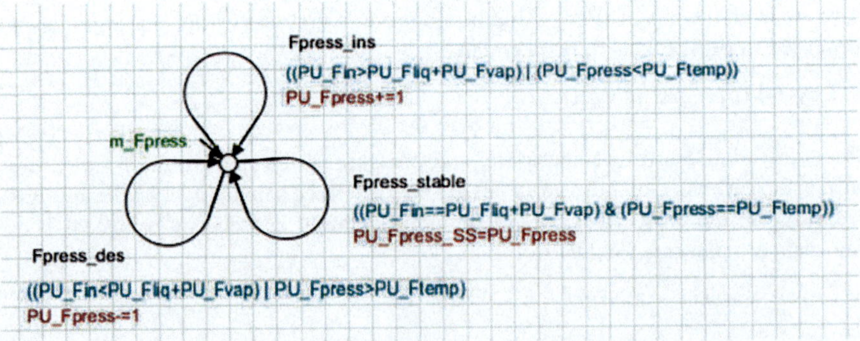

(c)

Fig. 6.19 Component models in the fourth level: **a** liquid level in flash drum; **b** temperature in flash drum; **c** pressure in flash drum (Reprinted with permission from Chen and Chang 2019. Copyright 2019 American Chemical Society)

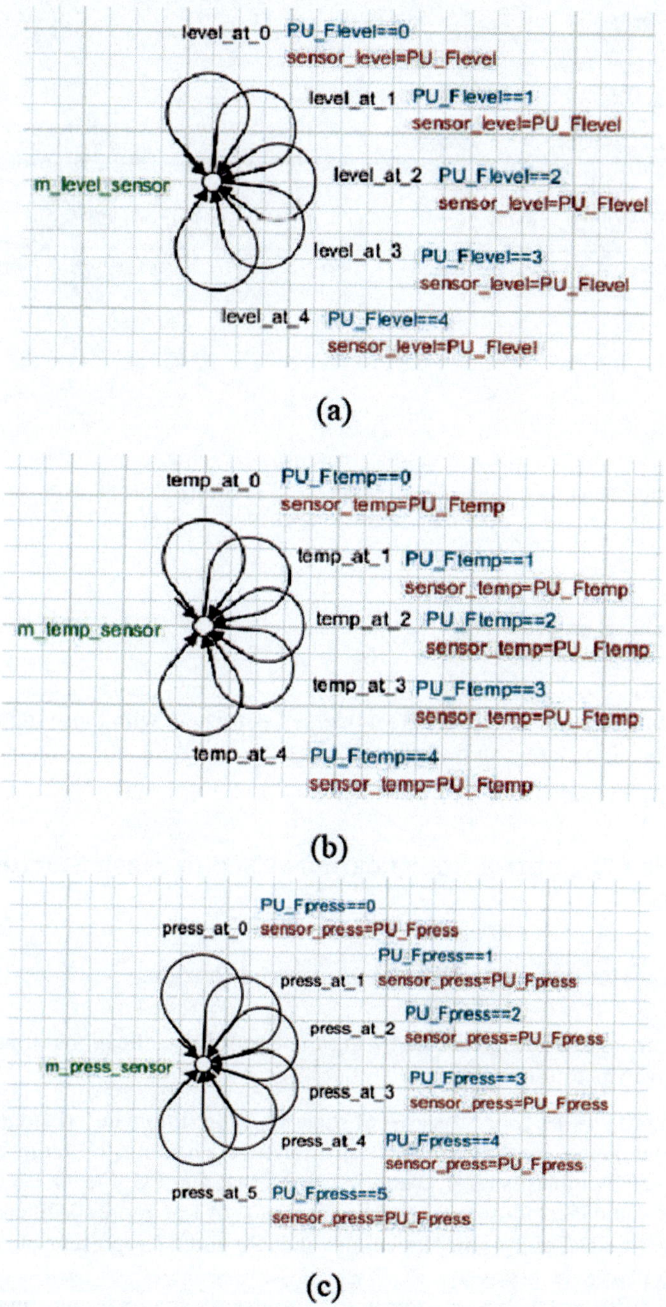

Fig. 6.20 Component models in the fifth level: **a** level measurement; **b** temperature measurement; **c** pressure measurement (Reprinted with permission from Chen and Chang 2019. Copyright 2019 American Chemical Society)

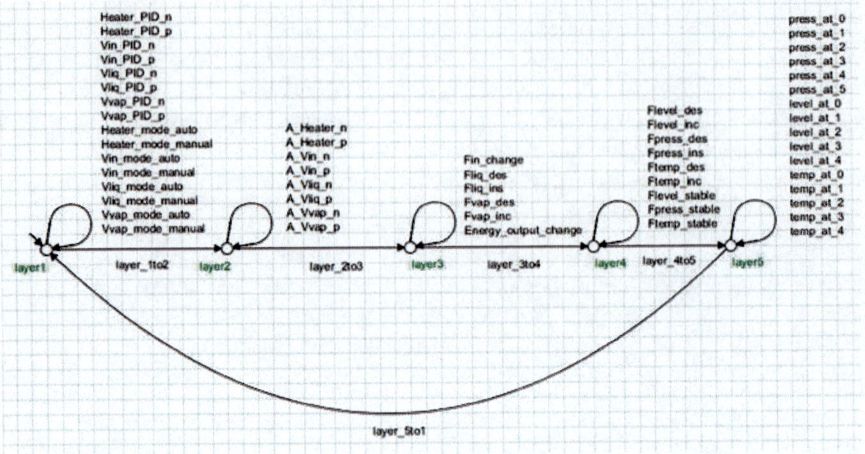

Fig. 6.21 Automaton used to represent the hierarchical structure of the flash startup process (Reprinted with permission from Chen and Chang 2019. Copyright 2019 American Chemical Society)

- **Hierarchical structure**:

An automaton should be built according to Fig. 6.2 to make sure that the cause-and-effect relationships embedded in the model are consistent with those in the realistic system. This automaton can be found in Fig. 6.21.

Appendix 6.2 Control Specifications Used in Flash Startup Example

- **Stage 1—condition adjustment and material charging**

Three automata are needed for depicting the control specifications of stage 1 and they are given as follows:

- **Spec 1.1 (Type A)**: Since the goal of stage 1 can be simply described as placing a certain amount of feedstock in the flash drum before heating, this target is specified in Fig. 6.22 as the guard of event Step 1. Notice that PU_Flevel == 1 denotes that the liquid level must reach the discrete value of 1 or, more specifically, around an actual height of 1.45m. Notice also that S0 and S1 represent the initial and marked places, respectively.
- **Spec 1.2 (Type C)**: It is also desired to minimize the charging time by maximizing the feed rate in this initial stage. The corresponding control specification can be modeled by the automaton in Fig. 6.23. Notice that the guard of event

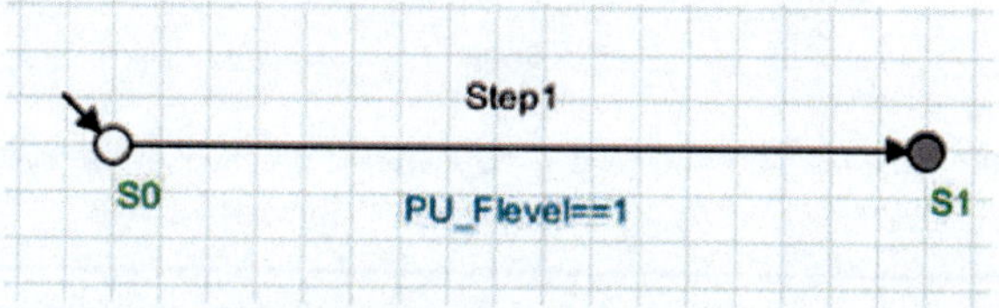

Fig. 6.22 Spec 1.1 in flash startup example (Reprinted with permission from Chen and Chang 2019. Copyright 2019 American Chemical Society)

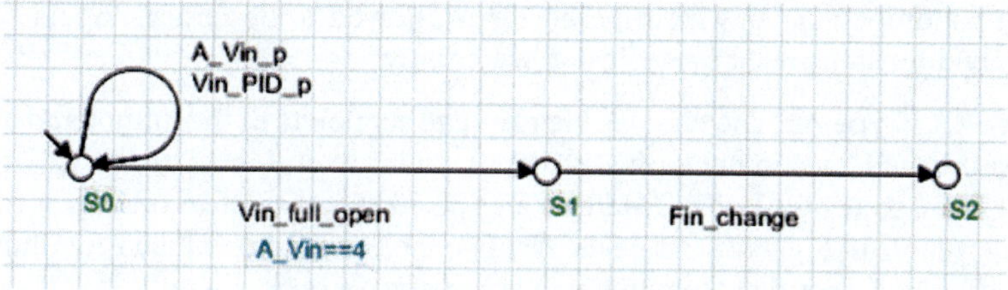

Fig. 6.23 Spec 1.2 in flash startup example (Reprinted with permission from Chen and Chang 2019. Copyright 2019 American Chemical Society)

Vin_full_open is A_Vin == 4 (the largest opening of valve Vin is at the discrete value of 4). In this example, this is the condition when Vin is fully open.

– **Spec 1.3 (Type D)**: The illegal strings specified in Fig. 6.24 are adopted to prevent the inlet valve (Vin) closing and the heater (H) activating accidentally. Notice that, in Supremica, the guard 1 == 0 is used to forbid the looping events, i.e., Vin_PID_n and H_PID_p, on S1.

- **Stage 2—condition adjustments, material charging, and unloading**

During this stage, all inlet and outlet valves should be open at the targeted positions, and the heater should be switched on to raise the liquid temperature in the flash drum to its set point. Furthermore, options should be incorporated to facilitate identification

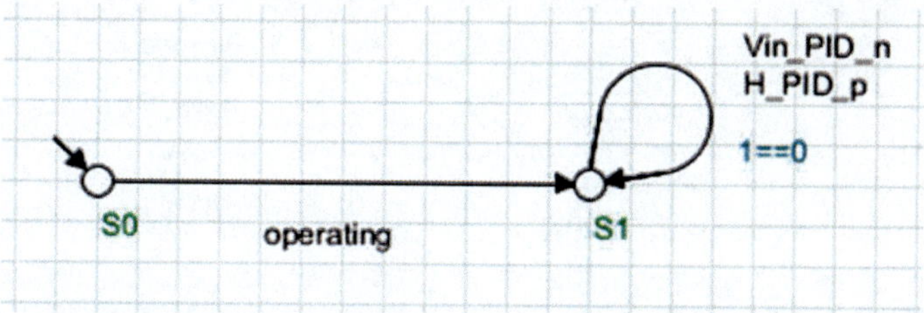

Fig. 6.24 Spec 1.3 in flash startup example (Reprinted with permission from Chen and Chang 2019. Copyright 2019 American Chemical Society)

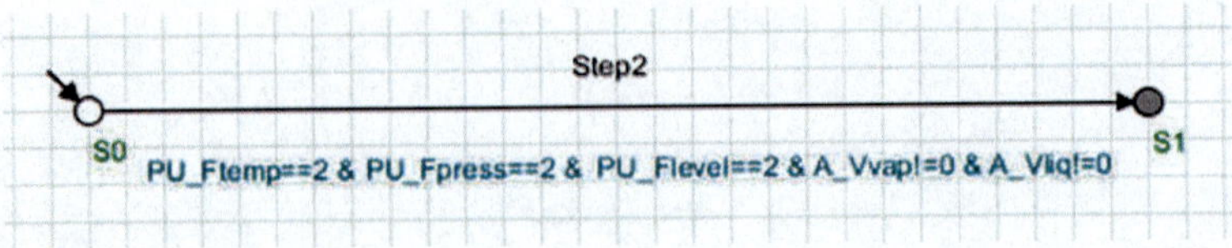

Fig. 6.25 Spec 2.1 in flash startup example (Reprinted with permission from Chen and Chang 2019. Copyright 2019 American Chemical Society)

of proper heater operation procedures. In particular, these options should include (1) the conditions under which the heater is allowed to be turned on and (2) the corresponding adjustment strategies of heater duty.

- **Spec 2.1 (Type A)**: The targeted steady-state set points of the drum temperature, pressure, and level in the flash startup operation are chosen to be 75 °C, 1.01bar, and 2.5m, respectively. These targets are specified in discretized values as the first three guards in Fig. 6.25, while the last two, i.e., A_Vvap! = 0 and A_Vliq! = 0, are used to ensure that the outlet valves are always open.
- **Spec 2.2 (Type B)**: As the liquid level continuously rises from 1.45m, the heater can be switched on and the inlet and outlet valves open at any time. Two options are provided with the specification model given in Fig. 6.26, i.e., the heating can start when the liquid level reaches either 1.45m or 2.5m. The former scenario is represented by event level_is_1, while the latter by event level_is_2.
- **Spec 2.3 (Type B)**: The energy transfer rate from the heater may be raised from zero directly to a level corresponding to the targeted steady-state temperature or to an intermediate temperature first and then to the final set point. In the specification presented in Fig. 6.27, these options are facilitated with the events once and twice, respectively. If event once is activated, all looping events on S2 must be repeated

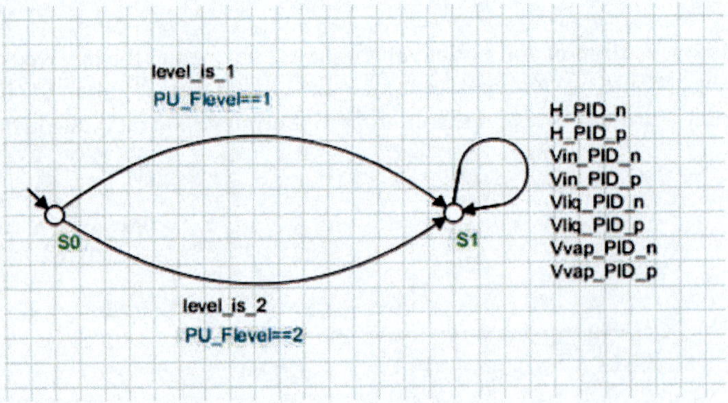

Fig. 6.26 Spec 2.2 in flash startup example (Reprinted with permission from Chen and Chang 2019. Copyright 2019 American Chemical Society)

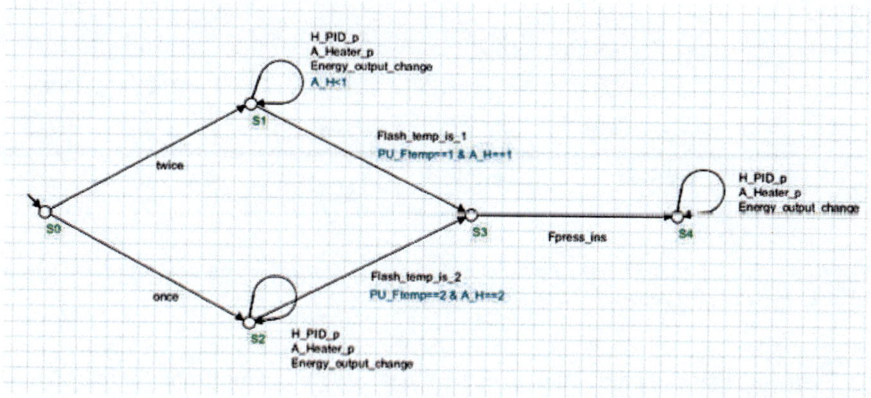

Fig. 6.27 Spec 2.3 in flash startup example (Reprinted with permission from Chen and Chang 2019. Copyright 2019 American Chemical Society)

two times in order to satisfy the guards of Flash_temp_is_2. As a result, the steady-state temperature should have already been reached at place S3. Since the pressure in the flash drum increases proportionally with temperature, the event Fpress_ins is introduced immediately after state S3. Since the steady-state conditions have been reached, the looping events on S4 must remain inactive in the scenario initiated by event once. On the other hand, if event twice is activated instead, it is only necessary to trigger all looping events on S2 one time for the purpose of satisfying the guards of Flash_temp_is_1. The same set of events should take place again on S4.

– **Spec 2.4 (Type C)**: Although the pressure in the flash drum increases with rising temperature after heating begins, this trend can be stopped by manipulating the outlet vapor valve Vvap at some instance so as to approach the set point smoothly. The corresponding specification can be found in Fig. 6.28. Notice that the guard of event Flash_press_is_2 is PU_Fpress == 2, which implies that the looping events on S1 can be triggered when the pressure is around 1.01bar and this pressure is also the initial pressure. Notice also that these looping events, i.e., A_Vvap_p and

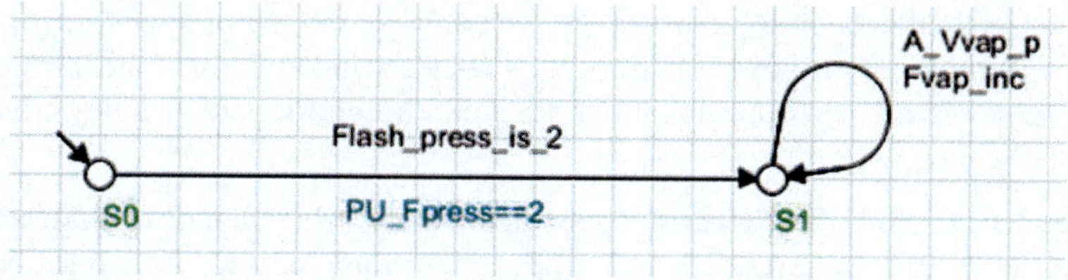

Fig. 6.28 Spec 2.4 in flash startup example (Reprinted with permission from Chen and Chang 2019. Copyright 2019 American Chemical Society)

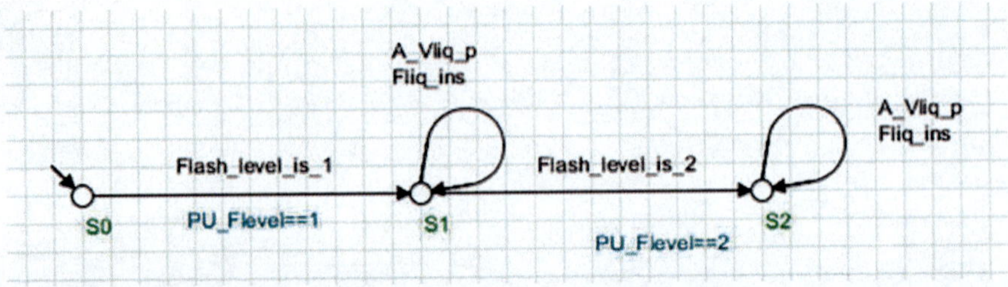

Fig. 6.29 Spec 2.5 in flash startup example (Reprinted with permission from Chen and Chang 2019. Copyright 2019 American Chemical Society)

Fvap_inc, represent the action to open outlet valve Vvap more and the subsequent event of increasing the corresponding vapor flow, respectively.

- **Spec 2.5 (Type C)**: In a similar fashion, the valve Vliq is also adjusted to drive the system toward a steady state. The corresponding specification is presented in Fig. 6.29. Notice first that the liquid level at the end of stage 1 is at 1.45m, i.e., the guard of event Flash_level_is_1 (PU_Flevel == 1) should be satisfied and, consequently, the looping event on S1 may be activated. Notice also that these looping events, i.e., A_Vliq_p and Fliq_inc, respectively, represent the action to open Vliq a little more and the subsequent event of increasing the corresponding liquid flow. On the other hand, if the above events are not triggered, the guard of Flash_level_is_2, i.e., PU_Flevels == 2, can be satisfied and the looping events on S2 should be activated around a liquid level of 2.5m. Finally, note that the looping events on S1 and S2 are the same.
- **Spec 2.6 (Type D)**: To maintain a stable feed rate, it is necessary to avoid significant disturbances from its steady-state level. The specification in Fig. 6.30 is utilized in the present work for this purpose. Notice first that the looping events on S1 are prohibited by the guard 1 == 0. Thus, the inlet valve can only be adjusted when the output of flow controller is at the nominal value, i.e., s_flow = 2.
- **Spec 2.7 (Type D)**: The specification in Fig. 6.31 is used to prevent the closing of the actuators accidentally in this stage.

- **Third stage of flash startup operation**

Since the desired set points have already been achieved in the previous stage, these operating conditions are supposed to be kept constant indefinitely in the third stage. Thus, the required control specifications are quite straightforward and they are listed as follows (Fig. 6.32):

- **Spec 3.1 (Type A)**: The guards of specification 3.1 below denote that the steady-state values of level, pressure, and temperature should be kept at their set points, respectively.
- **Spec 3.2 (Type D)**: As shown in Fig. 6.33, the controller outputs are not allowed to vary significantly in this stage so as to maintain a stable operation.

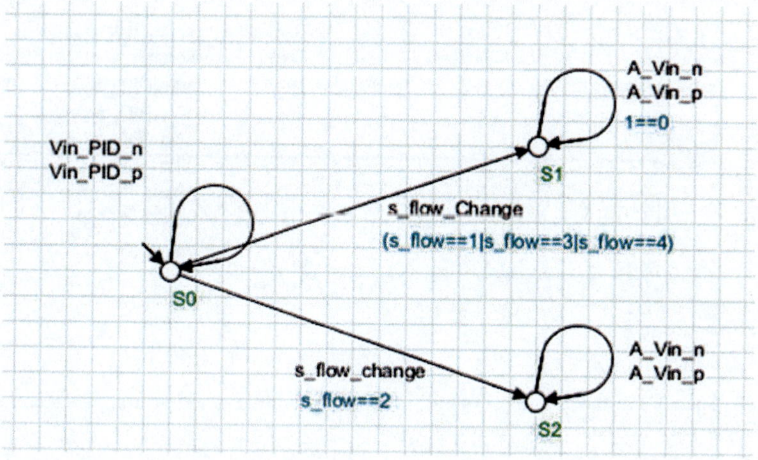

Fig. 6.30 Spec 2.6 in flash startup example (Reprinted with permission from Chen and Chang 2019. Copyright 2019 American Chemical Society)

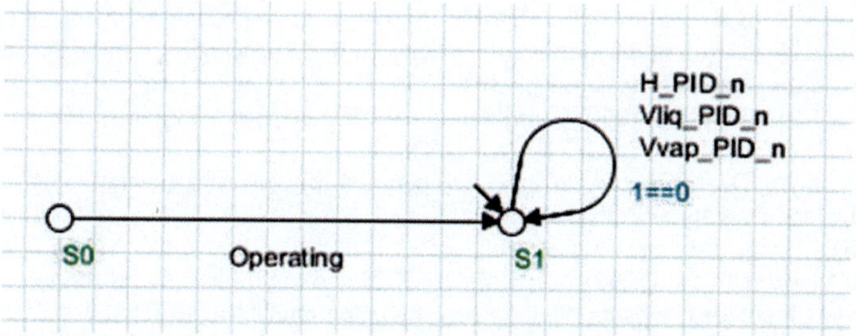

Fig. 6.31 Spec 2.7 in flash startup example (Reprinted with permission from Chen and Chang 2019. Copyright 2019 American Chemical Society)

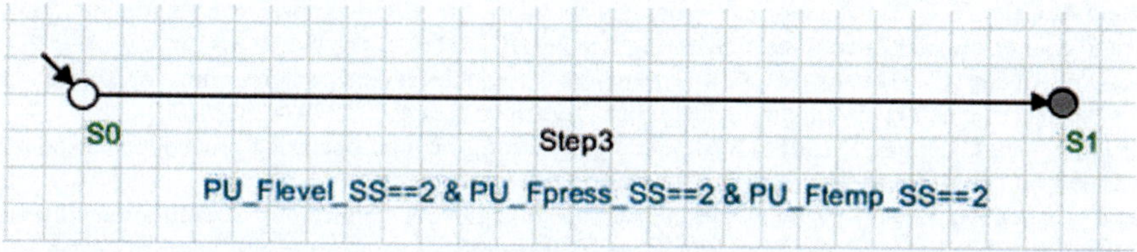

Fig. 6.32 Spec 3.1 in flash startup example (Reprinted with permission from Chen and Chang 2019. Copyright 2019 American Chemical Society)

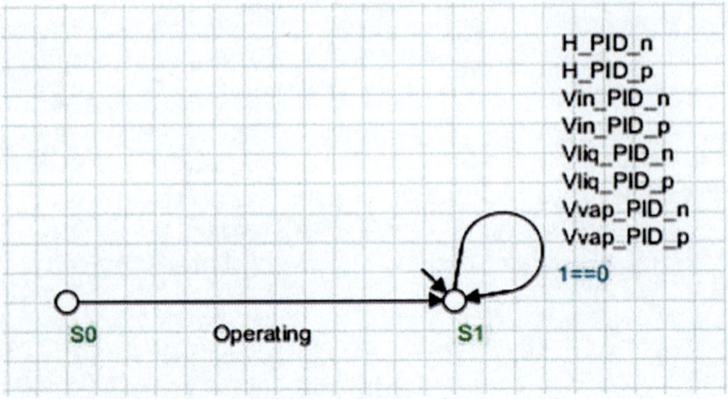

Fig. 6.33 Spec 3.2 in flash startup example (Reprinted with permission from Chen and Chang, 2019. Copyright 2019 American Chemical Society)

References

Åkesson K, Fabian M, Malik R (2006) SUPREMICA—an integrated environment for verification, synthesis and simulation of discrete event systems. In IEEE proceedings of the 8th international workshop on discrete event systems, pp 384–385

Cassandras CG, Lafortune S (2008) Introduction to discrete event systems, 2nd edn. Springer Science + Business Media, LLC, New York, NY, USA

Cochard T, Gouyon D, P´etin, J.-F., 2011. Safe operation sequences: a generation approach based on iterative refinements and abstractions of timed automata, IFAC Papers OnLine, 50-1, pp 6952–6957

Crooks CA, Macchietto SA (1992) A combined MILP and logic-based approach to the synthesis of operating proceduresfor batch plants. Chem Eng Commun 114:117–144

Ferrarini L, Piroddi L (2003) Modular design and implementation of a logic control system for a batch process. Comput Chem Eng 27:983–996

Foulkes NR, Walton MJ, Andow PK, Galluzzo M (1988) Computer-aided synthesis of complex pump and valve operations. Comput Chem Eng 12:1035–1044

Kang A, Chang CT (2014) Automata generated test plans for fault diagnosis insequential material-and energy-transfer operations. Chem Eng Sci 113:101–115

Lai JW, Chang CT, Hwang SH (2007) Petri-net based binary integer programs for automatic synthesis of batch operating procedures. Ind Eng Chem Res 46(9):2797–2813

Li JH, Chang CT, Jiang D (2014) Systematic generation of cyclic operating procedures based on timed automata. Chem Eng Res Des 92:139–155

Lu YC, Chen ZL, Lee HY (2017) Optimal start-up strategies for a conventional distillation column using simulated annealing. Chem Eng Trans 61:901–906

O'Shima E (1978) Safety supervision of valve operation. J Chem Eng Jpn 11:390–395

Rivas JR, Rudd DF (1974) Synthesis of failure-safe operation. AIChE J 20:320–325

Chen TY, Chang CT (2019) Design approach to synthesize, validate, and evaluate operating procedures based on untimed automata and dynamic simulation. Ind Eng Chem Res 58(19):8172–8183

Uthgenannt JA (1996) Path and equipment allocation for multiple, concurrent process on networked process plant units. Comput Chem Eng 20:1081–1087

Wang CJ, Chen YC, Feng ST, Chang CT (2017) Automata-based operating procedure for abnormal situation management in batch processes. Comput Chem Eng 97:220–241

Yang SH, Tan LS, He CH (2001) Automatic verification of safety interlock systems for industrial processes. J Loss Prev Process Ind 14:379–386

Yeh ML, Chang CT (2012a) An automata-based approach to synthesis untimed operating procedures in batch chemical processes. Korean J Chem Eng 29(5):583–594

Yeh ML, Chang CT (2012b) An automata based method for online synthesis of emergency response procedures in batch processes. Comput Chem Eng 38:151–170

Chapter 7
Normal Operating Procedures Obtained with Timed Automata

As mentioned in the previous chapter, an operating procedure should be synthesized according to the initial system state and also the ultimate goal. To overcome the difficulties caused by combinatorial explosion of all possible operation pathways, many published studies have focused on issues concerning systematic procedure synthesis. Since a comprehensive survey has already been given in Chap. 6, these studies are not enumerated here for the sake of brevity. Notice also that the common drawback of the earlier studies is that they emphasized only upon the procedure synthesis aspects and thus the resulting sequential function charts (SFCs) may not be implementable. In particular, these automatically synthesized standard operating procedures (SOPs) were not validated either in simulation studies with credible software or in the pilot plant experiments. Furthermore, if several candidates can be generated, it is necessary to evaluate them with reasonable criteria so as to identify the most suitable one. To fully address the above concerns, a design approach has been developed in Chap. 6 for synthesis, validation, and evaluation of alternative SFCs based on the *untimed* automata and dynamic simulation.

Although satisfactory results can be obtained with the untimed-automata (Alur and Dill 1994)-based approach suggested in Chap. 6, there is still an unsettled issue that may hinder practical applications. Specifically, due to budget constraints or technical difficulties, some of the online measurements required in the operating procedure may not be available. It is thus necessary to build timed automata to incorporate the elapsed times of various events (or actions) into the system model. By constraining the system model with the control specifications, both the shortest duration and fewest event traces can be extracted with software UPPAAL (Behrmann et al. 2004, 2006) and every such trace summarized with a SFC. Note that, in case when a needed online sensor is lacking, the elapsed time of an operation step can be stipulated in the corresponding activation condition of SFC according to the information embedded in the trace mentioned above. Of course, this SFC should also be simulated and verified with Aspen Plus Dynamics®. If the test results show that any SFC is unsafe and/or infeasible, one should discard/modify some of the control specifications and repeat the procedure synthesis steps.

C.-T. Chang et al., *Process Plant Operating Procedures*, Advances in Industrial Control,
https://doi.org/10.1007/978-3-030-70978-5_7

7.1 Timed Automata

A timed automaton is a finite-state machine equipped with one or more clocks (Cassandras and Lafortune 2008). All clocks progress synchronously, and the clock variable assumes a real positive value. To facilitate a clear description of the proposed modeling method, a brief summary of the automaton structure is given below. In particular, a timed automaton can be regarded as a six-tuple, i.e.,

$$TA = (L, \ell_0, C, A, I, E) \tag{7.1}$$

where L is a set of locations; $\ell_0 \in L$ is the initial location; C denotes the set of clock variables; A is a set of actions. In addition, $I : L \to B(C)$ denotes a function $I(\ell) = b(c)$ which assigns invariant(s) to location ℓ. Note that $B(C)$ denotes the set of conjunctions over simple conditions of the form $\{x \oplus c\}$ or $\{x - y \oplus c\}$, where $x, y \in C$, $c \in \mathbb{N}$, and $\oplus \in \{<, \leq, =, \geq, >\}$. Finally, the set $E \subseteq L \times A \times B(C) \times 2^C \times L$ contains all edges in the automaton. Each edge represents a transition process from one location to another, which is enabled by an action in the set A, constrained by a guard in the set $B(C)$ and timed according to a collection of clocks which belongs to the power set of C, i.e., 2^C.

7.2 Process Structure

To facilitate a clear illustration of the process structure, let us consider the startup operation of a modified version of the continuous flash process in Fig. 6.3 as an example. The corresponding process flow diagram (PFD) is given in Fig. 7.1. It is assumed that, at steady state, the feed is a mixture of 30 wt% water and 70 wt%

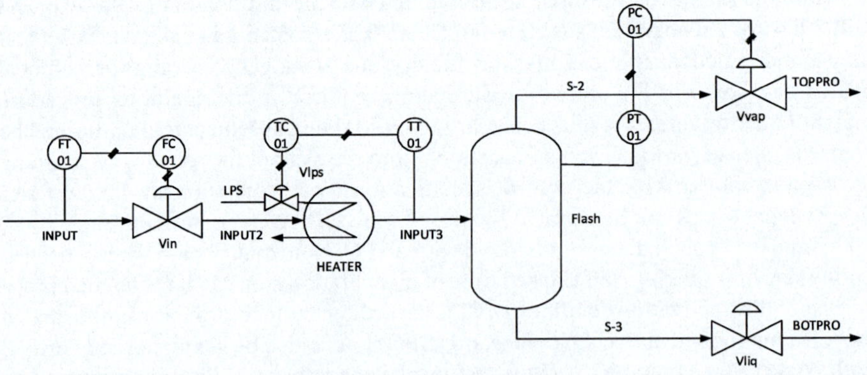

Fig. 7.1 PFD of a continuous flash process (Reprinted with permission from Zhang et al. 2020. Copyright 2020 American Chemical Society)

methanol and its flowrate, temperature, and pressure are kept at 1000 kmol/hr, 25°C, and 1.31 bar, respectively. The steady-state temperature and pressure of the top and bottom products are both set at 75°C and 1.01 bar, respectively, while the corresponding liquid level is 2.5 m. It is required that the concentration of methanol in the top product should not be lower than 87 wt%. The heating medium in the heater is assumed to be low steam pressure. In this system, there are three PID controllers (FC01, TC01, and PC01) for controlling the feed rate, the temperature, and the vapor pressure in the flash drum, respectively. The corresponding actuators are control valves, i.e.,Vin, Vlps, and Vvap. The level sensor and controller are not installed in this case for the purpose of demonstrating the use of elapsed time in startup operation, while Vliq is only a hand valve. It is assumed that, initially, all valves are closed, all controllers are on MANUAL and the flash drum is empty and at room temperature.

Every component in the above process is modeled with an automaton. Basically, every item in the PFD is treated as a component and they are again classified into a five-level hierarchy according to Fig. 3.1 in Chap. 3.

- Level 1: The top-level component is a human operator or a programmable logic controller (PLC).
- Level 2: The second-level components are the actuators, e.g., hand valves, control valves, and/or the corresponding PID controllers.
- Level 3: The material and energy flows among processing units in the given system are viewed as the components in the third level.
- Level 4: Every major unit operation in PFD, such as the flash drum itself in Fig. 7.1, is treated as a fourth-level component.
- Level 5: Every online sensor in PFD, such as the flow, temperature and pressure sensors, etc., is a component in the last level.

7.3 Construction of Component Models

The building principles of component models on the platform of UPPAAL can be summarized as follows:

All possible states of the component are first enumerated and denoted with circles (locations) in a graphic representation of the automaton. The location corresponding to the initial state is indicated with double concentric circles. All events that facilitate state transition should then be identified and each described with a directed edge between two locations. The guards (marked in green), the update variable values (marked in purple), and the synchronization mechanism(s) should be next added on to the corresponding edge. Every synchronization mechanism is incorporated via the event label (marked in blue). The "receiver" event is attached with a question mark (?) indicating that such event must also occur in other components at a prior instance. On the other hand, the exclamation mark (!) is used to specify an initiator or "sender" event that takes place in a component as long as all prerequisite conditions (guards) are satisfied. The corresponding receiver(s) should occur in another component later. Notice also that, if an event takes place almost instantaneously, then it is not necessary to specify the elapsed time as one of the guards on the corresponding edge. In this chapter, only the state-transition time of every event in the component model of each processing unit in the fourth level of the system hierarchy is estimated according to Aspen simulation results. For the sake of brevity, the required discretization and simulation procedures are provided in Appendix 7.1.

It is not possible to construct an automaton at this point to describe the level-1 component, i.e., PLC or human operator, since the operating procedure is not available a priori. For illustration conciseness, let us consider only the components in level 2, i.e., the actuators and the PID controllers, and level 4, i.e., the flash drum, as examples. All component models in the other levels and a layer model for integration on the basis of Fig. 3.1 can be found in Appendix 7.2.

7.3.1 *Actuators and PID Controllers*

Let us first construct a simple automaton to characterize the outlet valve Vliq in Fig. 7.1 by using the model building principles presented above. This model is given in Fig. 7.2. As mentioned in Appendix 7.1, the valve position is discretized into 5 levels (from 0 to 4) and, thus, the guards of the two events in this model, i.e., A_Vliq_p? (an increase in the valve opening) and A_Vliq_n? (a decrease in the valve opening), are specified as A_Vliq! $= 4$ and A_Vliq! $= 0$, i.e., the valve position is *not* fully open and *not* fully closed, respectively. Notice also that these two events are receivers and the corresponding senders can be found in the "layer" model which can be found in Appendix 7.2. Finally, note that the variable A_Vliq on either loop is updated with a C-like code, i.e., A_Vliq+ $= 1$ (or A_Vliq $=$ A_Vliq $+ 1$) and A_Vliq$-$ $= 1$ (or A_Vliq $=$ A_Vliq $- 1$),

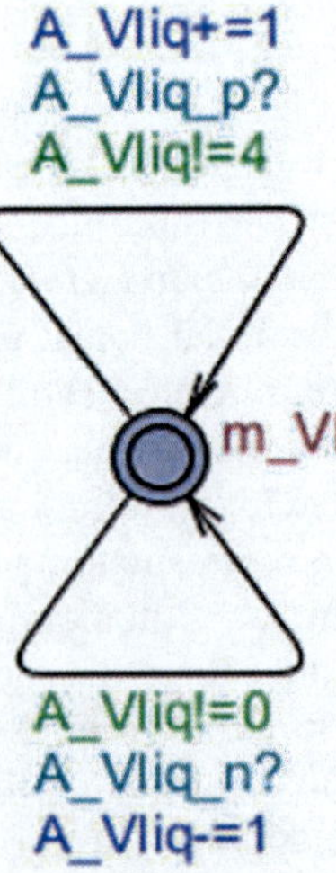

Fig. 7.2 Component model of hand valve Vliq in flash startup example (Reprinted with permission from Zhang et al. 2020. Copyright 2020 American Chemical Society)

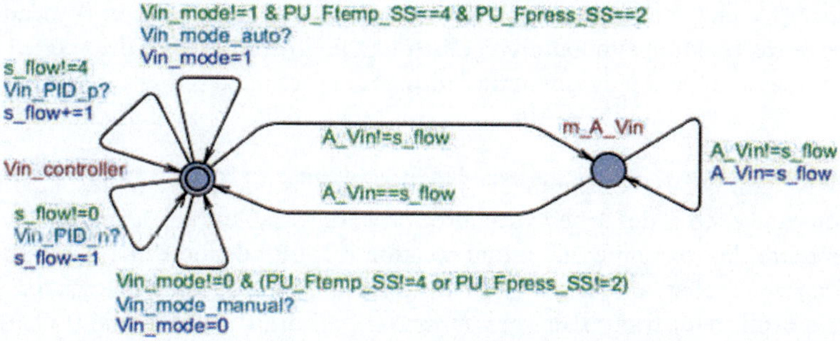

Fig. 7.3 Component model of flow controller FC01 and control valve Vin (Reprinted with permission from Zhang et al. 2020. Copyright 2020 American Chemical Society)

The flow control valve Vin and the corresponding controller FC01 are next modeled with the automaton in Fig. 7.3. The location Vin_controller and the edges attached to it in this model are used to describe the controller behavior of FC01, while location m_A_Vin and its corresponding edges are for characterizing valve Vin.

Note that there are four self-looping edges on location Vin_controller. The top and bottom ones are associated with receiver events Vin_mode_auto? and Vin_mode_manual?, respectively. These two events are actions to switch the controller mode from MANUAL (Vin_model = 0) to AUTO (Vin_model = 1) and vice versa. Note also that the guards of the former event are the targeted steady-state conditions reached when the controller is still on MANUAL and those of the latter are conditions when the steady state is not reached.

On the other hand, the two loops between the above two are associated with receiver events Vin_PID_p? and Vin_PID_n? and they are actions to adjust the controller output toward positive (s_flow+ = 1) and negative ($s_{\text{flow}}- =$ 1)directions, respectively. The guards of the two events are imposed to ensure the adjustments are feasible. If either event is triggered, then the guard of the edge from location Vin_controller to location m_A_Vin should be satisfied and, subsequently, the prerequisite on the self-looping edge of the latter location is also met. Note that the event associated with the loop on m_A_Vin is to reset the valve opening to that corresponding to the adjusted controller output, i.e., A_Vin = s_flow. As a result, the guard on the edge from location m_A_Vin to location Vin_controller should be satisfied and, thus, the component state should then return to Vin_controller waiting for the next request to alter the controller output.

7.3.2 Flash Drum

Since the elapsed time period of every state-transition event concerning the flash process is not negligible, it is necessary to estimate these time intervals via test runs using Aspen Plus Dynamics. The test procedure is detailed in Appendix 7.1 and the corresponding simulation results (i.e., the time profiles of discretized liquid level, temperature and pressure in flash drum) are presented in Figs. 7.30, 7.31, 7.32 and 7.33. By extracting the event times from these data, a timed automaton can be constructed (see Fig. 7.4).

In this automaton, the component states of the flash drum are represented by the locations and each is described with three process variables, i.e., level, temperature, and pressure. For example, the initial location L0T0P0 denotes that the flash drum is at a component state that can be characterized by the specific discretized values of liquid level, temperature and pressure, i.e., 0 (0 m), 0 $(20^{\circ}\mathrm{C})$, and 0 (1.01 bar). Note also that every directed arc in Fig. 7.4 may consist of more than one edge and a detailed listing of their guards and updated variables can be found in Appendix 7.3. To further illustrate its structure, let us consider arc 1 and arc 13 as examples. The specifications of the corresponding edges are given in Tables 7.1 and 7.2, respectively.

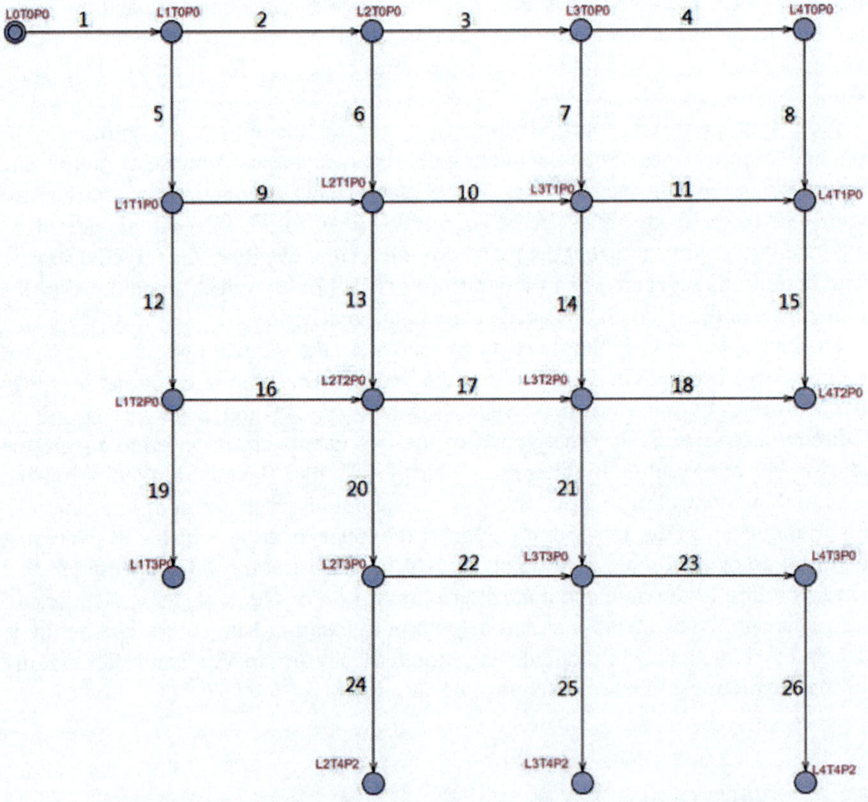

Fig. 7.4 Component model of the flash drum (Reprinted with permission from Zhang et al. 2020. Copyright 2020 American Chemical Society)

Table 7.1 Edge specifications of arc 1 in Fig. 7.4

Edge Label	Guards	Updates
1.1	PU_Flevel == 0 & PU_Fin == 4 && x >= 1	PU_Flevel = 1, x = 0

Table 7.2 Edge specifications of arc 13 in Fig. 7.4

Edge Label	Guards	Updates
13.1	PU_Ftemp == 1 & PU_Feng == 1 & A_Vvap == 2 & A_Vliq == 0 && x >= 8	PU_Ftemp = 2, x = 0
13.2	PU_Ftemp == 1 & PU_Feng == 2 & A_Vvap == 2 & PU_Fliq == 2 && x >= 6	PU_Ftemp = 2, x = 0

Notice first that every event label of each edge in Fig. 7.4 is numerically represented and any event in this model is neither a sender nor a receiver.

From Fig. 7.4 and Table 7.1, it can be observed that there is only one edge between L0T0P0 and L1T0P0, i.e., edge 1.1. Its guards imply that, to trigger the corresponding state transition, the liquid level should be at the lower bound (0) and inlet flowrate at the upper limit (4). Notice that these two prerequisites are connected by the AND operator (&), while the additional requirement on clock variable, i.e., $x \geq 1$, is incorporated with a double AND operator (&&). After all guard conditions are met, the liquid level should be updated to the discretized value of 1 and the clock variable reset to 0.

From Fig. 7.4 and Table 7.2, it can also be observed that there are two edges between L2T1P0 and L2T2P0, i.e., edge 13.1 and edge 13.2. The common guards of these two edges are PU_Ftemp == 1 (the temperature in flash drum is at the discretized value of 1) and A_{Vvap} == 2 (the opening of vapor valve is at the discretized value of 2). On the other hand, the net effect of energy input in the two scenarios should both be positive. Notice that, although the heating rate of the former (PU_Feng == 1) is slower than that of the latter (PU_Feng == 2), the energy output rate via liquid outlet flow for edge 13.1 (A_Vliq == 0) is also smaller than that for edge 13.2 (PU_Fliq == 2). Notice also that the event time of the former (x >= 8) is longer than that of the latter (x >= 6). After all guard conditions on either edge are met, the drum temperature can be transferred from the discretized value of 1 to 2 and the clock variable is again reset to 0.

7.4 Intrinsic Stages and Their Control Specifications

As indicated in Chap. 6, various operation paths can be extracted from a system automaton obtained by synchronizing all component models with an automaton that specifies the operation target. This target-setting automaton for the flash startup

Fig. 7.5 Ultimate target of flash startup operation (Reprinted with permission from Zhang et al. 2020. Copyright 2020 American Chemical Society)

operation is given in Fig. 7.5. However, since the system automaton is only loosely constrained by such a final goal, an overwhelmingly large number of unnecessary pathways may also be generated.

Although the ultimate goal of a specific operation can be easily given (e.g., Fig. 7.5), it can only be approached properly via a series of intermediate stages with interim goals which are often not explicitly stipulated. It is thus important to uncover these embedded subtasks and identify their features explicitly in advance. These features may be broadly classified as: (1) material charging, (2) material unloading, (3) reaction, (4) state adjustment, (5) phase change, (6) stable operation, etc. All such features of a stage are expressed as the "control specifications" first in natural language and then translated into automata.

For illustration purpose, let us revisit the flash startup process. Based on engineering knowledge and operational experience, it is clearly necessary to place a small quantity of raw material in the flash drum first and allow the liquid level reaching a height which is safe for heating. In the next stage, the temperature and pressure in the drum should be elevated to the set points and the input and output flow rates be raised to the steady-state levels. Finally, the stable operating conditions should be maintained for a relatively long period of time with the PID controllers.

The control specifications in all stages are detailed in the sequel.

7.4.1 Control Specifications for Stage 1

To save the operation time in stage 1, the inlet valve is opened fully before reaching the predetermined liquid level. At this designated level, the inlet valve opening is supposed to be reduced to 50%, which is the steady-state value. The above two requirements can be expressed by supervision 1 in Fig. 7.6 and specification 2 in Fig. 7.7, respectively. Also, the interim goals of stage 1 should be achieved if the liquid level in drum exceeds a discretized value of 1, i.e., between 0 and 0.8 m, and the inlet flowrate is adjusted to 2 (50%). This stage goal is represented by the automaton supervision 7_stage 1 in Fig. 7.8.

7.4.2 Control Specifications for Stage 2

Since the second stage starts immediately after completing stage 1, it is necessary to build an automaton to facilitate pathway connection between the two (see

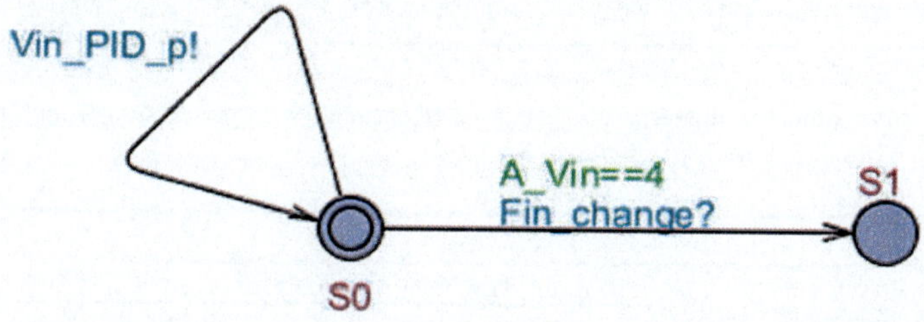

Fig. 7.6 Automaton for supervision 1 (Reprinted with permission from Zhang et al. 2020. Copyright 2020 American Chemical Society)

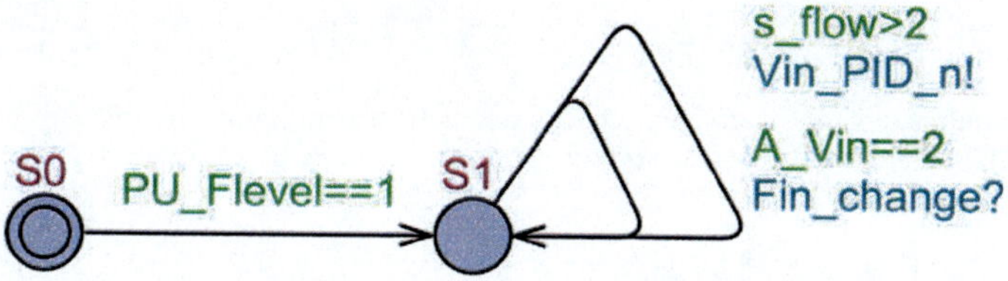

Fig. 7.7 Automaton for supervision 2 (Reprinted with permission from Zhang et al. 2020. Copyright 2020 American Chemical Society)

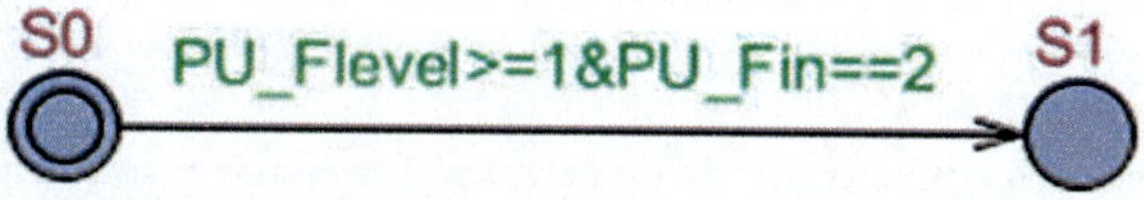

Fig. 7.8 Automaton for supervision 7_stage 1 (Reprinted with permission from Zhang et al. 2020. Copyright 2020 American Chemical Society)

Fig. 7.9).Note that the critical initial conditions of stage 2 are updated on the arc from S0 to S1 directly in this automaton. On the other hand, the interim goals of stage 2 are to drive all operating conditions to their steady-state values. These goals are also specified as the guards on the arc from S0 to S1 in the corresponding automaton in Fig. 7.10.

To achieve the above interim goals, Vlps, Vliq, and Vvap should be manipulated according to Figs. 7.11, 7.12, and 7.13, respectively. The heating is supposed to begin after liquid level reaching a discretized value greater than or equals to 1 and temperature is at 0 20 °C. From Appendix 7.1, it can be observed that the heating

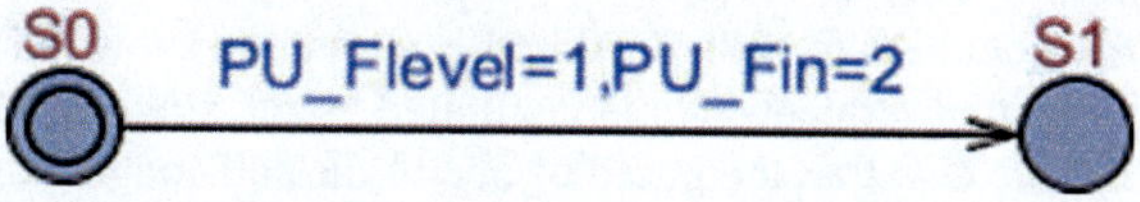

Fig. 7.9 Automaton for stage 1_stage 2 (Reprinted with permission from Zhang et al. 2020. Copyright 2020 American Chemical Society)

Fig. 7.10 Automaton for supervision 7_stage 2 (Reprinted with permission from Zhang et al. 2020. Copyright 2020 American Chemical Society)

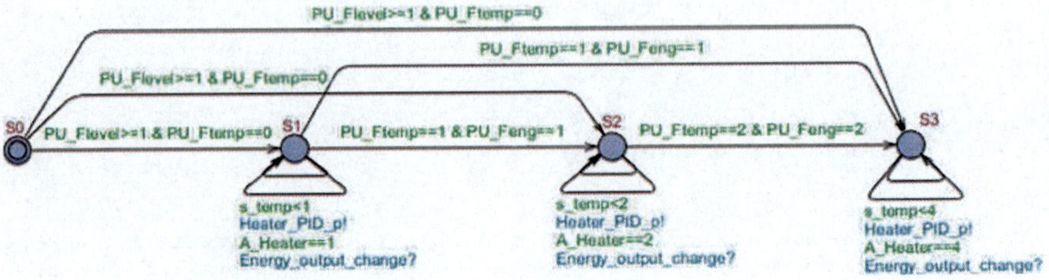

Fig. 7.11 Automaton for supervision 3 (Reprinted with permission from Zhang et al. 2020. Copyright 2020 American Chemical Society)

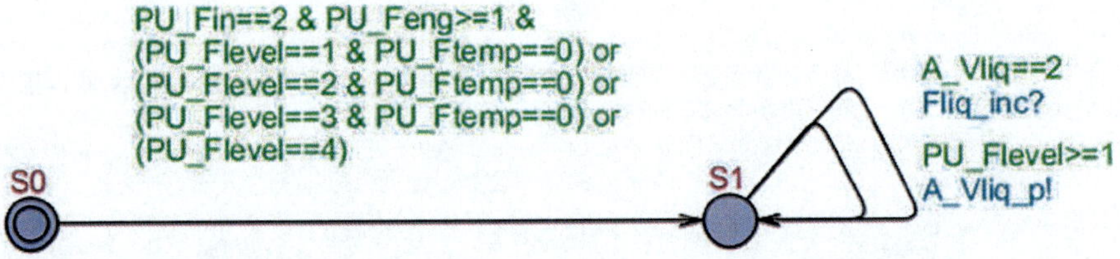

Fig. 7.12 Automaton for supervision 4 (Reprinted with permission from Zhang et al. 2020. Copyright 2020 American Chemical Society)

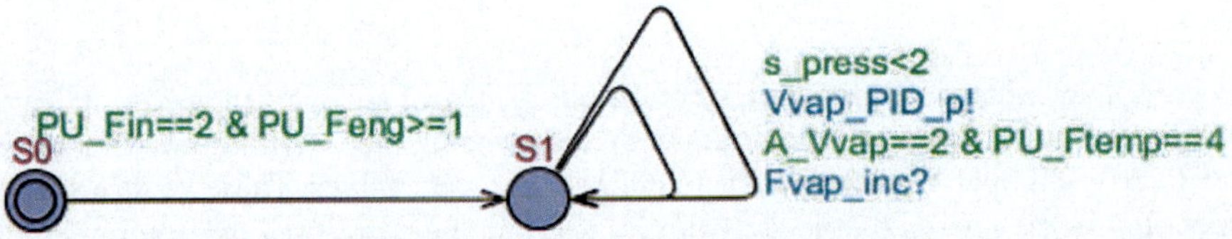

Fig. 7.13 Automaton for supervision 5 (Reprinted with permission from Zhang et al. 2020. Copyright 2020 American Chemical Society)

rate has been discretized into four values, i.e., 0 (0 MMkcal/hr), 1 (0.5 MMkcal/hr), 2 (0.9 MMkcal/hr), and 3 (1.68 MMkcal/hr). Thus, all heating sequences can be enumerated as follows: (1) $0 \rightarrow 3$; (2) $0 \rightarrow 2 \rightarrow 3$; (3) $0 \rightarrow 1 \rightarrow 3$; (4) $0 \rightarrow 1 \rightarrow 2 \rightarrow 3$. All sequences are embedded in the automaton shown in Fig. 7.11 (i.e., supervision 3). Note that sequence (1) is facilitated by the single arc between place S0 and place S3. Note also that the guard of one of the self-looping arcs on S3, that is, s_temp < 4 (which means the output of temperature controller is lower than the discretized value of 4 or 94%), is satisfied initially, and the corresponding sender event Heater_PID_p! is then triggered repeatedly to raise the controller output signal until the guard of the other self-looping arc, i.e., A_Heater == 4 (which means the opening of control valve of the heating medium equals the discretized value of 4 or 94%), is satisfied. At this point, the receiver event Energy_output_change? can be activated so as to complete the heating sequence0 → 3. Sequence (2) is facilitated in Fig. 7.11 first by the arc pointing from place S0 to place S2 and then by another from place S2 to place S3. Note that the guards of the former arc are the same as those of initial arc in sequence (1) and the guards of the latter are PU_Ftemp == 2 and PU_Feng == 2, which imply that the discretized values of both temperature and heat input must be raised to 2 before changing the heating rate. Since the self-looping arcs on S2 can be interpreted in the same way as those on S3, their descriptions are not repeated for the sake of brevity. Finally, since sequences (3) and (4) in Fig. 7.11 can be characterized in a similar fashion as (1) or (2), their explanations are also omitted.

7.4.3 *Control Specifications for Stage 3*

Because stage 3 immediately follows stage 2, it is necessary to use an automaton to forge a link between the two (see Fig. 7.14). Since stage 3 is also the final stage, its goal should be completing the startup operation (see Fig. 7.15). To achieve this goal, all controllers should be switched from MANUAL mode to AUTO mode after the steady-state conditions are reached (see Fig. 7.16).

7.5 Procedure Synthesis

The verification tool of UPPAAL is used here to search for the proper operation path within the real-time system. Specifically, the optimal pathway in every stage is synthesized in four distinct steps:

Fig. 7.14 Automaton for stage 1_stage 2 (Reprinted with permission from Zhang et al. 2020. Copyright 2020 American Chemical Society)

Fig. 7.15 Automaton for supervision 7_stage 3 (Reprinted with permission from Zhang et al. 2020. Copyright 2020 American Chemical Society)

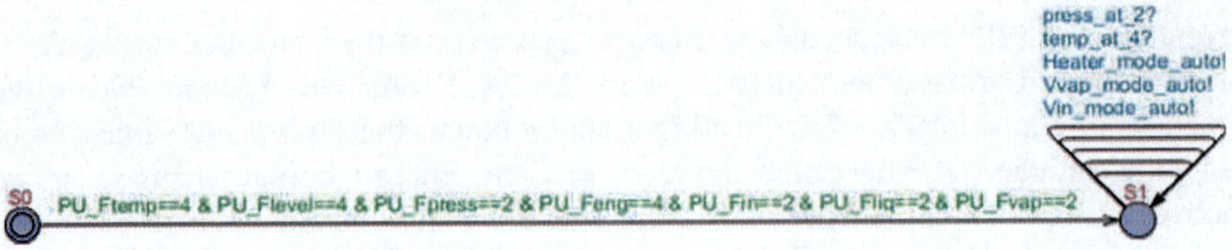

Fig. 7.16 Automaton for supervision 6 (Reprinted with permission from Zhang et al. 2020. Copyright 2020 American Chemical Society)

i. Build the automaton models of all components in the uncontrolled plant;
ii. Construct automata to represent the control specifications in every stage;
iii. Synchronize all automata created in the above two steps for each stage;
iv. Execute suitable property verification function in UPPAAL so as to locate the best operation pathway in each stage.

The operation pathways of all stages can then be pieced together to produce a procedure to facilitate the shortest operation duration or fewest state transitions. The former procedure is summarized in the sequential function chart (SFC-1) in Fig. 7.17, while the latter expressed with SFC-2 in Fig. 7.18. The procedures in SFC-7.1 and SFC-7.2 can be summarized as follows: The operator/PLC first opens

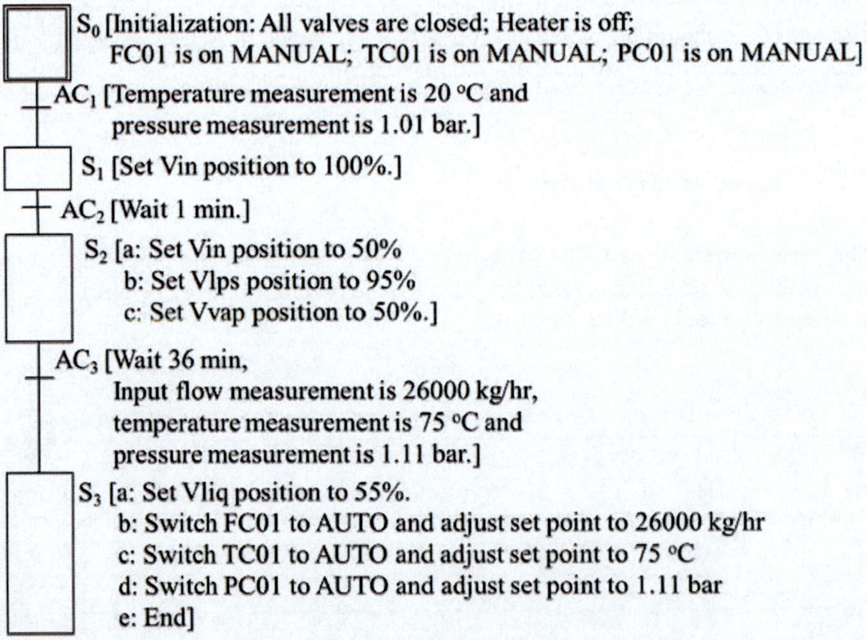

Fig. 7.17 Procedure to facilitate flash startup within shortest duration (SFC-7.1) (Reprinted with permission from Zhang et al. 2020. Copyright 2020 American Chemical Society)

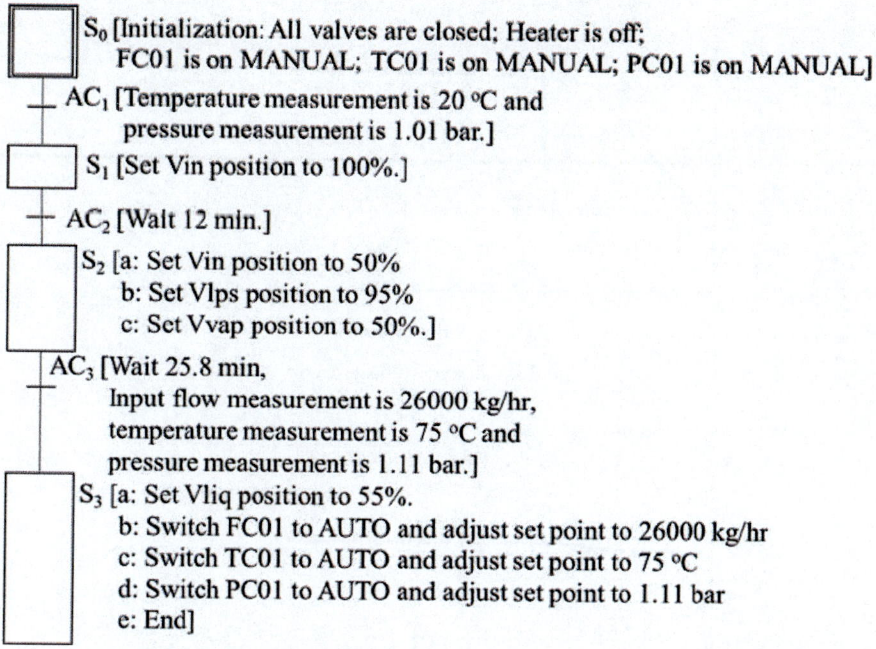

Fig. 7.18 Procedure to facilitate flash startup via fewest state transitions (SFC-7.2) (Reprinted with permission from Zhang et al. 2020. Copyright 2020 American Chemical Society)

the inlet valve Vin fully, waits for $x(x = 1$ or $12)$ minutes and then adjusts the inlet valve Vin to half-open position, the steam valve Vlps to 95% position, and the overhead vapor valve to 50% position. After waiting for another $y(y = 36$ or $25.8)$ minutes, the operator/PLC opens the bottom liquid valve to 55% position, switches all PID controllers from MANUAL to AUTO mode, and fixes their set points at the steady-state operating conditions.

7.6 Dynamic Simulation Studies

SFC-7.1 and SCF-7.2 have been converted to the Task files for simulation runs in Aspen Plus Dynamics, and the simulation results are presented in Figs. 7.19 and 7.20, respectively. Note that the target concentration in the overhead stream (87 wt%) is reached at 0.24 hr in the former case and 0.89 hr in the latter. The above two procedures are also compared on the basis of several performance indices in Table 7.3. It can be observed that, although the total operation time of SFC-7.1 is shorter than that of SFC-7.2, the total amounts of off-spec products and energy consumed in the former case are both greater than those in the latter.

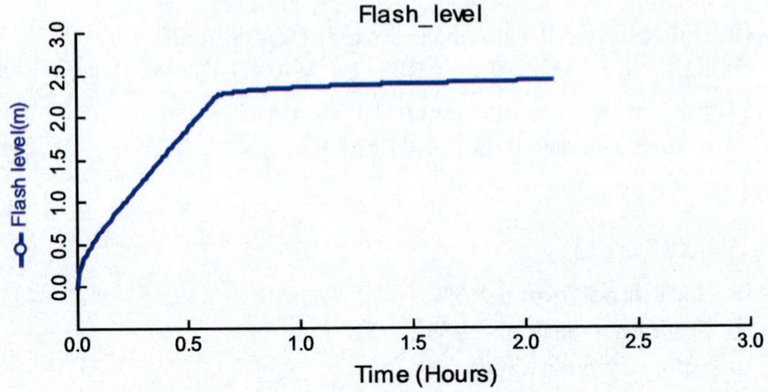

(a) Liquid level in flash drum

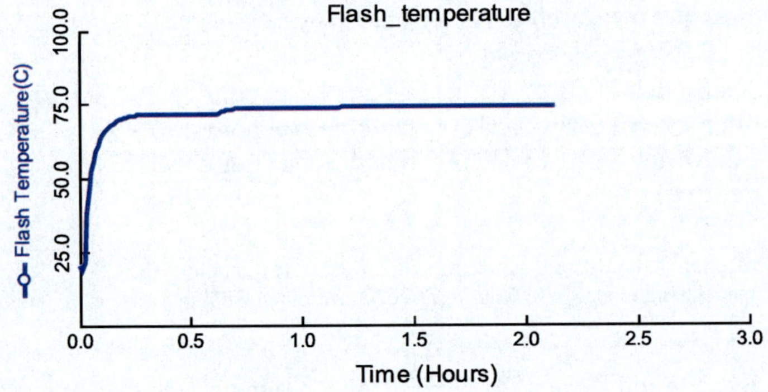

(b) Temperature in flash drum

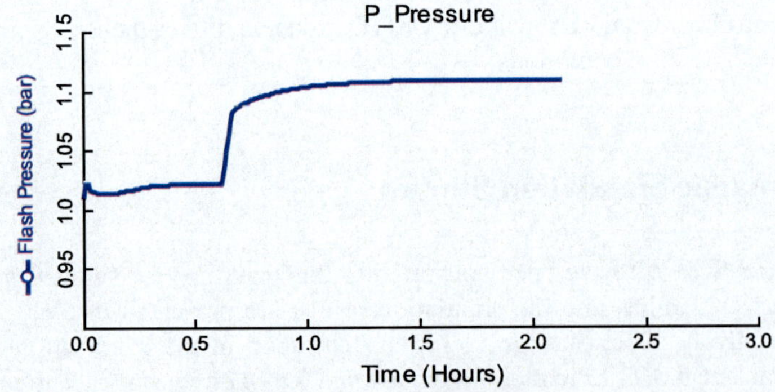

(c) Pressure in flash drum

Fig. 7.19 Time profiles of state and manipulated variables simulated for SFC-7.1 (Reprinted with permission from Zhang et al. 2020. Copyright 2020 American Chemical Society)

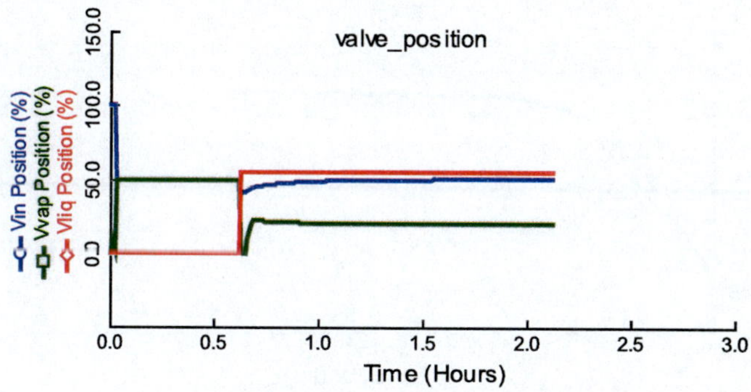

(d) Inlet and outlet valve openings

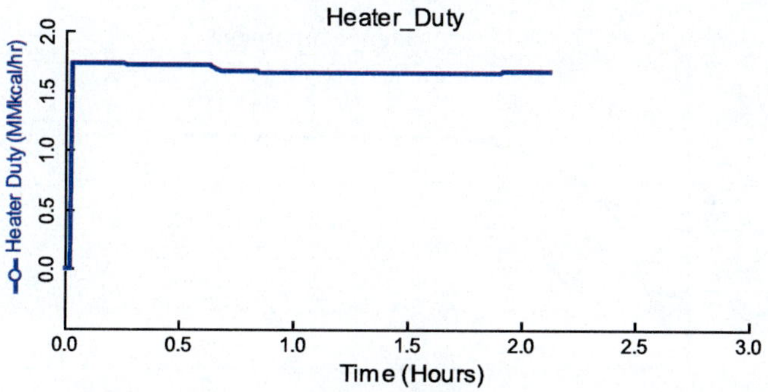

(e) Heater duty

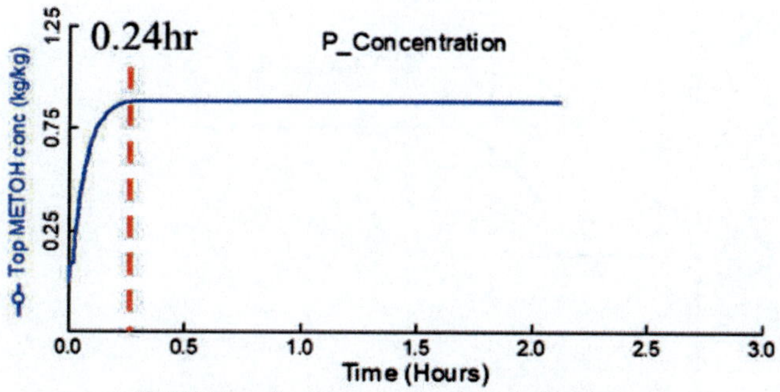

(f) Methanol concentration in overhead stream

Fig. 7.19 (continued)

7.7 Another Example

To demonstrate the feasibility of the proposed approach in more practical applications, a realistic example is presented in this section. Let us consider the startup operation of the continuous distillation process described in Fig. 7.21. It is assumed

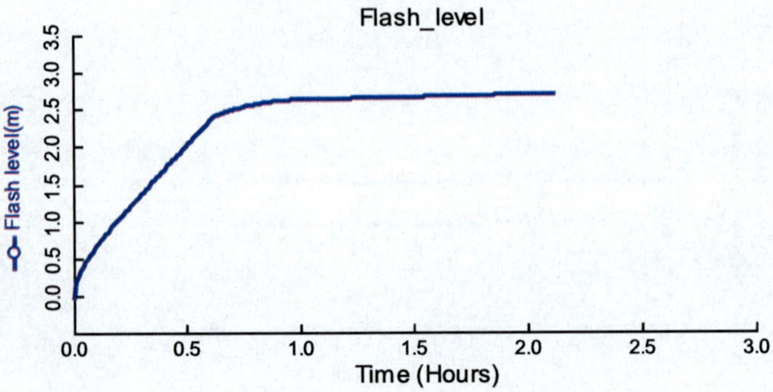

(a) Liquid level in flash drum

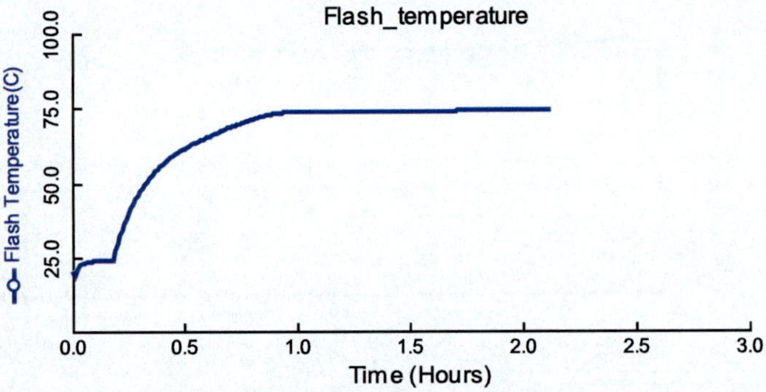

(b) Temperature in flash drum

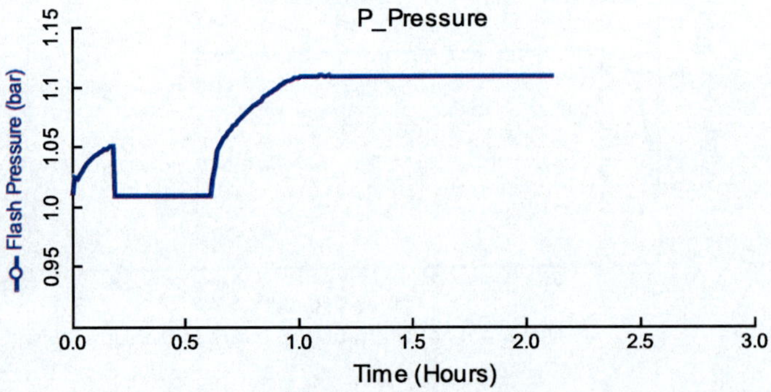

(c) Pressure in flash drum

Fig. 7.20 Time profiles of state and manipulated variables simulated for SFC-7.2 (Reprinted with permission from Zhang et al. 2020. Copyright 2020 American Chemical Society)

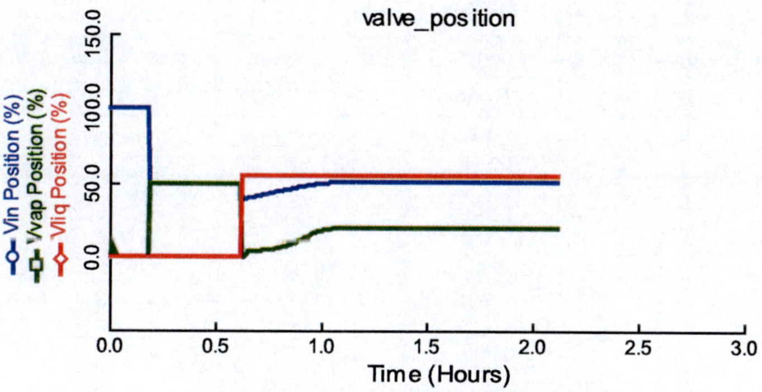

(d) Inlet and outlet valve openings

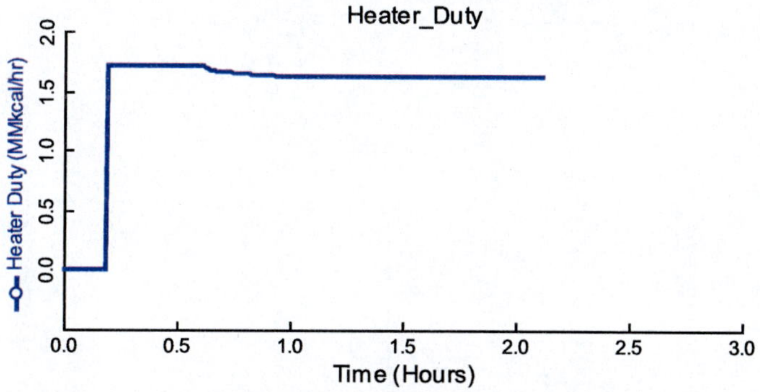

(e) Heater duty

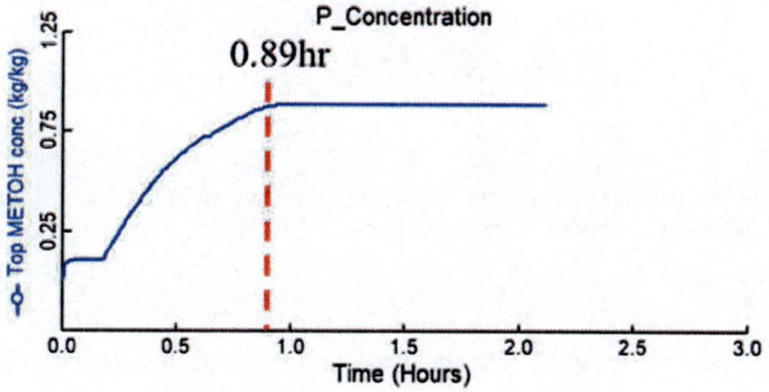

(f) Methanol concentration in overhead stream

Fig. 7.20 (continued)

Table 7.3 Performance indices of SFC-7.1 and SFC-7.2

SFC No	Amounts of off-spec products (kg)		Total amount of energy consumed (MMkcal)	Total operation time (min)
	Top	Bottom		
SFC-7.1	3490.89	34,445.95	3.51	37
SFC-7.2	2286.27	33,917.80	3.18	51

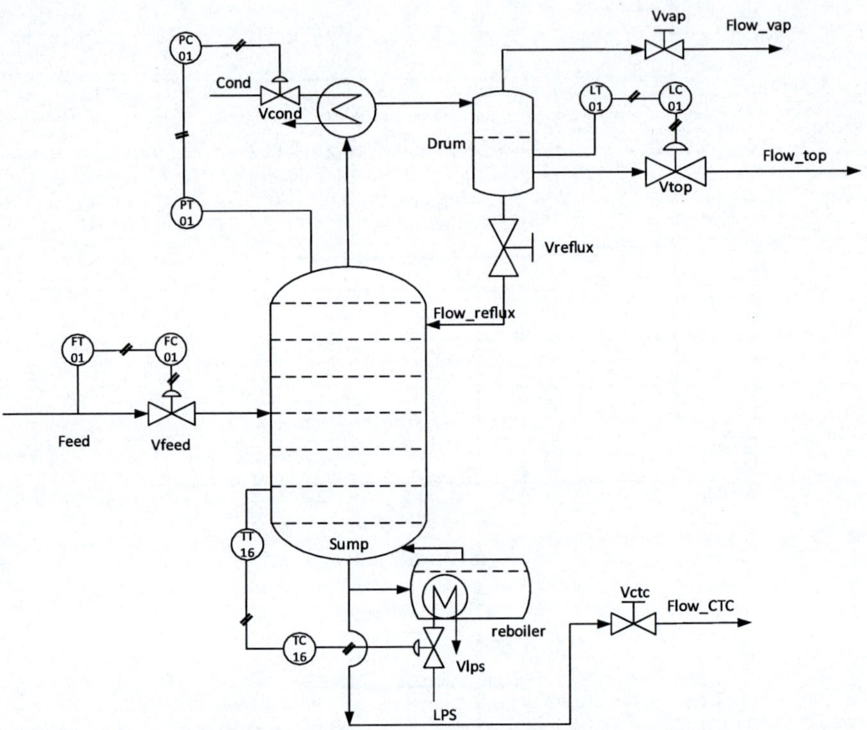

Fig. 7.21 PFD of a continuous distillation process (Reprinted with permission from Zhang et al. 2020. Copyright 2020 American Chemical Society)

that, at steady state, the feed is a mixture of 6 wt% CH_2Cl_2, 54 wt% $CHCl_3$, and 40 wt% CCl_4 and its flowrate, temperature, and pressure are kept at 10,000 kg/hr, 20°C, and 6 bar, respectively. The total number of plates in the distillation column is 20, while the feed is directed to the 10th plate. The steady-state set-point temperatures at the condenser and reboiler are at 80.3 °C and 105.0 °C, respectively. The steady-state reflux ratio is 5 mol/mole. The steady-state pressure settings at plate 1/condenser and plate 2 are chosen to be 2.00 and 2.02 bar, respectively, while the column pressure drop from bottom is 0.235 bar. It is also required that the concentration of light key ($CHCl_3$) in the top product should be greater than 81 mol% and that of heavy key (CCl_4) in the bottom product should not be lower than 98 mol%.

In this system, there are four PID controllers (FC01, TC16, PC01, and LC01) for controlling the feed rate, the temperature on the 16th plate, the top-plate pressure and the liquid level in reflux drum, respectively. The corresponding control valves are: Vfeed, Vlps, Vcond, and Vtop. It is assumed that, before the startup operation, all valves are closed, all controllers are on MANUAL and the reflux drum and column sump are both empty and at room temperature.

Finally, it should be noted that startup operation of this system has also been adopted by Aspen Plus Dynamics as a built-in example. Since the primary objective of the present case studies is to demonstrate the usefulness of the time-automata-based modeling approach, the sump level controller adopted in the Aspen built-in example has been deliberately excluded (see Fig. 7.21) for illustration purpose.

7.7.1 Operation Stages

Similar to the startup of flash drum, it is also necessary to place a small quantity of raw material in the column sump first and make sure that the liquid level reaches a height that is safe for heating. To allow this inlet flow, the outlet vapor valve (Vvap) should be partially open in advance. During the second stage, the heat input into the reboiler and heat output from the condenser should both be started to facilitate countercurrent vapor and liquid flows in the column. The product flows at the top and bottom should then be drawn consecutively from the column in the third stage to initiate the continuous operation. Finally, after all set points are reached, the stable operating conditions should be maintained for a relatively long period of time with the PID controllers.

7.7.2 Feasible Operating Procedures

By piecing together the automata-generated operation pathways obtained for achieving the interim goals of the aforementioned four stages, 33 different operating procedures can be created. Among them, 4 were considered to be unsafe for distillation startup on the basis of simulation results. The remaining feasible procedures were compared according to five performance indices. i.e., the total amounts of off-spec top and bottom products, the total amounts of heating and cooling utilities, and the total operation time. The best one is summarized by SFC-7.3 in Fig. 7.22, while the corresponding Aspen simulation results can be found in Fig. 7.23. The corresponding Task file can be found in Appendix 7.4.

Notice from both Figs. 7.22 and 7.23 that all valves in this system are closed initially except Vvap is at the 50% position. After confirming the designated initial conditions in AC_1, the inlet valve Vfeed is set to the 50% position to fill the column sump with liquid. The next "activation condition" is the elapsed time of the subsequent waiting period (26.6 min). It can be observed from Fig. 7.23a that, since the

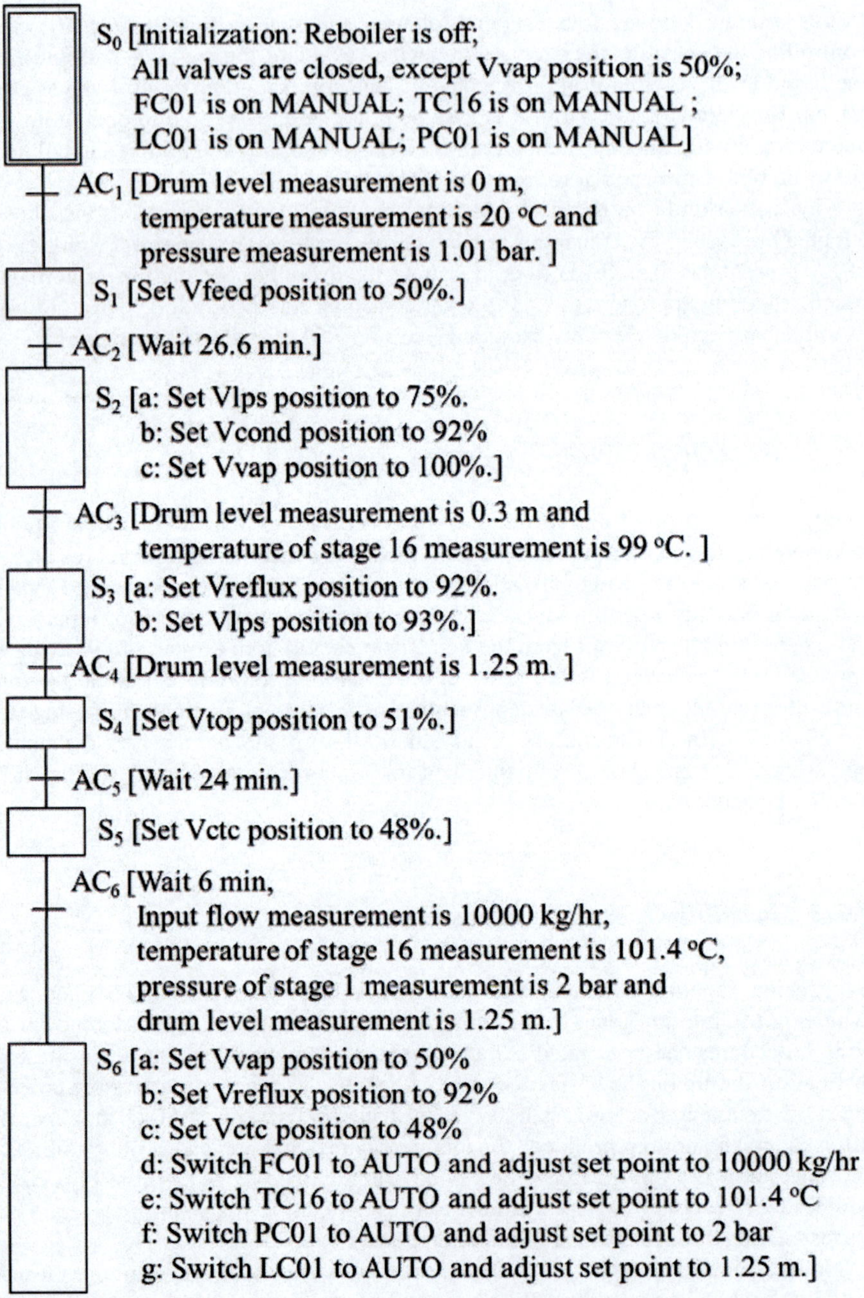

Fig. 7.22 SFC-7.3: The best operating procedure for distillation startup (Reprinted with permission from Zhang et al. 2020. Copyright 2020 American Chemical Society)

Fig. 7.23 Simulation results of distillation startup operated according to SFC-7.3 (Reprinted with permission from Zhang et al. 2020. Copyright 2020 American Chemical Society)

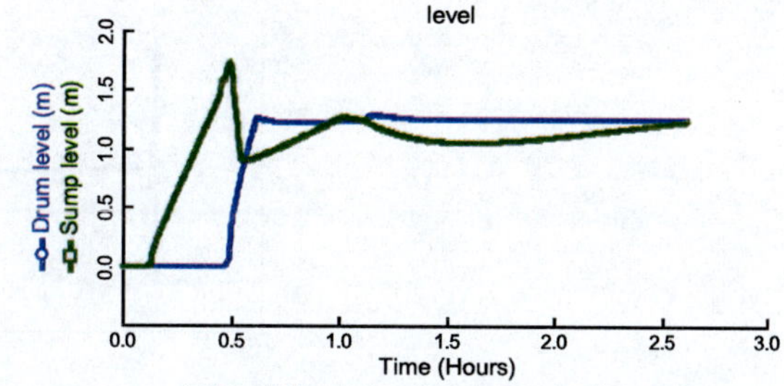

(a) Liquid levels in reflux drum and column sump

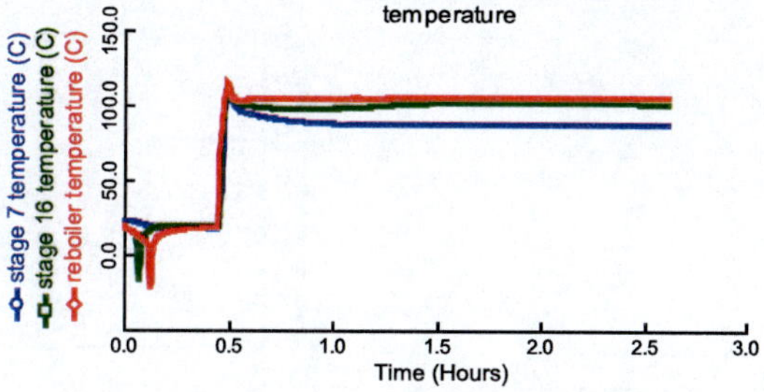

(b) Temperatures at plates 7, 16 and reboiler

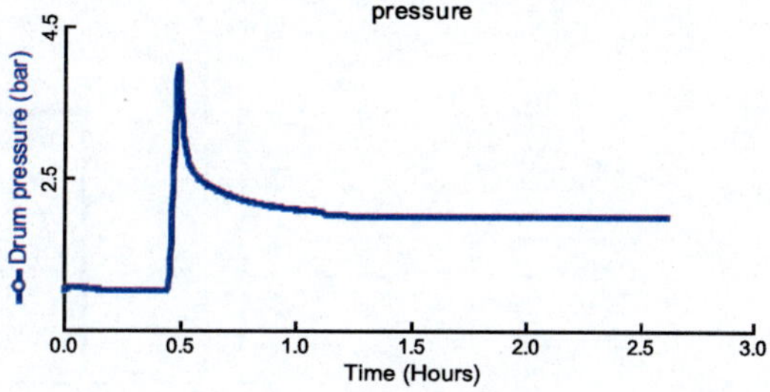

(c) Pressure in reflux drum.

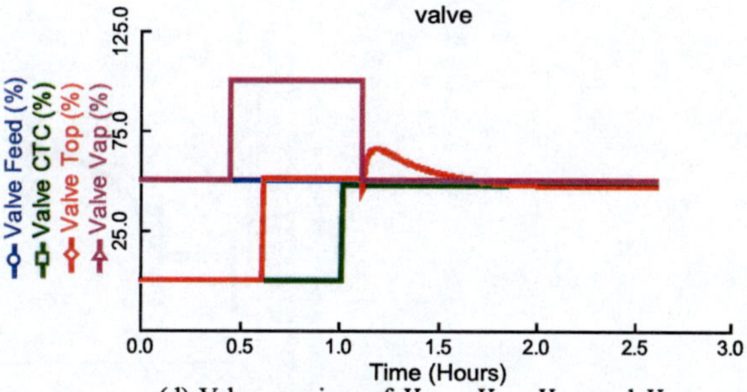

(d) Valve openings of V_{feed}, V_{ctc}, V_{top} and V_{vap}

Fig. 7.23 (continued)

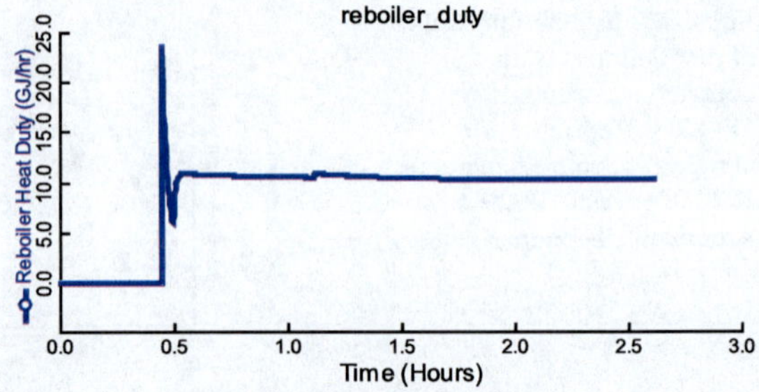

(e) Heat duty of reboiler

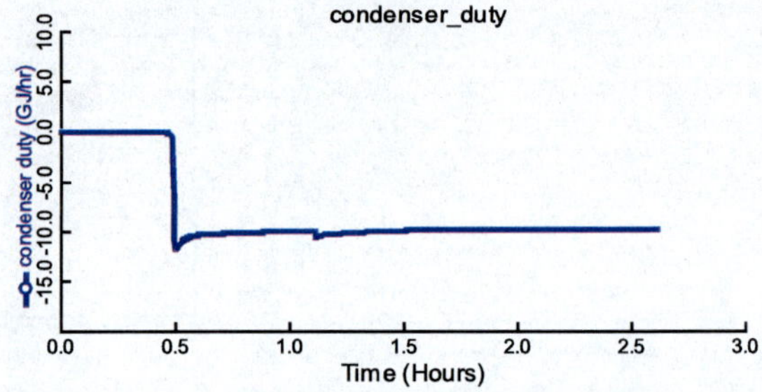

(f) Heat duty of condenser

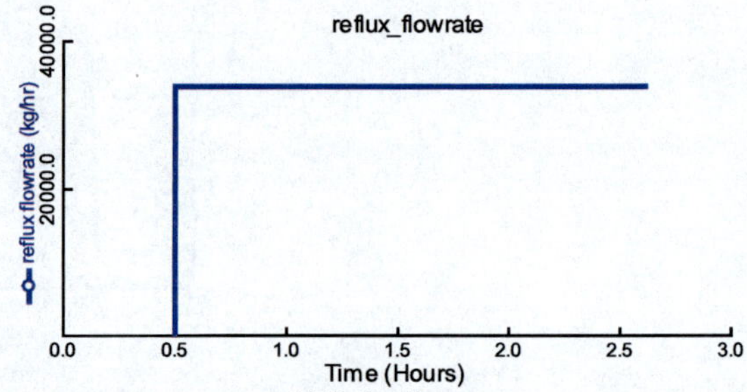

(g) Reflux flow rate

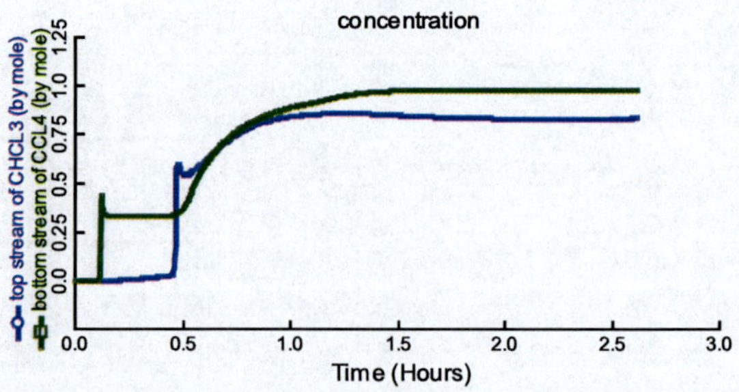

(h) Concentrations of light and heavy keys in the top and bottom product feed enters the column at the 10th plate, the height of liquid level in sump remains at 0 m initially for a period of approximately 10 min. In other words, it takes about the same amount of time for feed to travel from inlet to the sump. Notice also from Fig. 7.23b that the temperatures at plate 16 and in reboiler dip consecutively to below 0 °C during this 10-min period and then immediately recover to around 20 °C. These dips occur at instances when the simulated downward feed flow reaches the corresponding locations, and it should be noted that they may not be real in actual operation. During the next period from 10 min to 26.6 min, the height of liquid level in sump rises continuously before heating in reboiler and cooling in condenser begin, i.e., the flows of heating and cooling media are started at this instance by opening V_{lps} (75%) and V_{cond} (92%), respectively. At the same instance, V_{vap} is adjusted to 100% to avoid drastic overpressure in the reflux drum. Notice also that the height of liquid level in the column sump drops shortly afterwards, while that in the reflux drum begins to rise simultaneously. Next, when there is enough liquid accumulated in the reflux drum (i.e., its level reaches 0.3 m) and temperature of plate 16 is at 99 °C, V_{reflux} is opened to start the reflux flow going down to the column sump and, at the same time, V_{lps} is adjusted to 93% to increase the heat input to the reboiler and also the resulting upward vapor flow. Since the liquid outflow of the reflux drum is still considerably lower than inflow, the height of liquid level continues rising quickly until it reaches 1.25 m. At this time, the overhead product is drawn by opening V_{top} (51%) so as to hold the liquid level roughly at a constant height. After opening V_{top}, the operating procedure given in SFC-7.3 calls for a waiting period of 24 min. This is because the feed and reflux liquids slowly accumulate in the column sump during this period. Although the liquid in the sump is partially lost in the rising vapor due to heat input to reboiler, the downward trend of the liquid level can be eventually reversed in the waiting period and raised to about 1.25 m. At this time, V_{ctc} should be opened (48%) to draw the bottom product. It is predicted, after 6 min, all online measurements should reach their set-point values. At this point, all controllers should be switched to AUTO mode and all hand valves should be adjusted to their steady-state opening.

7.7.3 Unsafe Operating Procedures

An automata-generated unsafe operating procedure is shown in SFC-7.4 in Fig. 7.24, and the corresponding Aspen simulated level heights of the liquids in the reflux drum and column sump are presented in Fig. 7.25. In this procedure, after confirming the designated initial conditions in AC_1, the inlet valve Vfeed is set to the 50% position to fill the column sump with liquid. The subsequent waiting period is 18.9 min. At the end of this period, heating and cooling are started by directly opening V_{lps} and V_{cond} to their steady-state positions, i.e., 93% and 92%, respectively. Since the increase of heat input is too drastic in this case, the column sump becomes empty for a short period at around 27 min (see Fig. 7.25). Based on this observation, SFC-7.4 should be regarded as unsafe and, thus, excluded from further consideration.

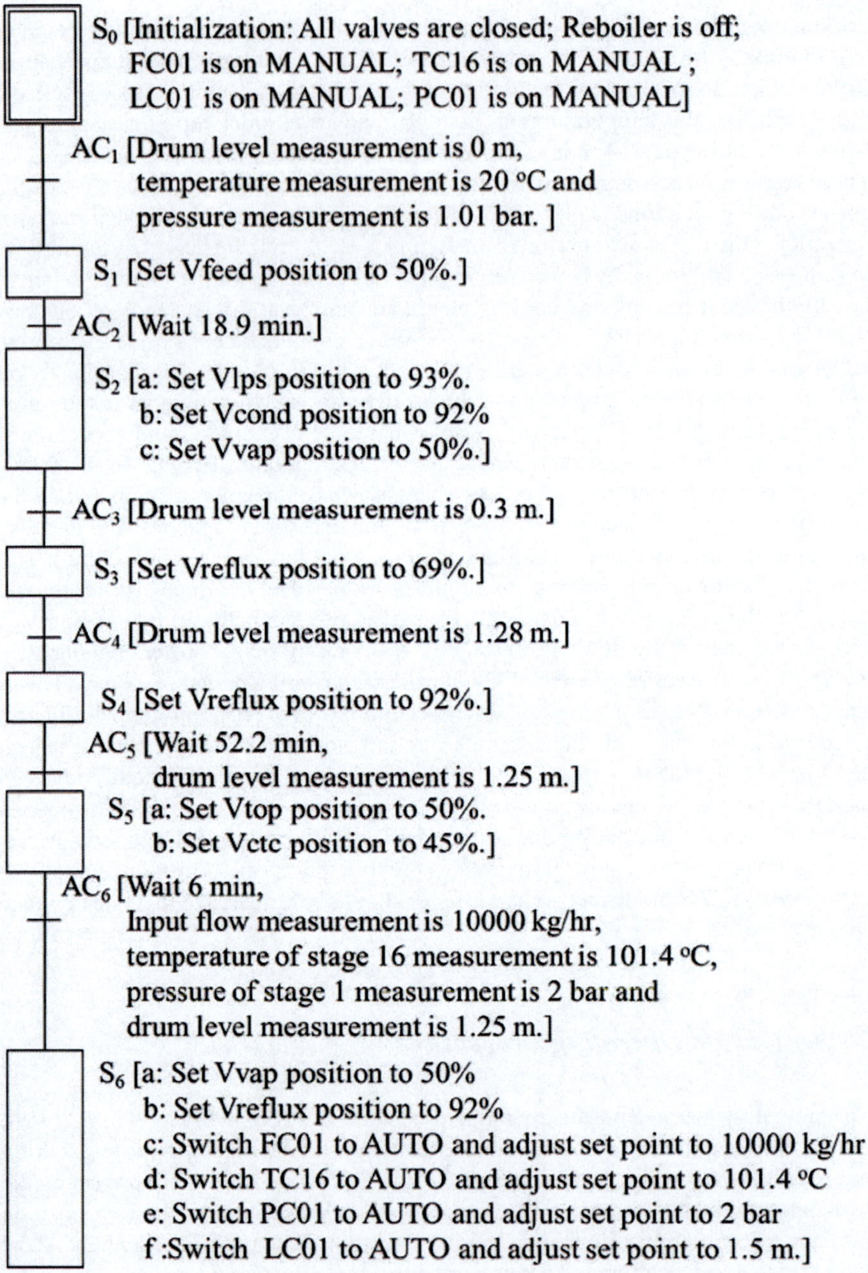

Fig. 7.24 SFC-7.4: An unsafe operating procedure for distillation startup (Reprinted with permission from Zhang et al. 2020. Copyright 2020 American Chemical Society)

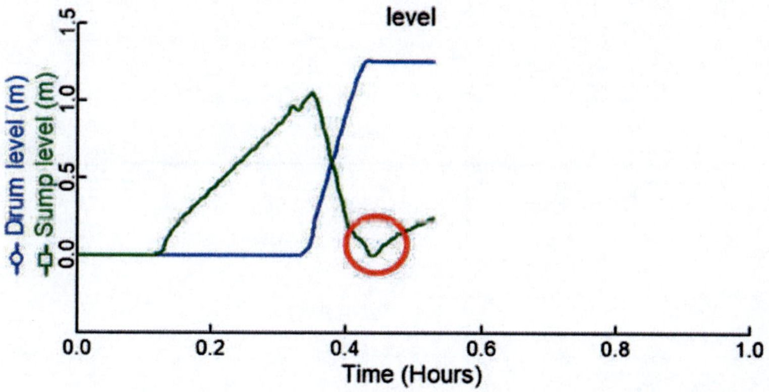

Fig. 7.25 Simulated liquid levels in reflux drum and column sump during distillation startup operated according to SFC-7.4 (Reprinted with permission from Zhang et al. 2020. Copyright 2020 American Chemical Society)

7.7.4 Comparison with Aspen Built-in Procedure

As mentioned before, the sump level controller adopted in the Aspen built-in example has been deliberately excluded for generating SFC-7.3 and SFC-7.4 in the above case studies. To be able to compare the Aspen procedure used for distillation startup (see Fig. 6.15 in the previous chapter) and the procedures generated with the proposed approach on a consistent basis, additional studies have been performed according to Fig. 7.26, which can be produced by adding back the sump level control loop to Fig. 7.21. By following the proposed procedure synthesis method, the SFC-7.5 in Fig. 7.27 can be obtained. Notice that, due to the extra level sensor and the corresponding PID controller in column sump, the waiting periods required in SFC-3 is replaced with the new features in SFC-7.5 which are marked by red rectangles. The resulting simulation data can be found in Fig. 7.28, while those obtained by executing the Aspen built-in procedure are presented in Fig. 7.29. It should be noted first that, by comparing Figs. 7.23 and 7.28, the dynamic behaviors of distillation startup operations that are executed according to SFC-7.3 and SFC-7.5 are actually quite similar. On the other hand, by comparing Figs. 7.28a and 7.29a, it can be observed that the liquid levels of drum and sump in the latter case fluctuates more frequently and in a larger range. This is because of the fact that there is an extra total-reflux stage (after the second stage) in the Aspen built-in procedure. This is essentially a more conservative measure to eliminate the possibility of emptying column sump. Furthermore, by comparing Figs. 7.28c and 7.29c, it can be seen that the drum pressure variation caused by implementing the Aspen-built-in procedure is less drastic than that by SFC-7.5. This is again due to the aforementioned less aggressive practice taken by the Aspen procedure. Finally, by comparing Figs. 7.28h and 7.29h, one can see that product concentrations change more before reaching steady state in the latter case, while those in the former are smoother and stabilize quicker.

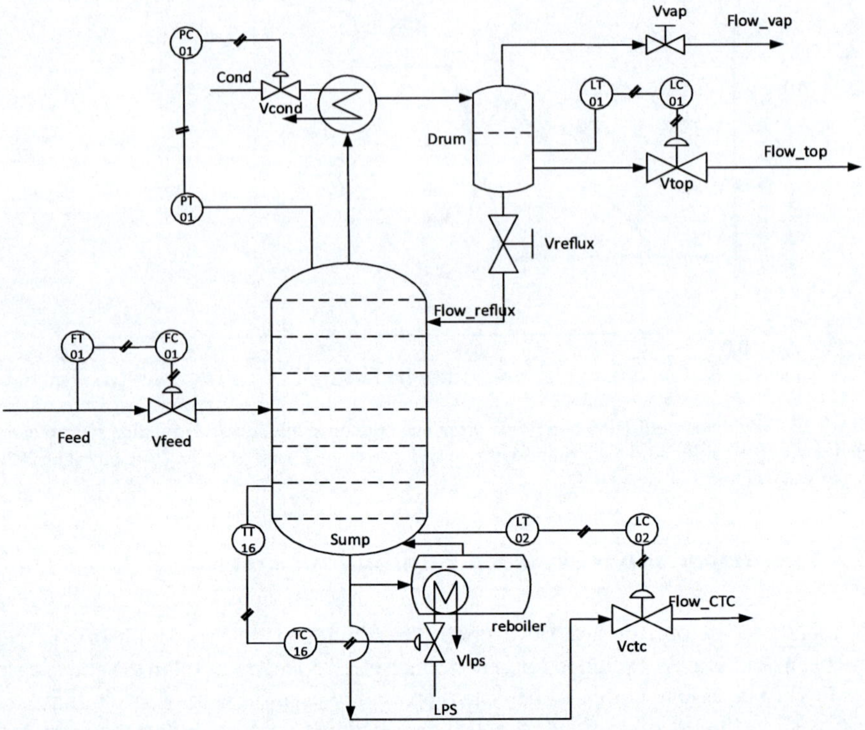

Fig. 7.26 PFD of a continuous distillation process used in Aspen built-in example (Reprinted with permission from Zhang et al. 2020. Copyright 2020 American Chemical Society)

Finally, the above procedures have also been compared on the basis of several performance indices calculated according to the simulation results (see Table 7.4). It can be found that the proposed SFC-7.5 (or SFC-7.3) outperforms the Aspen built-in procedure in almost every aspect.

7.8 Concluding Remarks

A generic approach has been presented in this chapter for systematically generating operating procedures based on the timed automata. The implementation steps include: (1) constructing automata for modeling the basic components and processing units according to engineering knowledge and with preliminary dynamics simulation results, respectively, (2) dividing the given operation into several intermediate stages and stipulating their control specifications in natural language and then converting them into automata, (3) synthesizing the operating procedures after synchronization of all automata mentioned above, and (4) verifying the procedures with Aspen Plus

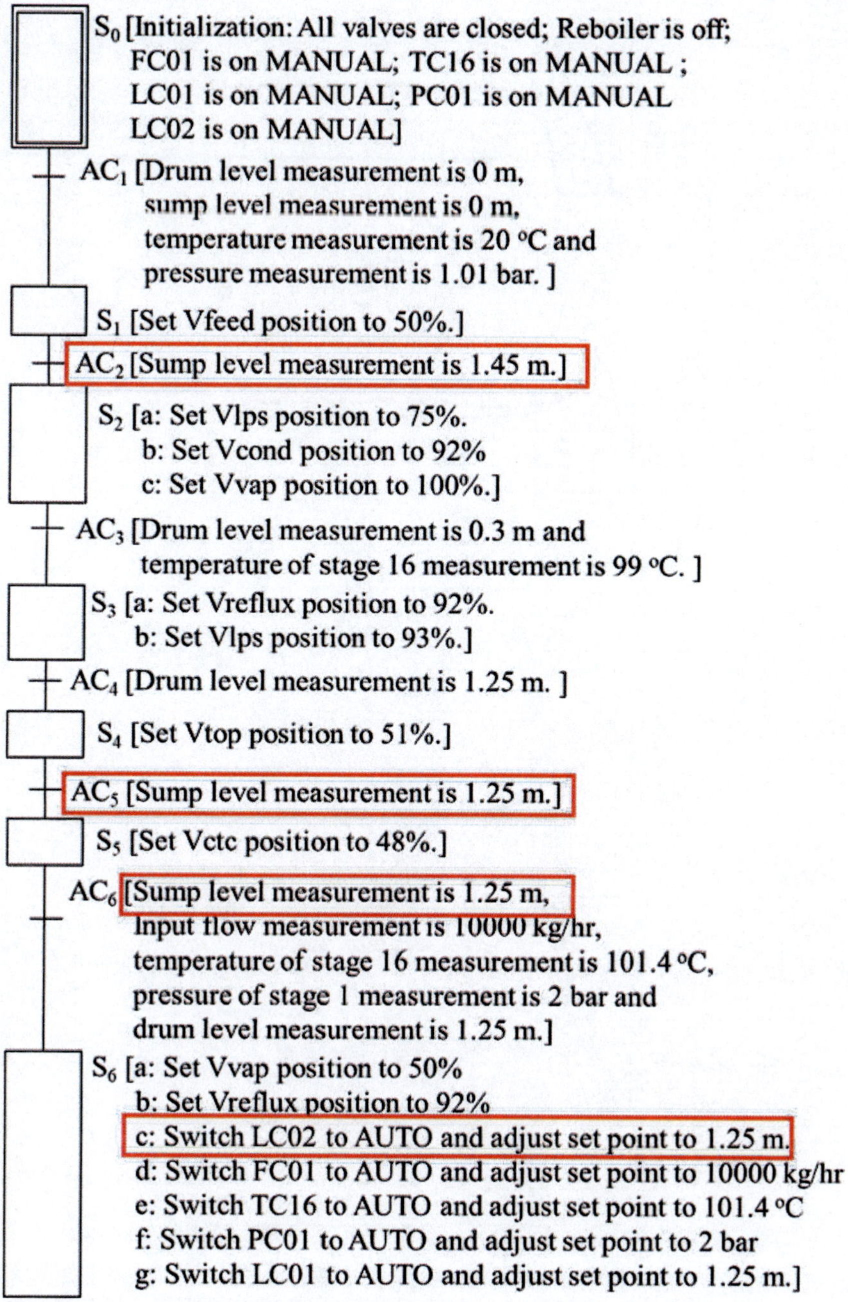

Fig. 7.27 SFC-7.5: The best operating procedure for distillation startup with extra level control loop in column sump (Reprinted with permission from Zhang et al. 2020. Copyright 2020 American Chemical Society)

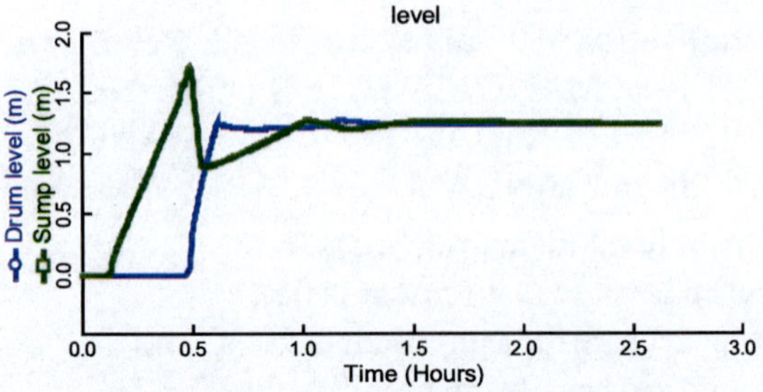

(a) Liquid levels in reflux drum and column sump

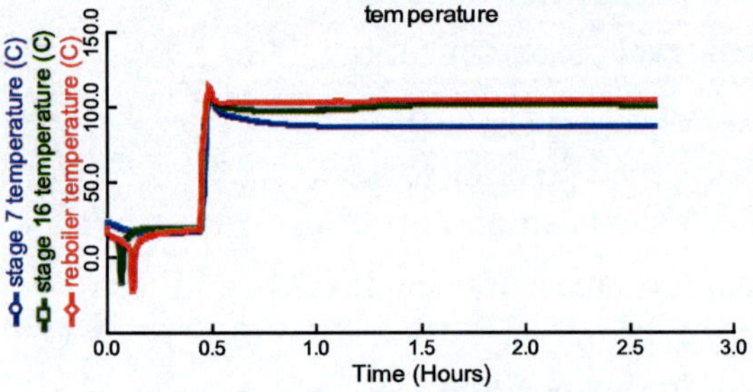

(b) Temperatures at plates 7, 16 and reboiler

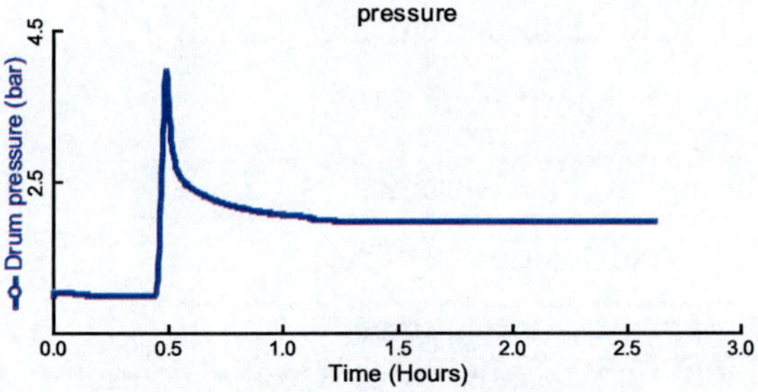

(c) Pressure in reflux drum.

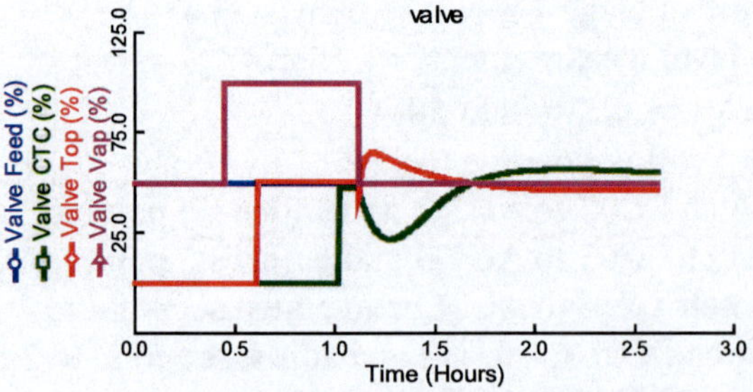

(d) Valve openings of V_{feed}, V_{ctc}, V_{top} and V_{vap}

Fig. 7.28 Simulation results of distillation startup operated according to SFC-7.5 (Reprinted with permission from Zhang et al. 2020. Copyright 2020 American Chemical Society)

Fig. 7.28 (continued)

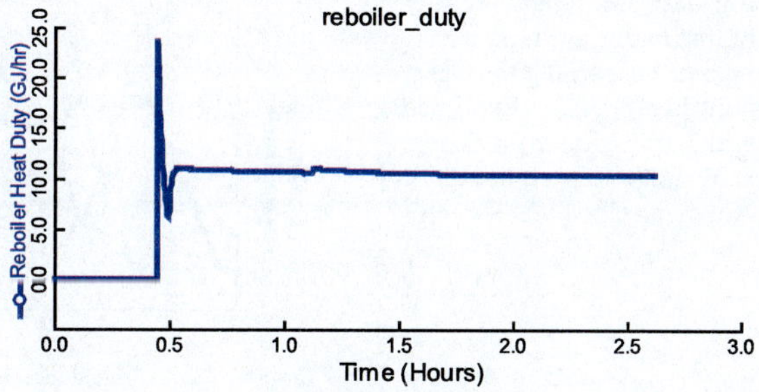

(e) Heat duty of reboiler

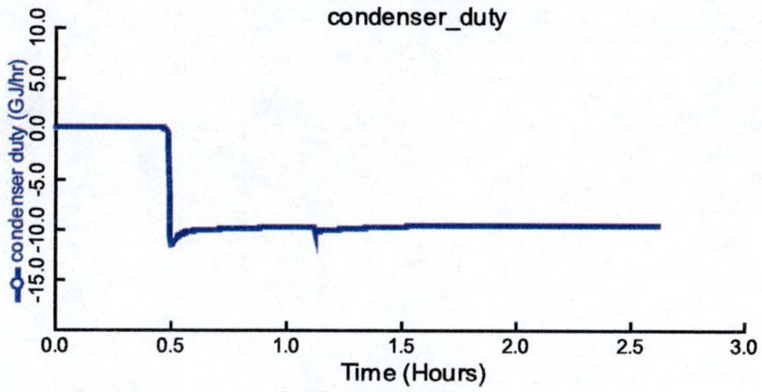

(f) Heat duty of condenser

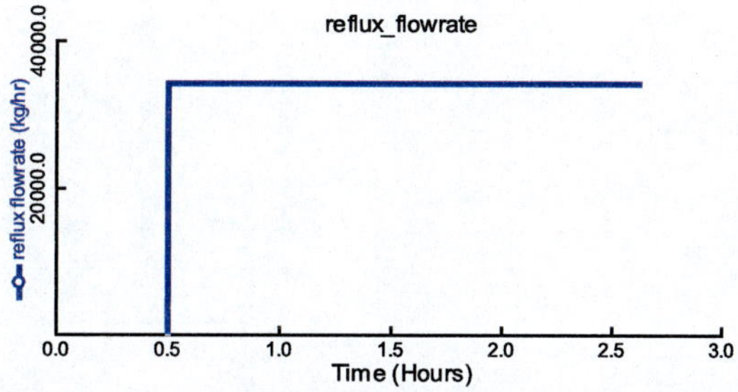

(g) Reflux flow rate

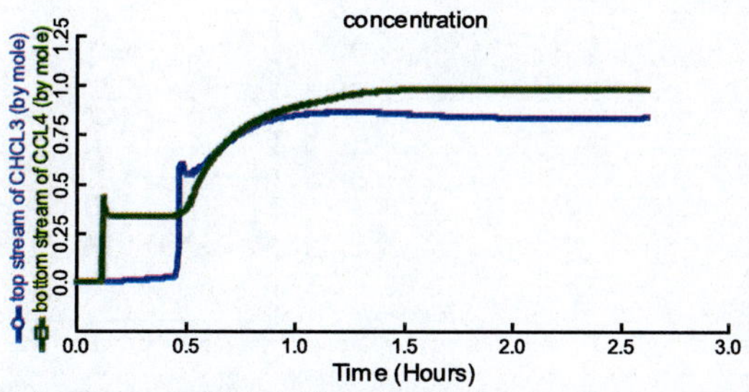

(h) Concentrations of light and heavy keys in the top and bottom products

Fig. 7.29 Simulation results of distillation startup operated according to Aspen built-in procedure (Reprinted with permission from Zhang et al. 2020. Copyright 2020 American Chemical Society)

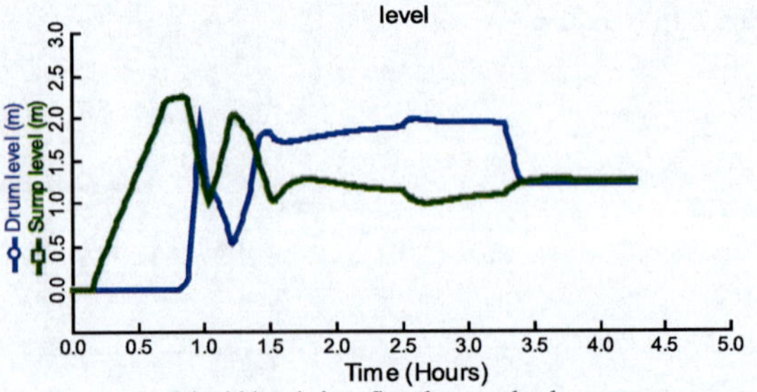

(a) Liquid levels in reflux drum and column sump

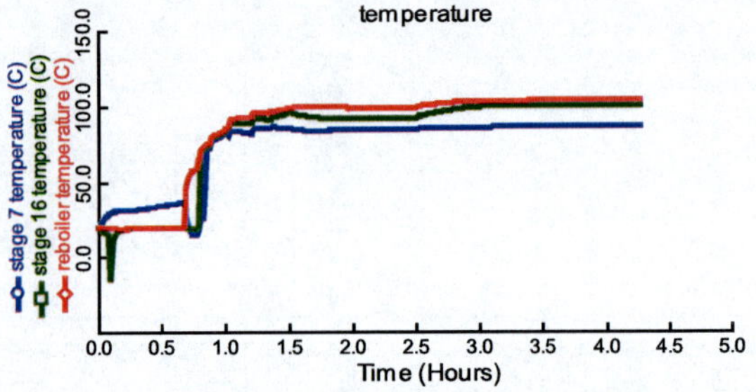

(b) Temperatures at plates 7, 16 and reboiler

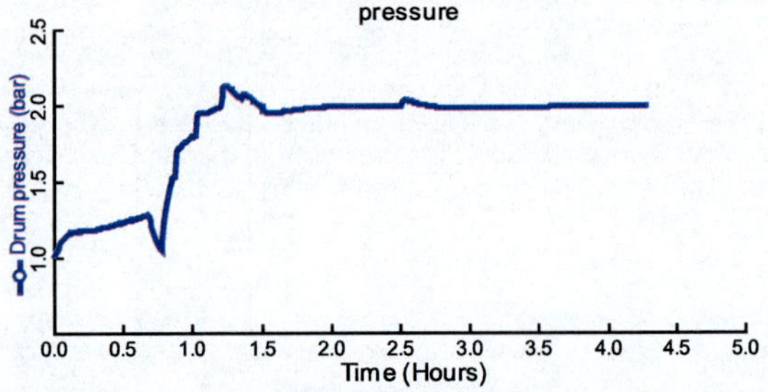

(c) Pressure in reflux drum

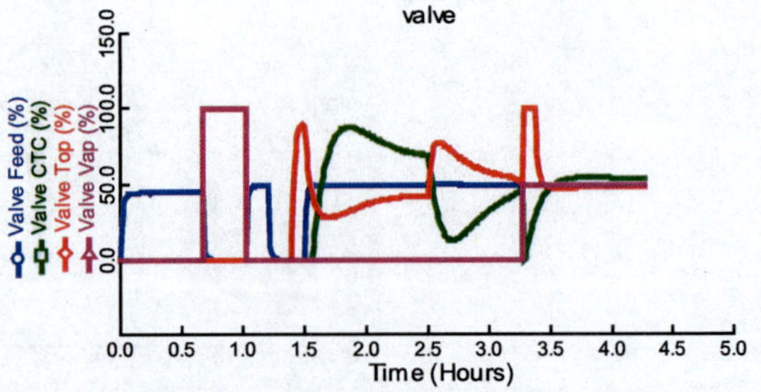

(d) Valve openings of V_{feed}, V_{ctc}, V_{top} and V_{vap}

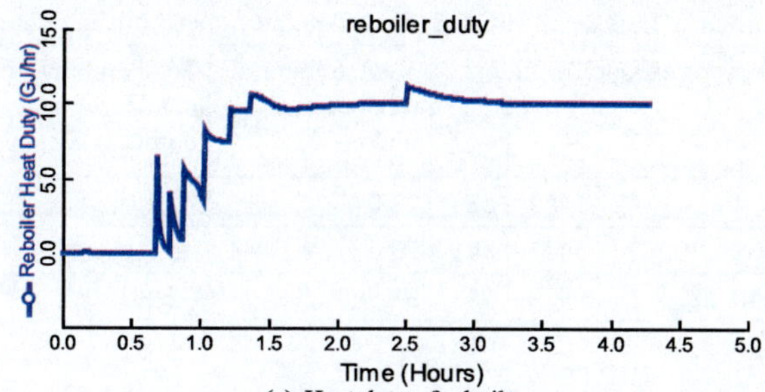

(e) Heat duty of reboiler

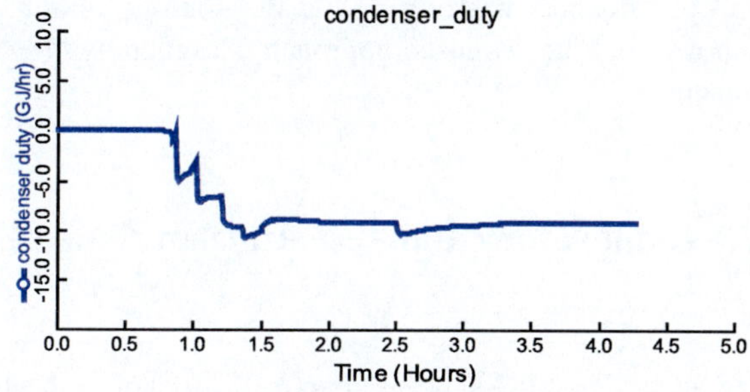

(f) Heat duty of condenser

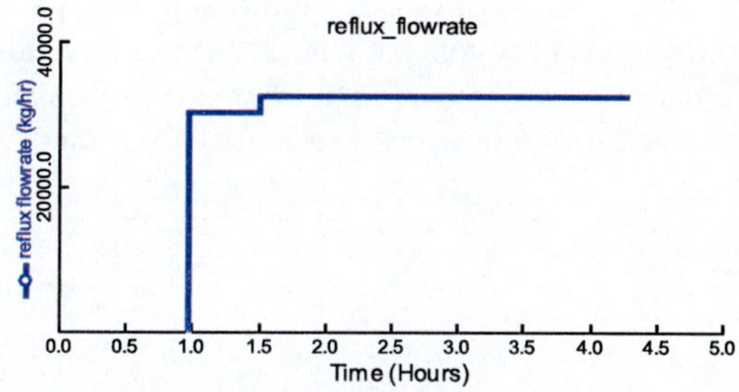

(g) Reflux flow rate

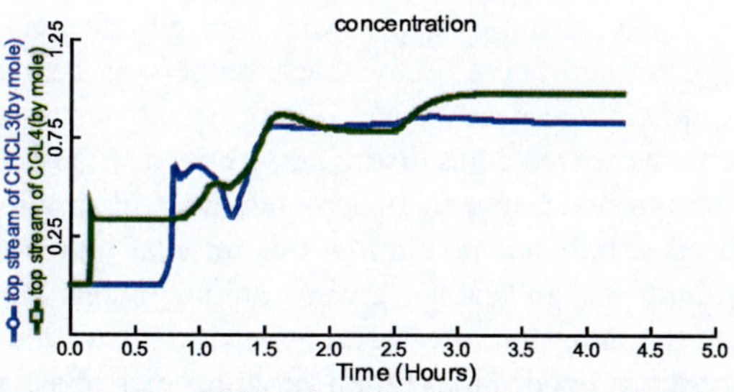

(h) Concentrations of light and heavy keys in the top and bottom products

Fig. 7.29 (continued)

Table 7.4 Performance indices of SFC-7.3, SFC-7.5, and Aspen built-in procedure

SFC #	Amount of off-spec product (kg)			Total amount of reboiler heat input (GJ)	Total amount of condenser heat output (GJ)	Total operation time (hr)
	Top	Ctc	Vap			
SFC-7.3	12,834	5467	838	23.0	−21.0	1.11
SFC-7.5	12,882	5415	843	23.0	−21.1	1.12
Aspen	18,640	9933	92	33.3	−31.0	3.27

Dynamics. This approach has been successfully applied to two realistic examples, i.e., the startup operations of flash drum and distillation column. Furthermore, it has also been shown that the proposed approach is especially effective for systems without critical sensors.

Appendix 7.1 Exploratory and Test Runs on Aspen Plus Dynamics

As mentioned in Sects. 7.2 and 7.4 of this chapter, the locations in a timed automaton denote system states (operating conditions) and each edge represents an event triggering state transition which may last a finite amount of time. To facilitate constructing such discrete-event system models, it is necessary to discretize the state variables and to estimate the state-transition times on the basis of simulation results obtained by simple exploratory runs using Aspen Plus Dynamics.

State Variables

Before discretizing the state variables, it is necessary to first determine the largest possible ranges of their variations. For this purpose, a fictitious operating procedure (see Fig. 7.30 for the flash startup example) was devised and then an exploratory simulation run was performed accordingly. Each range was next discretized into several intervals. The anticipated steady-state condition itself and the lower and upper bounds of each variable were treated as distinct discretized values. A separate value is also assigned to the interval between steady-state and maximum (or minimum) levels. If there is a need to enhance resolution, this interval may be further divided into more than one subinterval and each is given a unique integer value. It should be noted that partitioning too many subintervals may lead to an unnecessarily complex pathway network. The discretized values for all variables in the flash startup example are listed in Tables 7.5, 7.6, 7.7 and 7.8.

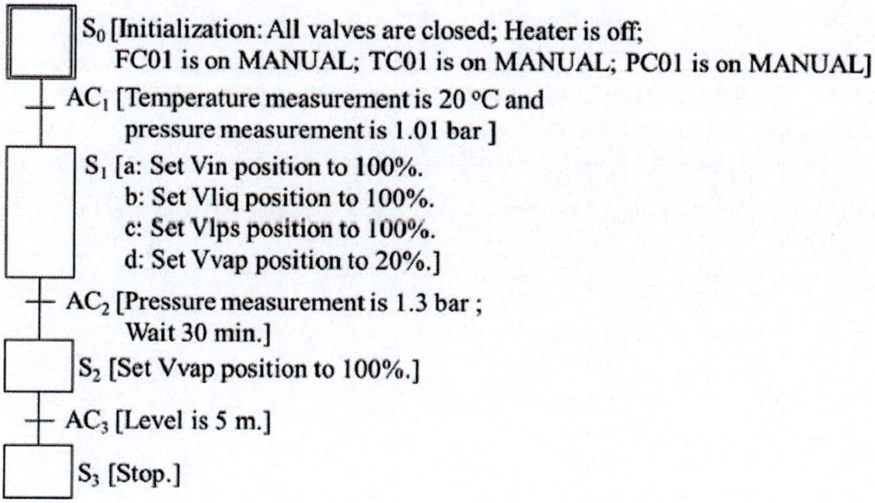

Fig. 7.30 Fictitious operating procedure for exploratory run (Reprinted with permission from Zhang et al. 2020. Copyright 2020 American Chemical Society)

Table 7.5 Intervals of control signals and their discretized values

Interval	s_flow	s_press	s_temp
0	(0, 0 + ε)%	(0, 0 + ε)%	(0, 0 + ε)%
1	(0 + ε, 50–ε)%	(0 + ε, 50 –ε)%	(0 + ε, 60)%
2	50 ± ε%	50 ± ε%	(60 + ε, 90)%
3	(50 + ε, 100–ε)%	(50 + ε, 100–ε)%	(90 + ε, 95–ε)%
4	(100–ε, 100)%	(100–ε, 100)%	95 + ε%
5			(95 + ε, 100–ε)%
6			(100-ε, 100)%

Table 7.6 Intervals of actuator states and their discretized values

Interval	A_Vin	A_Vliq	A_Vvap	A_Heater
0	(0, 0 + ε) %	(0, 0 + ε) %	(0, 0 + ε) %	(0, 0 + ε) %
1	(0 + ε, 50–ε) %	(0 + ε, 50–ε) %	(0 + ε, 50–ε) %	(0 + ε, 60)%
2	50 ± ε %	50 ± ε %	50 ± ε %	(60 + ε, 90)%
3	(50 + ε, 100–ε)%	(50 + ε, 100–ε)%	(50 + ε, 100–ε)%	(90 + ε, 95–ε)%
4	(100–ε, 100)%	(100–ε, 100)%	(100–ε, 100)%	95 ± ε %
5				(95 + ε, 100–ε)%
6				(100–ε, 100) %

Table 7.7 Intervals of input and output mass/energy flows and their discretized values

Interval	PU_Fin	PU_Fliq	PU_Fvap	PU_Feng
0	(0, 0 + ε) kg/hr	(0,0 + ε) kg/hr	(0, 0 + ε) kg/hr	(0, 0 + ε) MMkcal/hr
1	(0 + ε, 26,000–ε) kg/hr	(0 + ε, 22,000–ε) kg/hr	(0 + ε, 4000–ε) kg/hr	(0 + ε, 0.8) MMkcal/hr
2	26,000 ± ε kg/hr	22,000 ± ε kg/hr	4000 ± ε kg/hr	(0.8 + ε, 1.6) MMkcal/hr
3	(26,000 + ε, 42,000–ε) kg/hr	(22,000 + ε, 30,000–ε) kg/hr	(4000 + ε, 5000–ε) kg/hr	(1.6 + ε, 1,7-ε) MMkcal/hr
4	(40,000–ε, 40,000) kg/hr	(30,000–ε, 30,000) kg/hr	(5000–ε, 5000) kg/hr	1.7 ± ε MMkcal/hr
5				(1.7 + ε,2.2–ε) MMkcal/hr
6				(2.2 + ε, 2.2) MMkcal/hr

Table 7.8 Intervals of process variables and their discretized values

Interval	PU_Ftemp	PU_Flevel	PU_Fpress
0	(20, 20 + ε) °C	(0, 0 + ε) m	(1.01,1.01 + ε) bar
1	(20 + ε, 45) °C	(0 + ε, 0.8) m	(1.01 + ε, 1.11 –ε) bar
2	(45 + ε, 65) °C	(0.8 + ε, 1.6) m	1.11 ± ε bar
3	(65 + ε, 75–ε) °C	(1.6 + ε, 2.5–ε) m	(1.11 + ε,1.3–ε) bar
4	7 5 ± ε °C	2.5 ± ε m	(1.3–ε, 1.3) bar
5	(75 + ε, 83–ε) °C	(2.5 + ε, 5–ε) m	
6	(83–ε, 83) °C	(5–ε, 5) m	

State Transition Times

From Table 7.8, it can be observed that four discretized levels (i.e., from 1 to 4) are reachable from the initial level (i.e., 0) to the steady-state level (i.e., 4). Any of them can be chosen as the activation condition for heating to start. From Table 7.7, it can be seen that there are three discretized heating rates, i.e., 0.5 (1), 0.9 (2), and 1.68 (3) MMkcal/hr, for driving the temperature and pressure to the set-point conditions. Finally, two valve openings of V_{liq}, i.e., 0 and 50%, were selected to manipulate the liquid level to its steady state. Figure 7.31 is the fictitious SFC for carrying out test simulation runs to generate the time profiles of discretized state variables (see Figs. 7.32, 7.33, 7.34 and 7.35). It should be noted that the state transition times can then be estimated from these profiles.

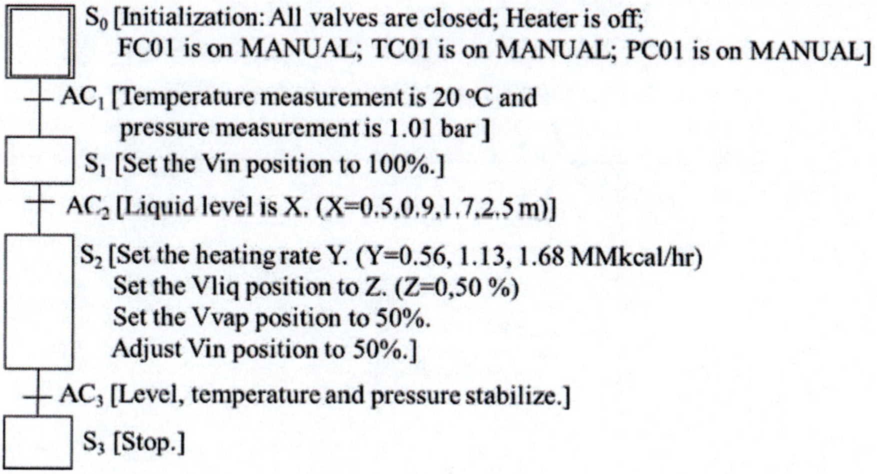

Fig. 7.31 Fictitious operating procedure for test runs (Reprinted with permission from Zhang et al. 2020. Copyright 2020 American Chemical Society)

Appendix 7.2 Component Models and Layer Model Used in the Flash Startup Example

Since the component model in level 1 is unavailable a priori and those in levels 2 and 4 have already been described in Sect. 7.4, only the component models in levels 3 and 5 and also the layer model are presented in the sequel.

Component Models in Level 3

As mentioned in Sect. 7.2, level 3 (process configuration) in the system hierarchy contains the material and energy flows among processing units. Since there is only one unit, i.e., the flash drum, in the flash startup example, it is only necessary to build models for the feed flowF_{in}, the vapor product flow F_{vap}, the liquid product flow F_{liq}, and the heat input flow F_{eng} (see Fig. 7.36). Let us use the component model of F_{in} in Fig. 7.36a as an example to illustrate the model-building principles. The guard $PU\_F_{in}! = A\_V_{in}$ is used to denote the condition that the feed flow (PU_F_{in}) does not match the inlet valve position (A_V_{in}). Since this inconsistent condition implies that the latter has been adjusted, the sender event F_{in}_change! should then be triggered. Subsequently, the feed state is varied to match the inlet valve state. After PU_F_{in} being altered to another value, the liquid level in the flash drum should also be affected accordingly.

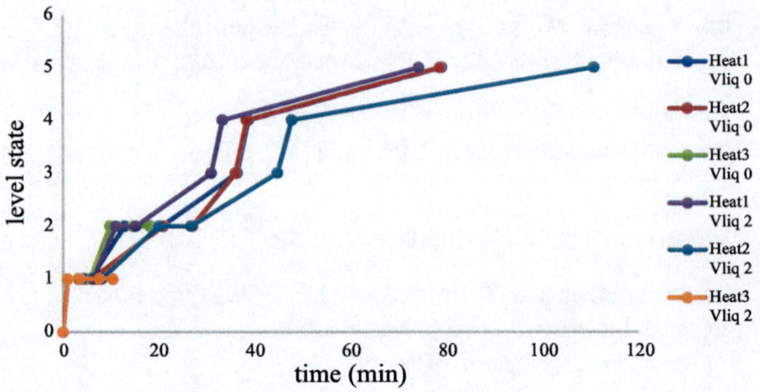

(a) Level

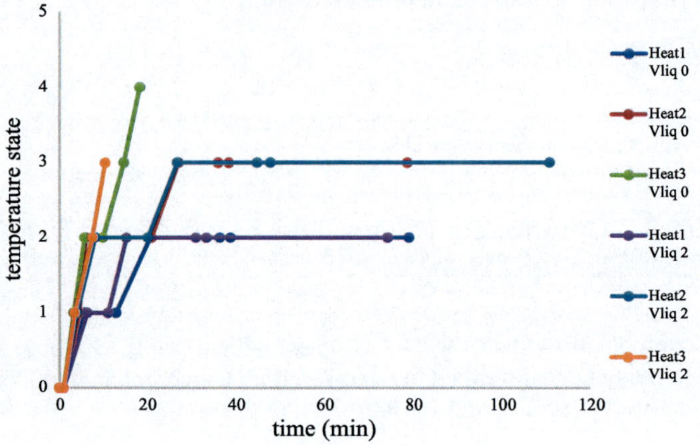

(b) Temperature

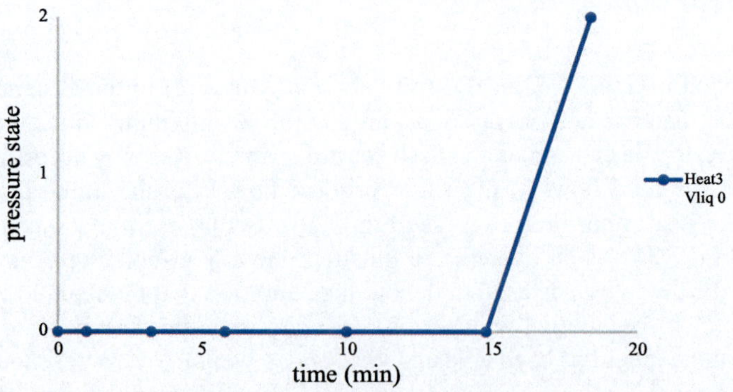

(c) Pressure

Fig. 7.32 Time profiles of discretized state variables for which heating starts when the liquid level reaches the discretized value of 1 (Reprinted with permission from Zhang et al. 2020. Copyright 2020 American Chemical Society)

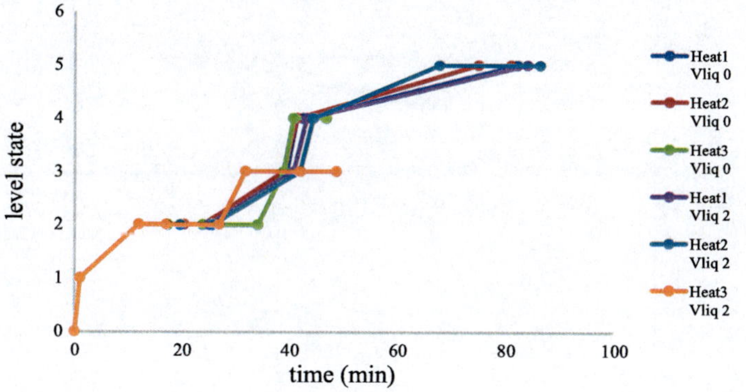

(a) Level

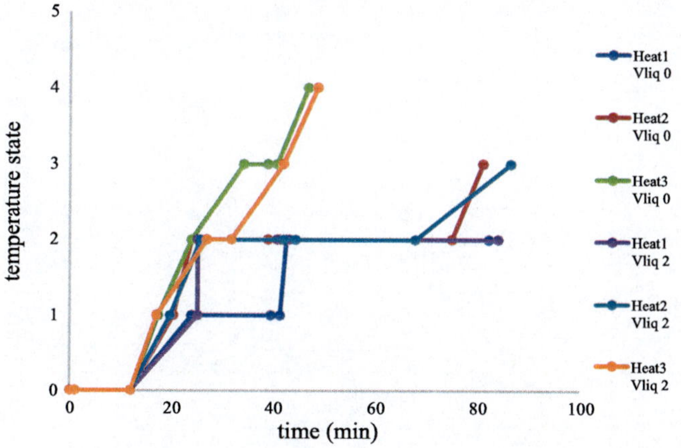

(b) temperature

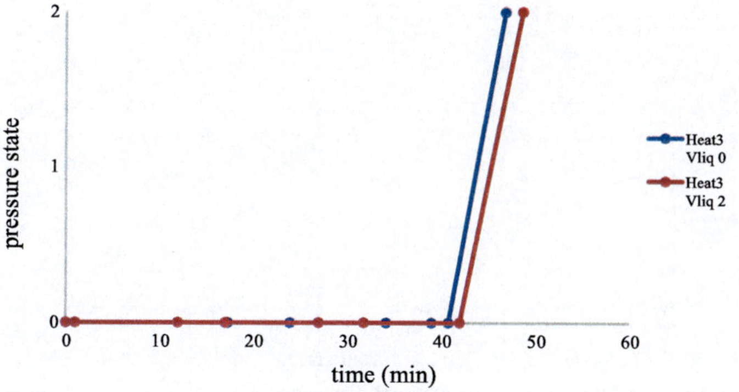

(c) pressure

Fig. 7.33 Time profiles of discretized state variables for which heating starts when the liquid level reaches the discretized value of 2 (Reprinted with permission from Zhang et al. 2020. Copyright 2020 American Chemical Society)

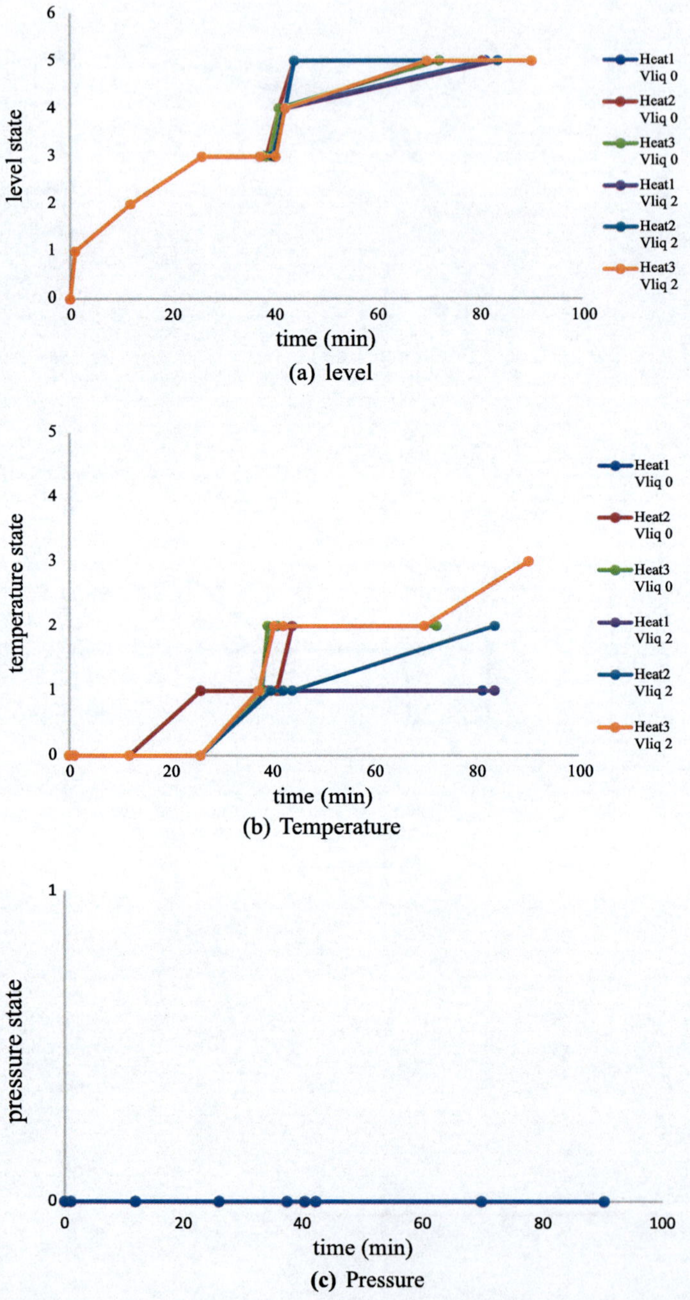

Fig. 7.34 Time profiles of discretized state variables for which heating starts when the liquid level reaches the discretized value of 3 (Reprinted with permission from Zhang et al. 2020. Copyright 2020 American Chemical Society)

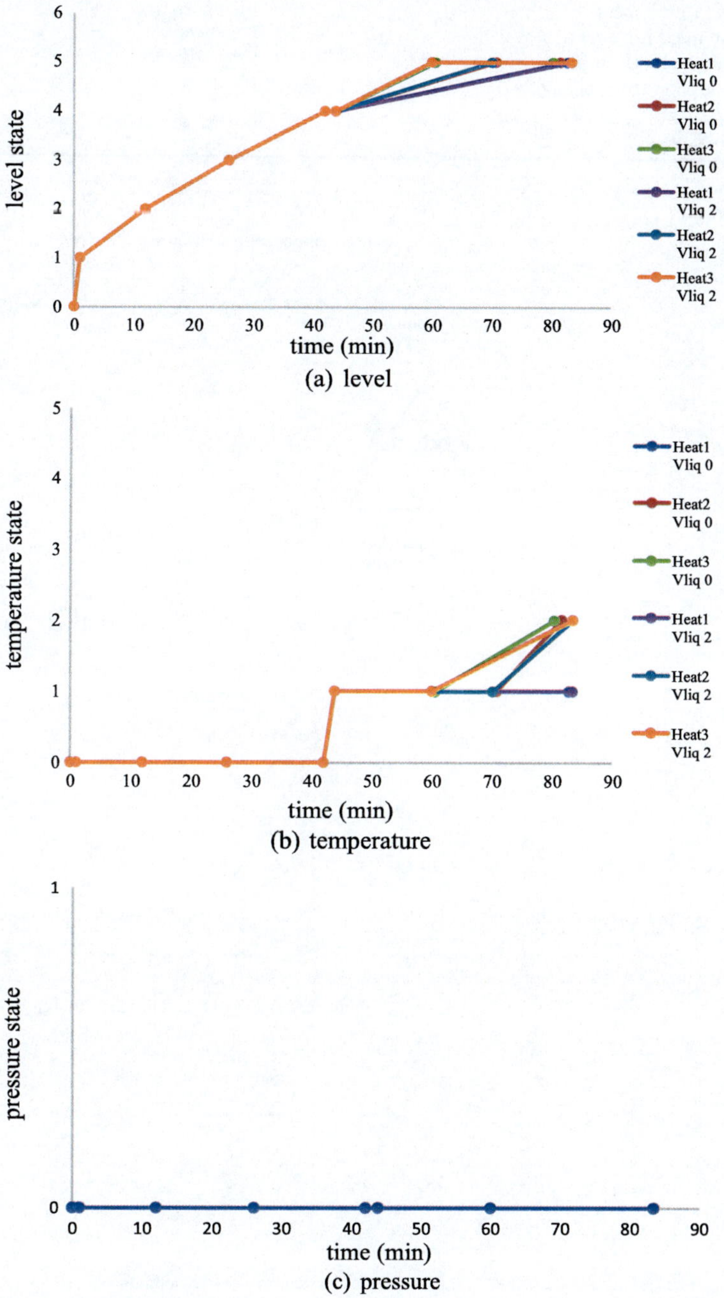

Fig. 7.35 Time profiles of discretized state variables for which heating starts when the liquid level reaches the discretized value of 4 (Reprinted with permission from Zhang et al. 2020. Copyright 2020 American Chemical Society)

Fig. 7.36 Components models in the third level of system hierarchy (Reprinted with permission from Zhang et al. 2020. Copyright 2020 American Chemical Society)

(a) Component model of feed flow (F_{in})

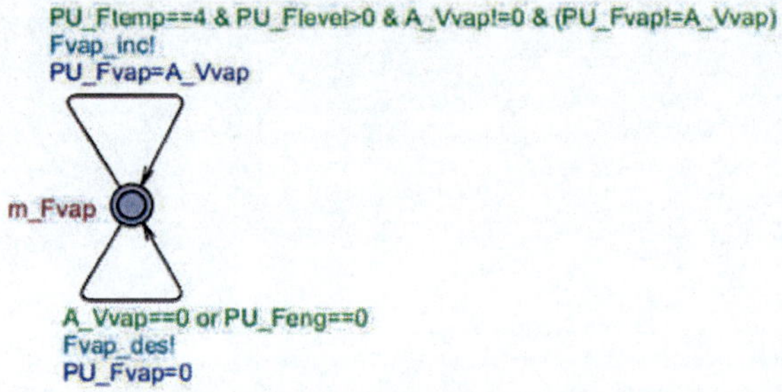

(b) Component model of vapor product flow (F_{vap})

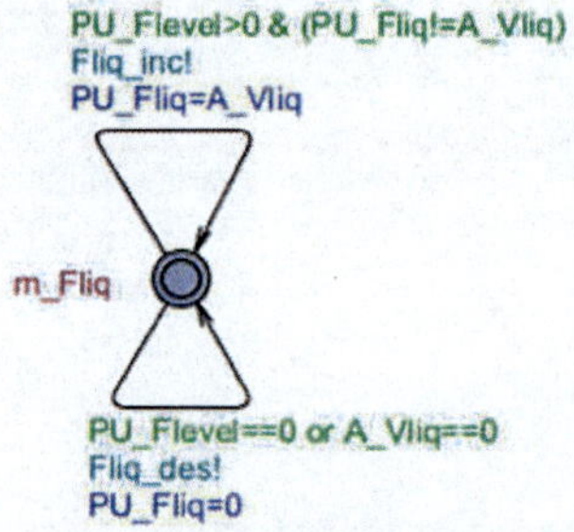

(c) Component model of liquid product flow (F_{liq})

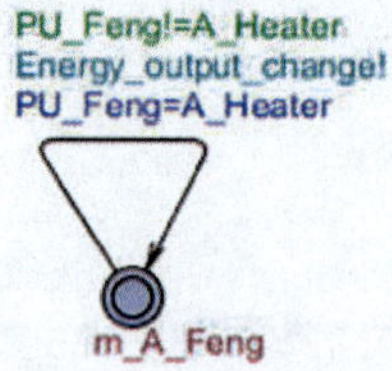

(d) Component model of heat input flow (F_{eng})

Component Models in Level 5

The liquid level is unmeasured in the flash startup example, while the models of the components of temperature and pressure sensors in the flash drum are presented in Fig. 7.37. It can be observed from Table 7.8 that the temperature and pressure are discretized into seven (7) and five (5) intervals, respectively, and the same numbers of self-looping edges can be found in the corresponding Fig. 7.37a and b. Notice also that each loop in Fig. 7.37 is used to represent a unique measurement-taking action.

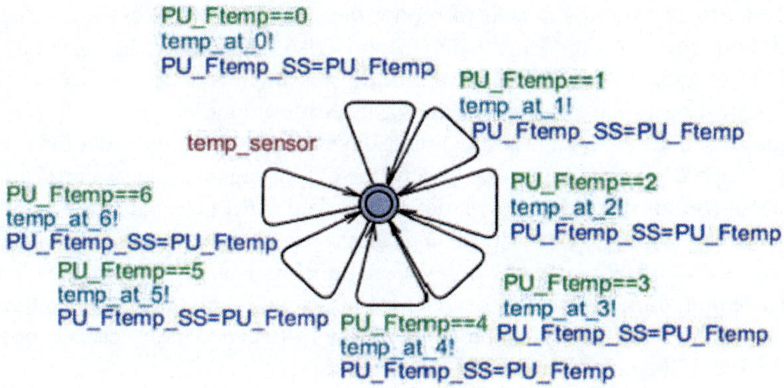

(a) Component model of temperature sensor

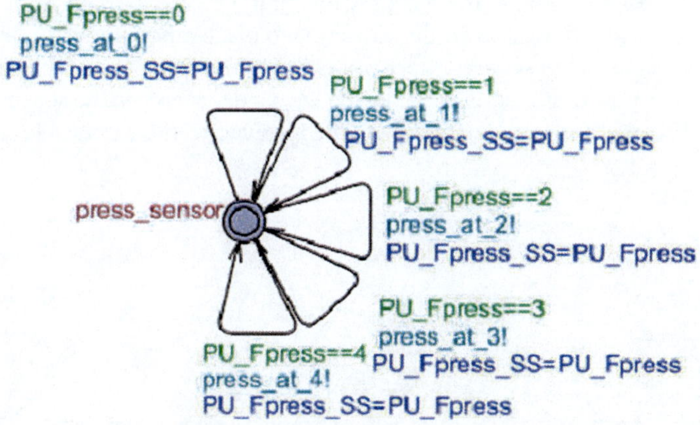

(b) Component model of pressure sensor

Fig. 7.37 Components models in the fifth level of system hierarchy (Reprinted with permission from Zhang et al. 2020. Copyright 2020 American Chemical Society)

Let us consider Fig. 7.37a as an example for illustration convenience. The guards of the seven self- loops are simply the discretized temperatures, i.e., PU_Ftemp == i (where i = 0, 1, ··· , 6). The corresponding sender events are: temp_at_i and i = 0, 1, ··· , 6. Finally, all sensor measurements are updated to the actual temperature values, i.e., PU_Ftemp_SS = PU_Ftemp.

Layer Model

In addition to the automata used to model the components in levels 2 – 5 of the system hierarchy (see Fig. 3.1 in Chap. 3), it is also necessary to construct a "layer model" to specify the precedence order of all possible events in the given system. Figure 7.38 shows such a model for the flash startup example.

Since all self-looping edges on the place layer1are used to represent the operator actions adopted to manipulate the controller output signals, every corresponding event should be considered to be a sender and marked with the exclamation symbol(!).

All self-looping edges on the place layer2represent the actuator adjustment actions taken by operator. Notice that only V_{liq} at the liquid outlet line in Fig. 7.1 is a hand valve and, thus, classified as sender events in the second layer. Since the other actuators in the flash startup example are all control valves, their actions should be consistent with the controller outputs in the first layer.

The self-looping edges on layer3denote the input and/or output flows of mass and energy. Since these flows are affected by sender events in layers 1 and 2, they are all considered to be receiver events that are marked by symbol(?).

Place layer4corresponds to the processing unit(s) in level 4 of the system hierarchy. In the present example, there is only one unit, i.e., the flash drum. As shown in Fig. 7.4, the state transitions in flash drum take place autonomously. Therefore, the are no needs to attach sender or receiver events onlayer4.

The states of processing units are reflected via sender events in the sensor models (see Fig. 7.37) and then synchronized with the receiver events on place layer5.

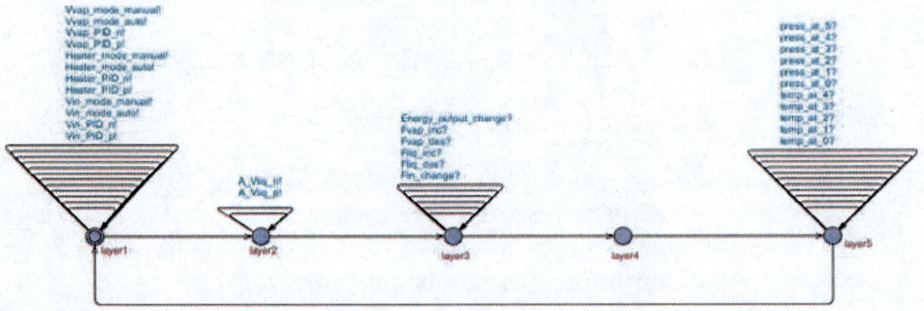

Fig. 7.38 Layer model (Reprinted with permission from Zhang et al. 2020. Copyright 2020 American Chemical Society)

Appendix 7.3 A Detailed Listing of the Guards and Updates of Every Edge in the Component Model of Flash Drum

(Tables 7.9, 7.10, 7.11, 7.12, 7.13, 7.14, 7.15, 7.16, 7.17, 7.18, 7.19, 7.20, 7.21, 7.22, 7.23, 7.24, 7.25, 7.26, 7.27, 7.28, 7.29, 7.30, 7.31, 7.32, 7.33 and 7.34)

Table 7.9 Edge specifications of arc 1 in Fig. 7.4

Edge Label	Guards	Updates
1.1	PU_Flevel = = 0 & PU_Fin = = 4 && x > = 1	PU_Flevel = 1, x = 0

Table 7.10 Edge specifications of arc 2 in Fig. 7.4

Edge Label	Guards	Updates
2.1	PU_Flevel = = 1 & PU_Fin = = 4 && x > = 11	PU_Flevel = 2, x = 0

Table 7.11 Edge specifications of arc 3 in Fig. 7.4

Edge Label	Guards	Updates
3.1	PU_Flevel = = 2 & PU_Fin = = 4 && x > = 14	PU_Flevel = 3, x = 0

Table 7.12 Edge specifications of arc 4 in Fig. 7.4

Edge Label	Guards	Updates
4.1	PU_Flevel = = 3 & PU_Fin = = 4 && x > = 16	PU_Flevel = 4, x = 0

Table 7.13 Edge specifications of arc 5 in Fig. 7.4

Edge Label	Guards	Updates
5.1	PU_Ftemp = = 0 & PU_Feng = = 1 &A_Vvap = = 2 & A_Vliq = = 0 & PU_Fin = = 2 && x > = 5	PU_Ftemp = 1, x = 0
5.2	PU_Ftemp = = 0 & PU_Feng = = 2 & A_Vvap = = 2 & A_Vliq = = 0 & PU_Fin = = 2 && x > = 4	PU_Ftemp = 1, x = 0
5.3	PU_Ftemp = = 0 & PU_Feng = = 4 & A_Vvap = = 2 & A_Vliq = = 0 & PU_Fin = = 2 && x > = 2	PU_Ftemp = 1, x = 0
5.4	PU_Ftemp = = 0 & PU_Feng = = 1 & A_Vvap = = 2 & PU_Fliq = = 2 & PU_Fin = = 2 && x > = 5	PU_Ftemp = 1, x = 0
5.5	PU_Ftemp = = 0 & PU_Feng = = 2 & A_Vvap = = 2 & PU_Fliq = = 2 & PU_Fin = = 2 && x > = 3	PU_Ftemp = 1, x = 0
5.6	PU_Ftemp = = 0 & PU_Feng = = 4 & A_Vvap = = 2 & PU_Fliq = = 2 & PU_Fin = = 2 && x > = 2	PU_Ftemp = 1, x = 0

Table 7.14 Edge specifications of arc 6 in Fig. 7.4

Edge Label	Guards	Updates
6.1	PU_Ftemp = = 0 & PU_Feng = = 1 & A_Vvap = = 2 & A_Vliq = = 0 & PU_Fin = = 2 && x > = 12	PU_Ftemp = 1, x = 0
6.2	PU_Ftemp = = 0 & PU_Feng = = 2 & A_Vvap = = 2 & A_Vliq = = 0 & PU_Fin = = 2 && x > = 8	PU_Ftemp = 1, x = 0
6.3	PU_Ftemp = = 0 & PU_Feng = = 4 & A_Vvap = = 2 & A_Vliq = = 0 & PU_Fin = = 2 && x > = 5	PU_Ftemp = 1, x = 0
6.4	PU_Ftemp = = 0 & PU_Feng = = 1 & A_Vvap = = 2 & PU_Fliq = = 2 & PU_Fin = = 2 && x > = 13	PU_Ftemp = 1, x = 0
6.5	PU_Ftemp = = 0 & PU_Feng = = 2 & A_Vvap = = 2 & PU_Fliq = = 2 & PU_Fin = = 2 && x > = 8	PU_Ftemp = 1, x = 0
6.6	PU_Ftemp = = 0 & PU_Feng = = 4 & A_Vvap = = 2 & PU_Fliq = = 2 & PU_Fin = = 2 && x > = 5	PU_Ftemp = 1, x = 0

Table 7.15 Edge specifications of arc 7 in Fig. 7.4

Edge Label	Guards	Updates
7.1	PU_Ftemp = = 0 & PU_Feng = = 1 & A_Vvap = = 2 & A_Vliq = = 0 & PU_Fin = = 2 && x > = 13	PU_Ftemp = 1, x = 0
7.2	PU_Ftemp = = 0 & PU_Feng = = 2 & A_Vvap = = 2 & A_Vliq = = 0 & PU_Fin = = 2 && x > = 13	PU_Ftemp = 1, x = 0
7.3	PU_Ftemp = = 0 & PU_Feng = = 4 & A_Vvap = = 2 & A_Vliq = = 0 & PU_Fin = = 2 && x > = 12	PU_Ftemp = 1, x = 0
7.4	PU_Ftemp = = 0 & PU_Feng = = 1 & A_Vvap = = 2 & PU_Fliq = = 2 & PU_Fin = = 2 && x > = 14	PU_Ftemp = 1, x = 0
7.5	PU_Ftemp = = 0 & PU_Feng = = 2 & A_Vvap = = 2 & PU_Fliq = = 2 & PU_Fin = = 2 && x > = 14	PU_Ftemp = 1, x = 0
7.6	PU_Ftemp = = 0 & PU_Feng = = 4 & A_Vvap = = 2 & PU_Fliq = = 2 & PU_Fin = = 2 && x > = 11	PU_Ftemp = 1, x = 0

Table 7.16 Edge specifications of arc 8 in Fig. 7.4

Edge Label	Guards	Updates
8.1	PU_Ftemp = = 0 & PU_Feng = = 1 & A_Vvap = = 2 & A_Vliq = = 0 & PU_Fin = = 2 && x > = 2	PU_Ftemp = 1, x = 0
8.2	PU_Ftemp = = 0 & PU_Feng = = 2 & A_Vvap = = 2 & A_Vliq = = 0 & PU_Fin = = 2 && x > = 2	PU_Ftemp = 1, x = 0
8.3	PU_Ftemp = = 0 & PU_Feng = = 4 & A_Vvap = = 2 & A_Vliq = = 0 & PU_Fin = = 2 && x > = 2	PU_Ftemp = 1, x = 0
8.4	PU_Ftemp = = 0 & PU_Feng = = 1 & A_Vvap = = 2 & PU_Fliq = = 2 & PU_Fin = = 2 && x > = 2	PU_Ftemp = 1, x = 0
8.5	PU_Ftemp = = 0 & PU_Feng = = 2 & A_Vvap = = 2 & PU_Fliq = = 2 & PU_Fin = = 2 && x > = 2	PU_Ftemp = 1, x = 0
8.6	PU_Ftemp = = 0 & PU_Feng = = 4 & A_Vvap = = 2 & PU_Fliq = = 2 & PU_Fin = = 2 && x > = 2	PU_Ftemp = 1, x = 0

Table 7.17 Edge specifications of arc 9 in Fig. 7.4

Edge Label	Guards	Updates
9.1	PU_Flevel = = 1 & PU_Feng = = 1 & A_Vvap = = 2 & A_Vliq = = 0 && x > = 7	PU_Flevel = 2, x = 0
9.2	PU_Feng = = 1 & PU_Flevel = = 1 & A_Vvap = = 2 & PU_Fliq = = 2 && x > = 5	PU_Flevel = 2, x = 0

Table 7.18 Edge specifications of arc 10 in Fig. 7.4

Edge Label	Guards	Updates
10.1	PU_Flevel = = 2 & PU_Feng = = 1 & A_Vvap = = 2 & A_Vliq = = 0 && x > = 16	PU_Flevel = 3, x = 0

Table 7.19 Edge specifications of arc 11 in Fig. 7.4

Edge Label	Guards	Updates
11.1	PU_Flevel = = 3 & PU_Feng = = 2 & A_Vvap = = 2 & A_Vliq = = 0 && x > = 2	PU_Flevel = 4, x = 0
11.2	PU_Flevel = = 3 & PU_Feng = = 1 & A_Vvap = = 2 & A_Vliq = = 0 && x > = 2	PU_Flevel = 4, x = 0
11.3	PU_Flevel = = 3 & PU_Feng = = 1 & A_Vvap = = 2 & PU_Fliq = = 2 && x > = 2	PU_Flevel = 4, x = 0
11.4	PU_Flevel = = 3 & PU_Feng = = 2 & A_Vvap = = 2 & PU_Fliq = = 2 && x > = 2	PU_Flevel = 4, x = 0

Table 7.20 Edge specifications of arc 12 in Fig. 7.4

Edge Label	Guards	Updates
12.1	PU_Ftemp = = 1 & PU_Feng = = 2 & A_Vvap = = 2 & A_Vliq = = 0 && x > = 2	PU_Ftemp = 2, x = 0
12.2	PU_Ftemp = = 1 & PU_Feng = = 4 & A_Vvap = = 2 & A_Vliq = = 0 && x > = 3	PU_Ftemp = 2, x = 0
12.3	PU_Ftemp = = 1 & PU_Feng = = 2 & A_Vvap = = 2 & PU_Fliq = = 2 && x > = 4	PU_Ftemp = 2, x = 0
12.4	PU_Ftemp = = 1 & PU_Feng = = 4 & A_Vvap = = 2 & PU_Fliq = = 2 && x > = 4	PU_Ftemp = 2, x = 0

Table 7.21 Edge specifications of arc 13 in Fig. 7.4

Edge Label	Guards	Updates
13.1	PU_Ftemp = = 1 & PU_Feng = = 1 & A_Vvap = = 2 & A_Vliq = = 0 && x > = 8	PU_Ftemp = 2, x = 0
13.2	PU_Ftemp = = 1 & PU_Feng = = 2 & A_Vvap = = 2 & A_Vliq = = 0 && x > = 4	PU_Ftemp = 2, x = 0
13.3	PU_Ftemp = = 1 & PU_Feng = = 4 & A_Vvap = = 2 & A_Vliq = = 0 && x > = 7	PU_Ftemp = 2, x = 0
13.4	PU_Ftemp = = 1 & PU_Feng = = 1 & A_Vvap = = 2 & PU_Fliq = = 2 && x > = 4	PU_Ftemp = 2, x = 0
13.5	PU_Ftemp = = 1 & PU_Feng = = 2 & A_Vvap = = 2 & PU_Fliq = = 2 && x > = 6	PU_Ftemp = 2, x = 0
13.6	PU_Ftemp = = 1 & PU_Feng = = 4 & A_Vvap = = 2 & PU_Fliq = = 2 && x > = 10	PU_Ftemp = 2, x = 0

Table 7.22 Edge specifications of arc 14 in Fig. 7.4

Edge Label	Guards	Updates
14.1	PU_Ftemp = = 1 & PU_Feng = = 4 & A_Vvap = = 2 & A_Vliq = = 0 && x > = 1	PU_Ftemp = 2, x = 0
14.2	PU_Ftemp = = 1 & PU_Feng = = 4 & A_Vvap = = 2 & PU_Fliq = = 2 && x > = 3	PU_Ftemp = 2, x = 0

Table 7.23 Edge specifications of arc 15 in Fig. 7.4

Edge Label	Guards	Updates
15.1	PU_Ftemp = = 1 & PU_Feng = = 1 & A_Vvap = = 2 & A_Vliq = = 0 && x > = 1	PU_Ftemp = 2, x = 0

Table 7.24 Edge specifications of arc 16 in Fig. 7.4

Edge Label	Guards	Updates
16.1	PU_Flevel = = 1 & PU_Feng = = 2 & A_Vvap = = 2 & A_Vliq = = 0 && x > = 15	PU_Ftemp = 2, x = 0
16.2	PU_Flevel = = 1 & PU_Feng = = 4 & A_Vvap = = 2 & A_Vliq = = 0 && x > = 4	PU_Ftemp = 2, x = 0
16.3	PU_Flevel = = 1 & PU_Feng = = 2 & A_Vvap = = 2 & PU_Fliq = = 2 && x > = 12	PU_Ftemp = 2, x = 0

Table 7.25 Edge specifications of arc 17 in Fig. 7.4

Edge Label	Guards	Updates
17.1	PU_Flevel = = 2 & PU_Feng = = 2 & A_Vvap = = 2 & A_Vliq = = 0 && x > = 15	PU_Flevel = 3, x = 0
17.2	PU_Flevel = = 2 & PU_Feng = = 1 & A_Vvap = = 2 & A_Vliq = = 0 && x > = 16	PU_Flevel = 3, x = 0
17.3	PU_Flevel = = 2 & PU_Feng = = 1 & A_Vvap = = 2 & PU_Fliq = = 2 && x = = 16	PU_Flevel = 3, x = 0
17.4	PU_Flevel = = 2 & PU_Feng = = 2 & A_Vvap = = 2 & PU_Fliq = = 2 && x > = 16	PU_Flevel = 3, x = 0
17.5	PU_Flevel = = 2 & PU_Feng = = 4 & A_Vvap = = 2 & PU_Fliq = = 2 && x > = 5	PU_Flevel = 3, x = 0

Table 7.26 Edge specifications of arc 18 in Fig. 7.4

Edge Label	Guards	Updates
18.1	PU_Flevel = = 3 & PU_Feng = = 4 & A_Vvap = = 2 & A_Vliq = = 0&& x > = 2	PU_Flevel = 4, x = 0
18.2	PU_Flevel = = 3 & PU_Feng = = 2 & A_Vvap = = 2 & A_Vliq = = 0 && x > = 2	PU_Flevel = 4, x = 0
18.3	PU_Flevel = = 3 & PU_Feng = = 1 & A_Vvap = = 2 & A_Vliq = = 0 && x > = 1	PU_Flevel = 4, x = 0
18.4	PU_Flevel = = 3 & PU_Feng = = 1 & A_Vvap = = 2 & A_Vliq = = 0 && x > = 2	PU_Flevel = 4, x = 0
18.5	PU_Flevel = = 3 & PU_Feng = = 1 & A_Vvap = = 2 & PU_Fliq = = 2 && x > = 2	PU_Flevel = 4, x = 0
18.6	PU_Flevel = = 3 & PU_Feng = = 1 & A_Vvap = = 2 & PU_Fliq = = 2 && x > = 2	PU_Flevel = 4, x = 0
18.7	PU_Flevel = = 3 & PU_Feng = = 2 & A_Vvap = = 2 & PU_Fliq = = 2 && x > = 2	PU_Flevel = 4, x = 0
18.8	PU_Flevel = = 3 & PU_Feng = = 4 & A_Vvap = = 2 & PU_Fliq = = 2 && x > = 2	PU_Flevel = 4, x = 0

Table 7.27 Edge specifications of arc 19 in Fig. 7.4

Edge Label	Guards	Updates
19.1	PU_Ftemp = = 2 & PU_Feng = = 4 & A_Vvap = = 2 & PU_Fliq = = 2 && x > = 3	PU_Ftemp = 3, x = 0

Table 7.28 Edge specifications of arc 20 in Fig. 7.4

Edge Label	Guards	Updates
20.1	PU_Ftemp = = 2 & PU_Feng = = 2 & A_Vvap = = 2 & A_Vliq = = 0 && x > = 6	PU_Ftemp = 3, x = 0
20.2	PU_Ftemp = = 2 & PU_Feng = = 4 & A_Vvap = = 2 & A_Vliq = = 0 && x > = 5	PU_Ftemp = 3, x = 0
20.3	PU_Ftemp = = 2 & PU_Feng = = 2 & A_Vvap = = 2 & PU_Fliq = = 2 && x > = 7	PU_Ftemp = 3, x = 0
20.4	PU_Ftemp = = 2 & PU_Feng = = 4 & A_Vvap = = 2& A_Vliq = = 0 && x > = 10	PU_Ftemp = 3, x = 0

Table 7.29 Edge specifications of arc 21 in Fig. 7.4

Edge Label	Guards	Updates
21.1	PU_Ftemp = = 2 & PU_Feng = = 4 & A_Vvap = = 2 & PU_Fliq = = 2 && x > = 10	PU_Ftemp = 3, x = 0

Table 7.30 Edge specifications of arc 22 in Fig. 7.4

Edge Label	Guards	Updates
22.1	PU_Flevel = = 2 & PU_Feng = = 2 & A_Vvap = = 2 & A_Vliq = = 0 && x > = 9	PU_Ftemp = 3, x = 0
22.2	PU_Flevel = = 2 & PU_Feng = = 4 & A_Vvap = = 2 & A_Vliq = = 0 && x > = 5	PU_Ftemp = 3, x = 0
22.3	PU_Flevel = = 2 & PU_Feng = = 2 & A_Vvap = = 2 & PU_Fliq = = 2 && x > = 18	PU_Ftemp = 3, x = 0

Table 7.31 Edge specifications of arc 23 in Fig. 7.4

Edge Label	Guards	Updates
23.1	PU_Flevel = = 3 & PU_Feng = = 2 & A_Vvap = = 2 & A_Vliq = = 0 && x > = 2	PU_Ftemp = 4, x = 0
23.2	PU_Flevel = = 3 & PU_Feng = = 4 & A_Vvap = = 2 & A_Vliq = = 0 && x > = 2	PU_Ftemp = 4, x = 0
23.3	PU_Flevel = = 3 & PU_Feng = = 2 & A_Vvap = = 2 & PU_Fliq = = 2 && x > = 3	PU_Flevel = 4, x = 0

Table 7.32 Edge specifications of arc 24 in Fig. 7.4

Edge Label	Guards	Updates
24.1	PU_Ftemp = = 3 & PU_Feng = = 4 & A_Vvap = = 2 & A_Vliq = = 0 && x > = 4	PU_Ftemp = 4, PU_Fpress = 2,x = 0

Table 7.33 Edge specifications of arc 25 in Fig. 7.4

Edge Label	Guards	Updates
25.1	PU_Ftemp = = 3 & PU_Feng = = 3 & A_Vvap = = 2 & PU_Fliq = = 2 && x > = 7	PU_Ftemp = 4, PU_Fpress = 2,x = 0

Table 7.34 Edge specifications of arc 26 in Fig. 7.4

Edge Label	Guards	Updates
26.1	PU_Ftemp = = 3 & PU_Feng = = 4 & A_Vvap = = 2 & A_Vliq = = 0 && x > = 6	PU_Ftemp = 4, PU_Fpress = 2,x = 0

Appendix 7.4 Task File Used for SFC-7.3 Simulation

```
runs at 0
Blocks("FC4").Automan:1;
Blocks("FC4").OPman:50;
wait 0.444;
Blocks("VVENT").pos:100;
Blocks("TC16").Automan:1;
Blocks("TC16").OPman:70,000;
Blocks("PC2").Automan:1;
Blocks("PC2").OPman:45,000;
wait for Blocks("LC1").PV > = 0.3;
Blocks("TC16").Automan:1;
Blocks("TC16").OPman:105,000;
Blocks("TC7").Automan:1;
Blocks("TC7").OPman:34,000;
wait for Blocks("LC1").PV > = 1.25;
Blocks("LC1").Automan:1;
Blocks("LC1").OPman:51;
wait 0.4;
Blocks("LC3").Automan:1;
```

```
Blocks("LC3").OPman:48;
wait 0.1;
Blocks("VVENT").pos:50;
Blocks("FC4").Automan:0;
Blocks("FC4").SP:10,000;
Blocks("TC16").Automan:0;
Blocks("TC16").SP:101.4;
Blocks("PC2").Automan:0;
Blocks("PC2").SP:2;
Blocks("LC1").Automan:0;
Blocks("LC1").SP:1.25;
wait 1.5;
End.
```

References

Alur R, Dill DL (1994) A theory of timed automata. Theor Comp Sci 126:183–235

Behrmann G, David A, Larsen KG (2004) A tutorial on UPPAAL. In Formal methods for the design of real-time systems (pp. 200–236). Springer, Berlin, Heidelberg

Behrmann G, David A, Larsen KG (2006) A Tutorial on Uppaal 4.0,.Department of Computer Science, Aalborg University, Denmark

Cassandras CG, Lafortune S (2008) Introduction to Discrete Event Systems, 2nd edn. Springer Science+Business Media, New York

Zhang CR, Yeh CY, Chang CT (2020) Synthesis, validation, and evaluation of operating procedures based on timed automata and dynamic simulation. Ind Eng Chem Res 59(18):8769–8782

Chapter 8
Generation of Test Plans for Fault Diagnosis with Untimed Automata

Unexpected failures in a chemical plant often result in undesirable consequences, e.g., deterioration in product quality, reduction in productivity and, in worse scenarios, fire, explosion, or toxic release, etc. Since the offline hazard assessment methods can limit the total expected loss of accidents only to a certain extent, online fault diagnosis is an alternative means for further improvement of operational safety.

According to Venkatasubramanian et al. (2003a, b, c), the available fault diagnosis methods could be classified into three types: (1) quantitative-model-based approaches; (2) qualitative-model-based approaches; (3) process-history-based approaches. These methods were developed primarily for the continuous steady-state processes, while considerably less effort has been devoted to the unsteady operations. Nomikos and MacGregor (1994; 1995) developed the multi-way principal component analysis for batch process monitoring, which has later been utilized in online diagnosis studies (Kourti and Macgregor 1995; Kourti et al. 1995; Undey et al. 2003; Lee et al. 2004). In addition, other fault identification techniques based on the artificial neural networks, the knowledge-based expert systems, and the observers (Ruiz et al. 2001a,b; Pierri et al. 2008) have also been proposed for sequential operations. Although satisfactory results were reported, these methods are mostly effective for fault diagnosis in a system with relatively few interconnected units and, also, the diagnostic resolution in cases of coexisting failures may not be acceptable.

In order to expand the scope of fault diagnosis, Chen et al. (2010) developed several Petri net-based algorithms to configure online identification systems for batch plants with many more units. However, since the event sequences (or traces) in multi-failure scenarios cannot be conveniently generated with the Petri-net models, this approach was limited to single-failure accidents. Generally speaking, such model deficiencies can be improved (or avoided) with automata (Sampath et al. 1995, 1996; Sampath et al. 1998; Baroni et al. 1999, 2000; Debouk et al. 2000; Benveniste et al. 2003; Zad et al. 2003; Qiu and Kumar 2006; Yeh and Chang 2011). With this alternative modeling approach, a so-called "diagnoser" can be constructed to predict all observable fault-propagation event sequences in the given system and to determine the corresponding fault origins. Since the root cause(s) of a trace may or may not

C.-T. Chang et al., *Process Plant Operating Procedures*, Advances in Industrial Control,
https://doi.org/10.1007/978-3-030-70978-5_8

be unique, there is still a need to enhance the diagnostic resolution with additional measures.

In general, the diagnostic performance can always be improved by capturing more process information and gaining deeper insights into the current plant status. These goals are traditionally achieved with more sensors so as to secure extra online measurement data under abnormal process conditions. However, since the execution of diagnostic test plans seems to be a feasible option that has not been systematically discussed in the chemical engineering literature, e.g., see Yeh and Chang (2011), it is the objective of this chapter to present an effective method to synthesize the required diagnostic procedures.

8.1 Model-Building Methods

It should first be noted that a generic automaton construction method has already described in Chap. 3. For the sake of illustration clarity, this method is repeated here with a simple example. Specifically, let us consider a fictitious liquid transfer system represented by the piping and instrumentation diagram (P&ID) in Fig. 8.1 and also the sequential function chart (SFC) in Fig. 8.2 and Table 8.1. If a three-way valve is "closed" in this liquid transfer system, the port connecting to the horizontal pipeline in Fig. 8.1, i.e., P-2 in the case of V-1 or P-3 in the case of V-3, is assumed to be blocked. Otherwise, its inlet flow(s) should be directed to every outlet pipeline. It is assumed that all valves *except* V-4 are placed at the "close" position initially. Thus, it clear from the given SFC that, during the normal operation, the buffer tank is filled

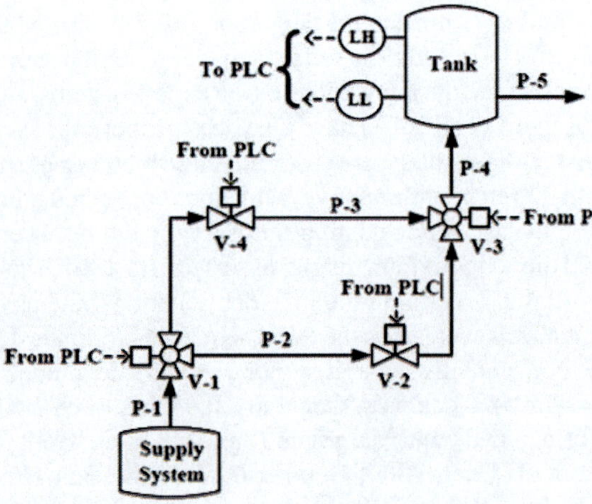

Fig. 8.1 P&ID of a liquid transfer system (Reprinted with permission from Kang and Chang 2014. Copyright 2014 Elsevier)

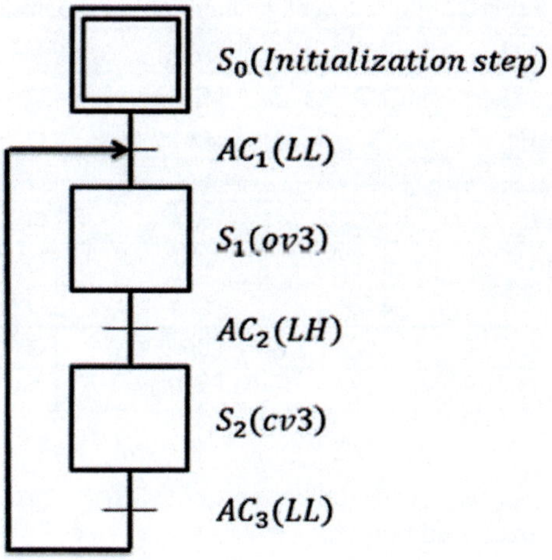

Fig. 8.2 Normal SFC of a liquid transfer operation (Reprinted with permission from Kang and Chang 2014. Copyright 2014 Elsevier)

Table 8.1 The normal transfer procedure: (a) operation steps; (b) activation conditions

(a)		(b)	
Operation Step	Control Actions	Symbol	Conditions
S_0	Initialization	AC_1	START
S_1	Open V–3	AC_2	LH
S_2	Close V–3	AC_3	LL

with liquid via P-1, P-3, and P-4 by manipulating V-3 and then drained via P-5 by gravity.

For the sake of brevity, only three failures are considered here:

- A large leak develops in tank (which is referred to as "T1leak" or F_1);
- V-3 fails at the "close" position (which is referred to as "v3s_c" or F_2);
- V-3 fails at the "open" position (which is referred to as "v3s_o" or F_3).

Based on the aforementioned assumptions, a total of 8 possible process configurations (pc01 - pc08) can be identified and they are listed in Table 8.2. Note that, for valve V-3, there are four possible states: (1) state O, i.e., it is at the normally open position; (2) state C, i.e., it is at the normally close position; (3) state SC, i.e., it sticks at the close position; (4) state SO, i.e., it sticks at the open position. On the other hand, only two tank states are adopted depending on whether or not a leak develops. Finally, notice that the states of V-1, V-2, and V-4 are unchanged since they are not used in normal operation and also their failures are not considered in the present example.

As described in Chap. 3, the plant model can be obtained by first building automata to model all components in the given process and then integrating them via the standard parallel decomposition operation (Cassandras and Lafortune 2008). In this example, the controller and the remaining components are modeled, respectively,

Table 8.2 Process configurations of liquid transfer system without diagnostic tests

V-1	V-2	V-3	V-4	*T1leak*	Symbol
C	C	C	O	N	pc01
C	C	O	O	N	pc02
C	C	SC	O	N	pc03
C	C	SO	O	N	pc04
C	C	C	O	Y	pc05
C	C	O	O	Y	pc06
C	C	SC	O	Y	pc07
C	C	SO	O	Y	pc08

with two different approaches. The corresponding model building strategies are illustrated below:

As a general rule, every component model in all except the first hierarchical level is used to represent a finite set of identifiable states of the hardware item under consideration and also the specific events facilitating the state transitions. A transition from one state to another is caused by the so-called *state-transition event*, while a self-looping transition results from the *state-maintaining event*. Notice that the latter event may (or may not) be bypassed if one or more former event is allowed at the originating state. To illustrate this model building principle, let us use the component model of V-3 (Fig. 8.3) as an example. Under the normal conditions, there are only two valve states (i.e., V3O and V3C) representing the open and close positions respectively.

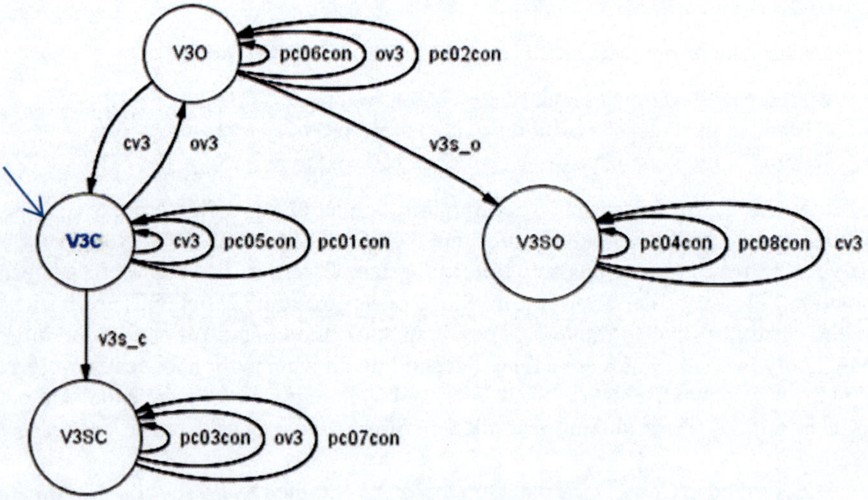

Fig. 8.3 Component model for valve V-3 in liquid transfer system (Reprinted with permission from Kang and Chang 2014. Copyright 2014 Elsevier)

Note that the action to close or open V-3 (denoted as cv3 and ov3 respectively) can be either a state-transition or a state-maintaining event depending on the starting valve state. The eight (8) additional state-maintaining events in this model, i.e., pc01con – pc08con, represent the scenarios that the corresponding configurations are maintained for a sufficiently long period of time. Finally, notice that the initial state in this model is marked by attaching an incoming arrow without origins.

Another important feature is that the state-maintaining event at a higher hierarchical level may be used as a lower-level state-transition event. To illustrate this model-building technique, let us use the tank model (Fig. 8.4) as an example. In this automaton, T1H and T1L denote the normal operating conditions when the liquid height reaches the designated maximum and minimum values respectively. The state-transition events between these two states, i.e., pc01con and pc02con, are the state-maintaining events in the automaton representing V-3. Note also that, since no sensor failures are considered here, it is assumed that the online measurements always accurately reflect the tank states and, thus, the sensor model can be omitted.

After building the automaton to model a normal component, additional mechanisms can be incorporated to describe its failures. Generally speaking, every failure can be modeled as a state-transition event, which triggers a change from one of the normal states to an abnormal one. Let us again consider the models of V-3 and

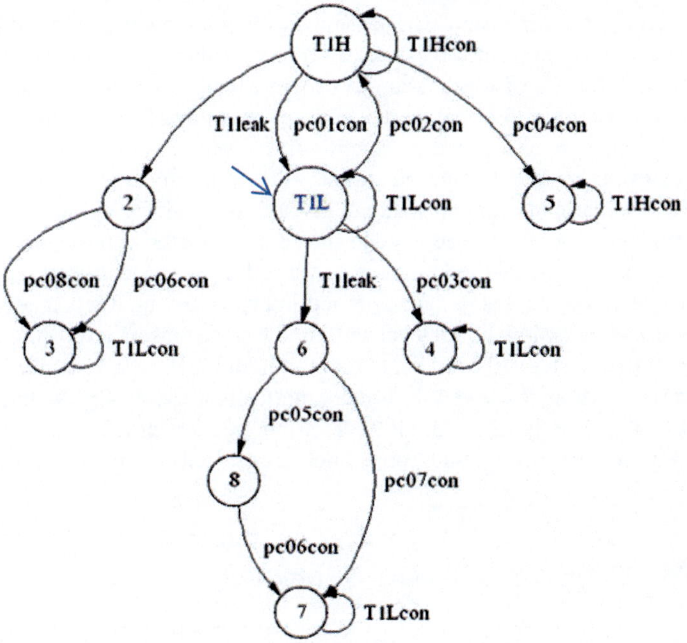

Fig. 8.4 Tank model in liquid transfer system (Reprinted with permission from Kang and Chang 2014. Copyright 2014 Elsevier)

Tank in Figs. 8.3 and 8.4. In the case of V-3, the failure v3s_c activates the transition from V3C (normal) to V3SC(abnormal) while v3s_o from V3O (normal) to V3SO(abnormal). In the model concerning Tank, the failure T1leak induces two different transitions from the normal states T1H and T1L to the abnormal states 2 and 6, respectively. Since the states of a lower level component are affected by the upper-level failures, the impacts of v3s_c and v3s_o should also appear in the tank model via the state-transition events pc03con – pc08con. Notice that each of these abnormal configurations can only be obtained after reaching a corresponding normal one, i.e., after the event pc01con or event pc02con. For illustration convenience, let us also assume that the effect of leak dominates that of inlet flow and, thus, the eventual liquid level is low after T1leak. On the other hand, due to the absence of sensor failures, the state-maintaining events in the tank model, i.e., T1Hcon and T1Lcon, should also be treated as the corresponding online level measurements. It should be emphasized that the practice of omitting the sensor models in this example is by no means restrictive. If a more comprehensive fault diagnosis is required, it is only necessary to build the neglected component models with the aforementioned techniques. The test-plan synthesis procedure described later is still applicable.

As mentioned before, the level-1 component should be modeled with a different approach. In particular, a subset of all events that are allowed in the remaining levels should be selected and assembled according to the operation steps and activation conditions specified in the given SFC. For example, the PLC model for the aforementioned liquid transfer system can be synthesized with this strategy (see Fig. 8.5). Notice first that the loop formed by states 0 – 5 represents the normal operation cycle (see Fig. 8.2). Since it is also necessary to stipulate the controller behavior after one or more failure occurs, additional 6 branches are attached to this loop to describe the fault propagation paths associated with configurations pc03 –pc08, respectively. The self-looping event at state 7 is placed simply to facilitate succinct representation of two different scenarios (with and without T1leak) after failure v3s_c occurs, while the same modeling approach is adopted at state 8 to characterize the event sequences after failure v3s_o. The self-looping events placed at state 6 and state 17 are used to describe the scenarios that the actuator action ov3 is executed repeatedly without changing the persisting low liquid level and, for the same reason, a self-looping event cv3 is introduced at state 10 to indicate that the liquid level is constantly high even after multiple attempts to close V-3. Finally, notice that the deadlock states 19 and 21 are adopted primarily to highlight the fact that no control actions are called for since the required activation condition cannot be met in these two scenarios.

8.2 Observable Event Traces in Diagnoser

After obtaining the system model, a diagnoser can then be produced by following the procedure given below (Cassandras and Lafortune 2008):

(1) Assign a distinct numerical label to a selected state in the system model.

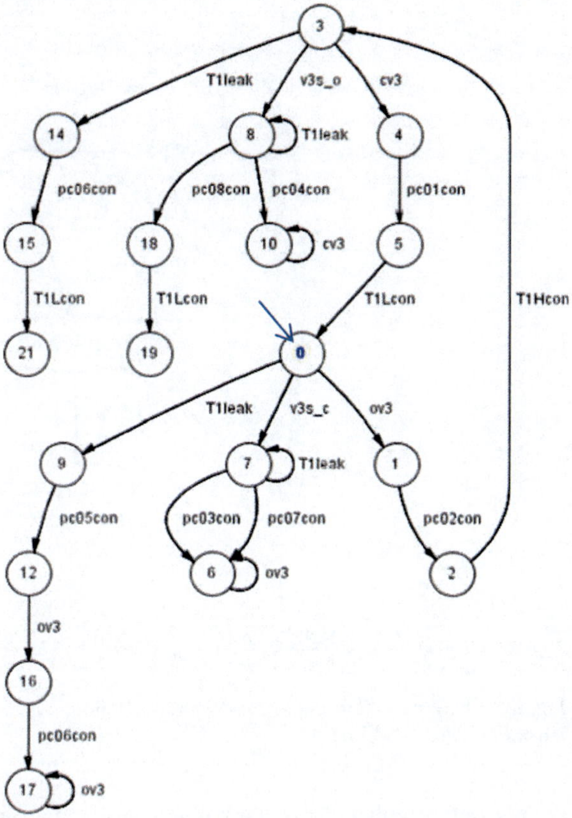

Fig. 8.5 PLC model in the liquid transfer system (Reprinted with permission from Kang and Chang 2014. Copyright 2014 Elsevier)

(2) Identify all paths between the initial state and the one selected in step (1). Consider these paths one at a time. If one or more failure is present on a path, then augment the numerical label with the corresponding failure label(s). Otherwise, augment the numerical label with the label "N".
(3) Repeat steps (1) and (2) until all states are exhausted.
(4) Under the assumption that only the actuator actions and sensor measurements are observable, hide every unobservable event in the system model by merging its input and output states.
(5) Ensure the liveliness of the resulting diagnoser by adding a fictitious self-looping event "STOP" at every deadlock state.

Notice that the system model and the corresponding diagnoser can be easily generated from the component models with existing free software DESUMA. The live diagnoser for the aforementioned liquid transfer system is presented in Fig. 8.6. In this case, the event "STOP" should be interpreted as "the sensor reading remains unchanged for a long time."

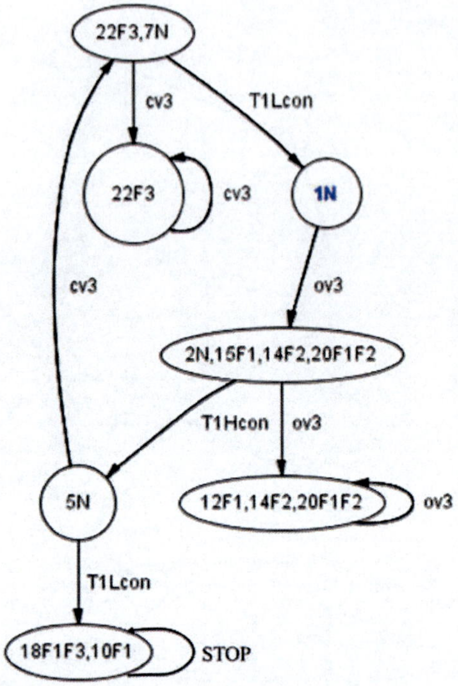

Fig. 8.6 Live diagnoser of liquid transfer system (Reprinted with permission from Kang and Chang 2014. Copyright 2014 Elsevier)

It is well established that fault diagnosis can be rigorously performed according to the observable traces in the diagnoser. In the liquid transfer example, three traces can be extracted from Fig. 8.6 and they are sketched in Fig. 8.7. Notice that every trace starts with an initial transition "i cycles", which denotes the event sequence in i complete normal cycles ($i = 0, 1, 2, \cdots$). A detailed analysis of the corresponding fault propagation sequences is given below:

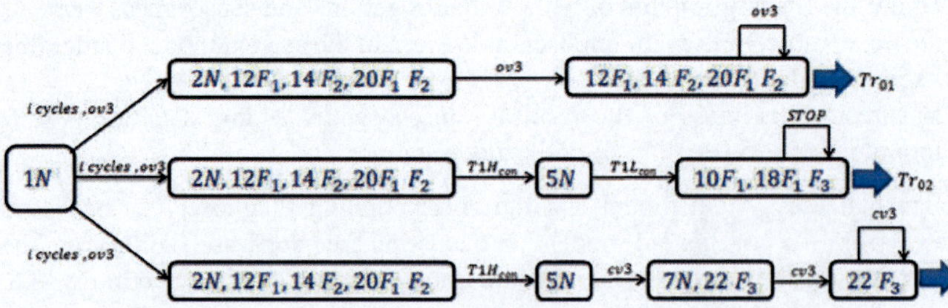

Fig. 8.7 Observable event traces in the diagnoser of liquid transfer system (Reprinted with permission from Kang and Chang 2014. Copyright 2014 Elsevier)

- **Trace 1** (Tr_{01}): If the actuator action ov3 is first performed at the initial state (which may be reached after completing any nonnegative number of normal cycles), the resulting liquid level should be T1Hcon in normal operation. Since the sensor reading stays unchanged at T1Lcon in this case, the action ov3 must be attempted again and again according to the SFC given in Fig. 8.2. After observing this trace, one should be able to deduce that there are three possible fault origins, i.e., (1) F_1, (2) F_2 and (3) $F_1 F_2$.
- **Trace 2** (Tr_{02}): After applying the actuator action *ov3* at the initial state (which may be reached after completing any nonnegative number of normal cycles), the sensor reading shows T1Hcon (normal). However, the liquid level then quickly drops to T1Lcon(abnormal) before the actuator action to close V-3 (cv3) can be executed. Observation of this trace indicates that the fault origins could be (1) F_1 or (2) $F_1 F_3$.
- **Trace 3** (Tr_{03}): By executing the actuator action ov3 at the initial state (which may be reached after completing any nonnegative number of normal cycles), the sensor reading can be raised to T1Hcon (normal). However, the next actuator action cv3 fails to bring down the liquid level even after repeated attempts. There is only one explanation for these phenomena, i.e., V-3 sticks at the open position (F_3).

Since the implied fault origins are not unique whenever Tr_{01} or Tr_{02} is observed online, there is a need to further enhance diagnostic resolution. For this purpose, Yeh and Chang (2011) suggested two complementary design options, i.e., installation of new sensors and implementation of extra operation steps. Let us focus on the latter in the present example. Notice also that, due to the model building conventions adopted in this example, every observable trace in a live diagnoser ends at a self-looping event. It is thus assumed that the diagnostic tests can only be applied at the corresponding states.

8.3 Control Specifications for Diagnostic Tests

In the supervisory control paradigm (Ramadge and Wonham, 1987; 1989), the plant to be operated is represented with an automaton P and the supervisor S is viewed as a mapping from the language generated by P to the power set of E, i.e., $S : \mathcal{L}(P) \rightarrow 2^E$, where $\mathcal{L}(P)$ denotes the set of all traces generated from P and E is the event set of P. If $t \in \mathcal{L}(P)$, then $S(t)$ should be interpreted as the allowed actuator actions after executing trace t. A sketch of this conceptual framework has already been given previously in Chap. 3 (see Fig. 3.27). In the traditional applications, P can be assembled with all component models under normal process conditions and its supervisor S is a model of the given SFC to be executed by an operator or a PLC.

Since multiple failure mechanisms are incorporated into automata and the diagnosability of the resulting system cannot always be guaranteed, it is, therefore, necessary to perform a diagnostic test to differentiate the fault origins implied by each

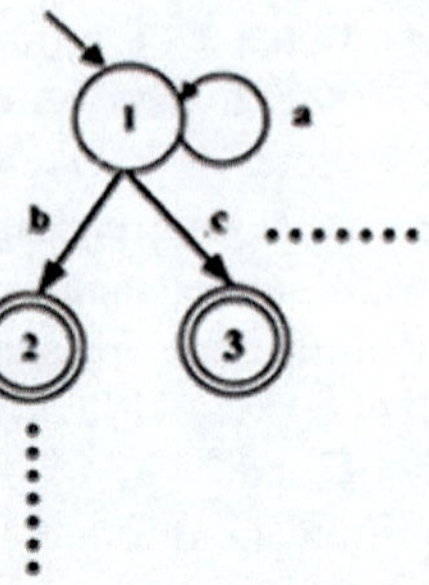

Fig. 8.8 Control specification for reaching the operational target(s) (Reprinted with permission from Kang and Chang 2014. Copyright 2014 Elsevier)

observable trace in the diagnoser. Every test plan is essentially viewed as a dedicated SFC for online implementation in the corresponding scenario. In order to synthesize such a SFC, a set of control specifications for the corresponding supervisor must be stipulated in advance to eliminate the unacceptable traces in $\mathcal{L}(P)$. The general structures of these specifications are summarized below:

8.3.1 *Test Target*

The primary function of a diagnostic test is to construct a unique observable pathway in the system automaton for each fault origin implied by an undiagnosable trace. The control specification to facilitate this goal can be characterized with the generalized automaton sketched in Fig. 8.8. The failures, actuator actions, online measurements, and/or process configurations (which last for a sufficiently long period of time) may all be utilized as the self-looping events in this model, e.g., event a, while all possible sensor outputs should be adopted as the transition-causing events, e.g., event b and event c. Notice that this structure may be repeated for more than one layer, which can be determined on a trial-and-error basis. Notice also that the total number of layers (M) is bounded from above, i.e., $M \leq F - 1$, where F is the number of fault origins implied by the undiagnosable trace under consideration.

8.3.2 *Auxiliary Constraint*

The auxiliary constraint is used mainly to facilitate the identification of a feasible SFC for implementing the diagnostic test with the fewest steps. The generalized model structure of this constraint can be found in Fig. 8.9. All sensor measurements and process configurations should be chosen as the self-looping events (α) in this automaton, while all actuator actions the transition-causing events (β). Notice that N is the number of assumed test steps and its lower limit is also identified with a trial-and-error approach.

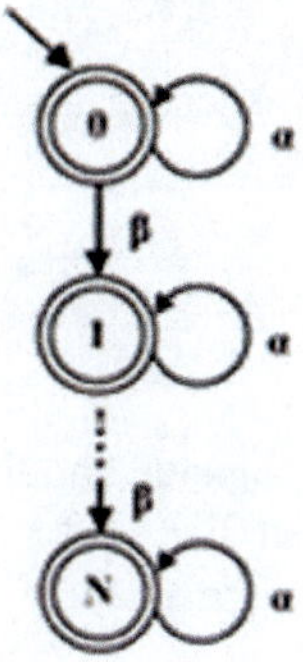

Fig. 8.9 Specification to facilitate identification of the optimal SFC (Reprinted with permission from Kang and Chang 2014. Copyright 2014 Elsevier)

8.4 Test-Plan Synthesis

By using the proposed control specifications, a set of new supervisors can be produced systematically for maximizing the diagnostic resolution of any given system. To every trace in diagnoser, the following test-plan synthesis procedure can be applied:

- **Step 1:** Set the initial state of every component in the last three levels in a system hierarchy by considering three scenarios:

(a) If a component failure is confirmed by the observable trace under consideration, then set the corresponding failed state as the initial condition.
(b) If a component failure can be neither confirmed nor rejected by observing the given trace, then set the normal state prior to this failure as the initial condition.
(c) If it can be certain that the component is normal, the initial component condition should be the final state normally achieved by the actuator actions in the trace.

- **Step 2:** Identify all process configurations allowed in the test plan.
- **Step 3:** Modify and simplify the component models according to the results obtained in Step 1 and Step 2.
- **Step 4:** Select the layer numbers (M and N) in the two control specifications, and then build the corresponding automata.
- **Step 5:** Produce the diagnostic supervisor by assembling the modified component models and the selected control specifications with parallel composition.
- **Step 6:** Repeat Step 4 and Step 5 in a trial-and-error fashion until the best candidate is identified.

To illustrate this synthesis procedure, let us first consider trace 1 (Tr_{01}) in the diagnoser of the liquid transfer system (see Fig. 8.7). The implementation steps are summarized below:

- **Step 1:**

Since the three implied fault origins, in this case, are (1) F_1, (2) F_2, and (3) F_1F_2, the component failures F_1 (T1leak) and F_2 (v3s_c) can be neither confirmed nor

Table 8.3 Allowed configurations of liquid transfer system for the test plan of Tr_{01}

V-1	V-2	V-3	V-4	T1leak	Symbol
O	O	SC	O	N	pc09
O	O	SC	O	Y	pc10
O	O	O	O	Y	pc11

rejected and thus the initial conditions of Tank and V-3 in the test plan should be set at T1L and V3C, respectively. Since the other components in the given system are normal, their starting states should be V1C, V2C and V4O, respectively.

- **Step 2:**

Since the failure of V-3 cannot be ruled out, there is a need to provide an alternative means to fill the surge tank by opening V-1 and V-2. Thus, the allowed process configurations in this scenario should at least include those listed in Table 8.3. Note that all three fault origins are incorporated while, for the purpose of simplifying test steps, V-1 and V-2 are not permitted to be closed again after they are opened.

- **Step 3:**

For the liquid transfer system, the original component models are obtained without considering the diagnostic tests. These models should now be modified according to the initial component states and process configurations identified in Step 1 and Step 2, and then simplified on the basis of specific process requirements (see Figs. 8.10, 8.11, 8.12, 8.13 and 8.14). For reasons already mentioned in the previous step, the actuator actions to close V-1 and V-2 are not included in Figs. 8.10 and 8.11. Similarly, for model reduction purpose, the action cv3 is also omitted in the automaton representing V-3 (Fig. 8.12). Since the state of V-4 is really irrelevant in the test plan, it is kept open in the test plan to avoid considering pointless options (see Fig. 8.13). Finally,

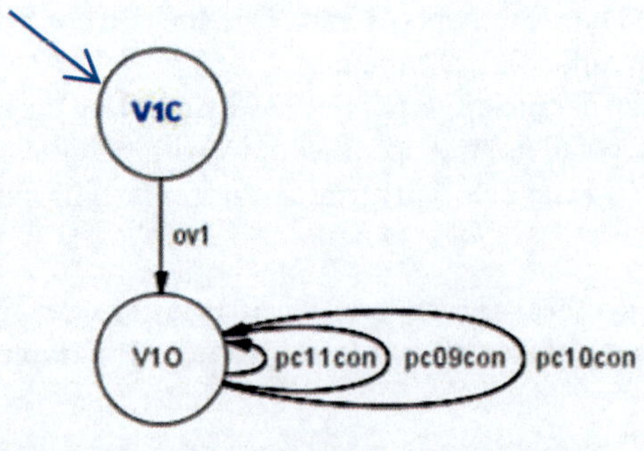

Fig. 8.10 Modified component model of V-1 for Tr_{01} in liquid transfer system (Reprinted with permission from Kang and Chang 2014. Copyright 2014 Elsevier)

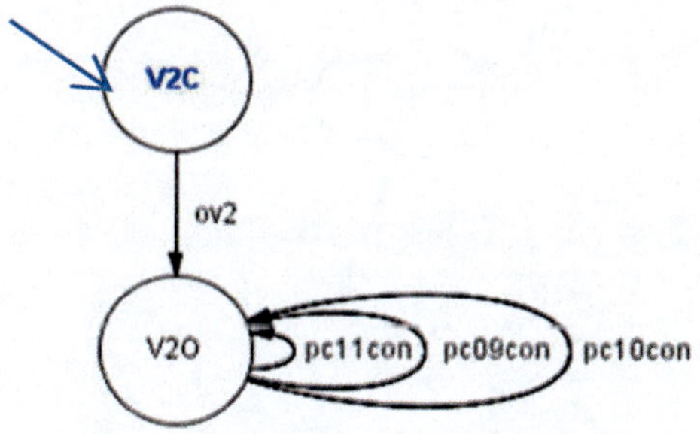

Fig. 8.11 Modified component model of V-2 for Tr_{01} in liquid transfer system (Reprinted with permission from Kang and Chang 2014. Copyright 2014 Elsevier)

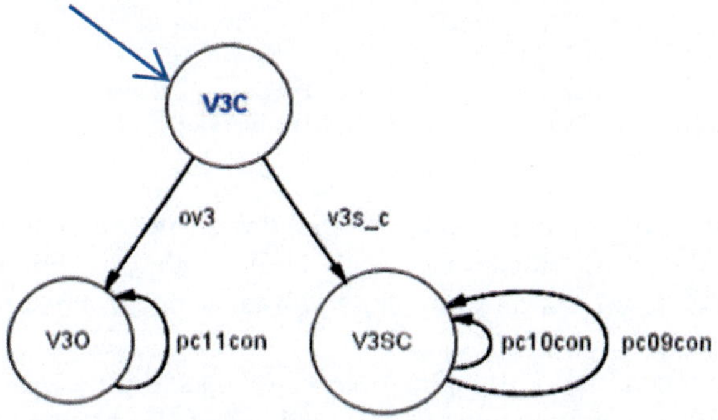

Fig. 8.12 Modified component model of V-3 for Tr_{01} in liquid transfer system (Reprinted with permission from Kang and Chang 2014. Copyright 2014 Elsevier)

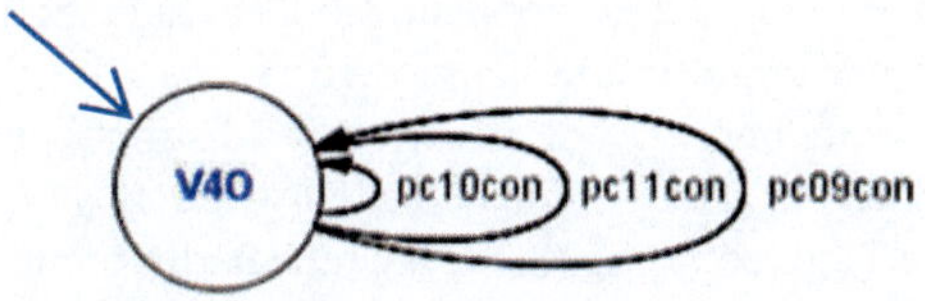

Fig. 8.13 Modified component model of V-4 for Tr_{01} in liquid transfer system (Reprinted with permission from Kang and Chang 2014. Copyright 2014 Elsevier)

the same approach can be taken to produce the tank model in Fig. 8.14 by modifying and simplifying the automaton in Fig. 8.4.

- **Step 4–Step 6:**

Since 3 possible fault origins are embedded in Tr_{01}, there should be at most 2 layers in the control specification for setting the target. The self-looping events in the corresponding automaton should be the actuator actions (ov1, ov2 and ov3), the

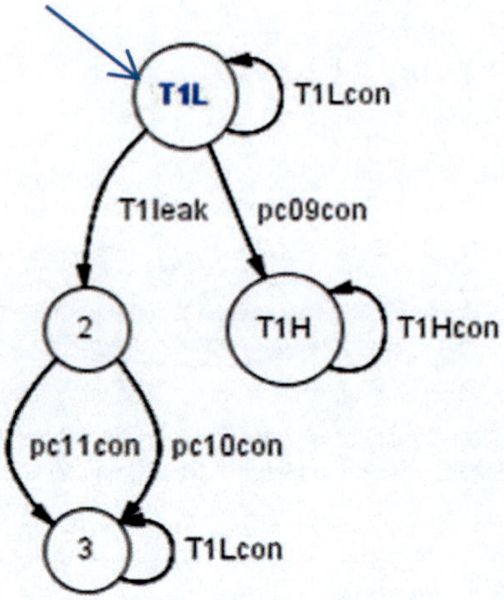

Fig. 8.14 Modified component model of Tank for Tr_{01} in liquid transfer system (Reprinted with permission from Kang and Chang 2014. Copyright 2014 Elsevier)

anticipated failures (v3s_c and T1leak), and the allowed process configurations (pc09 – pc11), while the transition-causing events should be the possible sensor readings (T1Hcon and T1Ccon). As an example, a two-layer target model is provided in Fig. 8.15.

On the other hand, the self-looping events in the auxiliary constraint should be the allowed process configurations (pc09 – pc11) and all possible sensor readings (T1Hcon and T1Ccon). The transition-causing events, in this case, should be the selected actuator actions (ov1, ov2 and ov3) and implied failures (v3s_c and T1leak). As an example, a 3-layer (i.e., $N = 3$) automaton is presented in Figure 8.16.

By repeatedly composing the component models (Figs. 8.10 , 8.10, 8.11, 8.12, 8.13 and 8.14) with the target model (Fig. 8.15) in parallel, it can be found that a two-layer target specification is not effective for distinguishing all implied fault origins and the best performance can be achieved when $M = 1$. The corresponding diagnostic supervisor is given in Fig. 8.17. By repeatedly composing the diagnostic supervisor (Fig. 8.17) with the auxiliary constraint (Fig. 8.16) in parallel, it can be

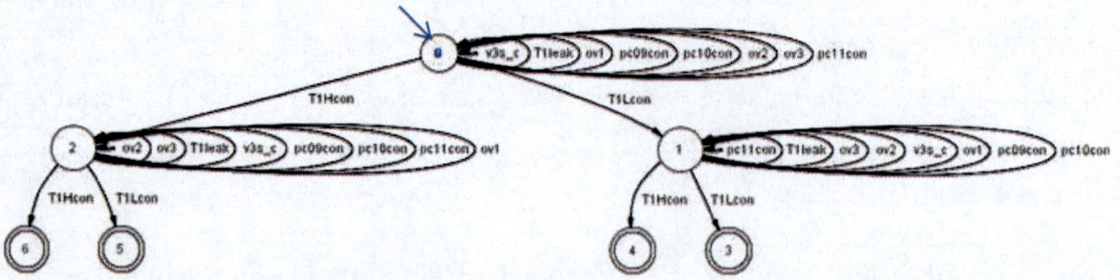

Fig. 8.15 Specification to achieve the operational target for Tr_{01} in liquid transfer system (Reprinted with permission from Kang and Chang 2014. Copyright 2014 Elsevier)

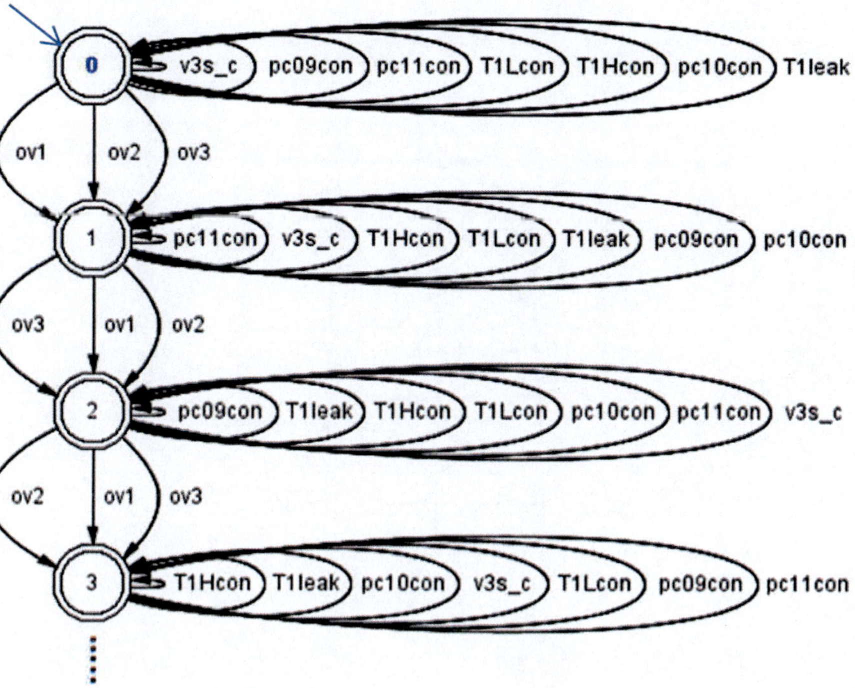

Fig. 8.16 Specification to synthesize SFC for Tr_{01} in liquid transfer system (Reprinted with permission from Kang and Chang 2014. Copyright 2014 Elsevier)

found that the smallest test plan can be identified at $N = 2$ (Fig. 8.18) and the corresponding SFC is given in Fig. 8.19.

On the basis of the above discussion, let us summarize the test plan of trace Tr_{01} as follows: Trace Tr_{01} in Fig. 8.7 represents the scenario that the abnormal sensor reading T1Lcon persists after multiple attempts to open V-3 (ov3). Three possible fault origins, i.e., F_1 (T1leak), F_2 (v3s_c), and F_1F_2 (T1leak and v3s_c), can be deduced by observing this sequence. According to Figs. 8.18 and 8.19, the required operation steps at this point should be opening both V-1 and V-2 (i.e., ov1 and ov2) to allow an alternative flow into the buffer tank. A resulting high liquid level indicates that the suspected leak (F_1) is not present and thus the correct fault origin should be F_2. Otherwise, the remaining two origins, i.e., F_1 and F_1F_2, should both be possible candidates but they are not distinguishable.

Finally, due to the assumption that the effect of leak dominates that of inlet flow, it can also be concluded that the two fault origins implied by Tr_{02}, i.e., F_1 (T1leak) and F_1F_3 (T1leak & v3s_o), cannot be made differentiable with any test plan.

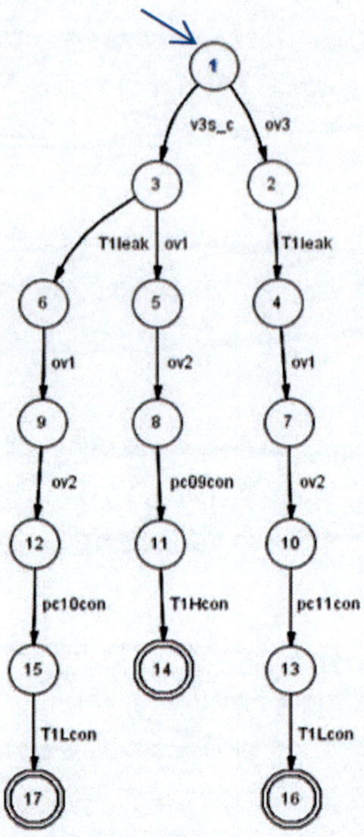

Fig. 8.17 Diagnostic supervisor for Tr_{01} in liquid transfer system (Reprinted with permission from Kang and Chang 2014. Copyright 2014 Elsevier)

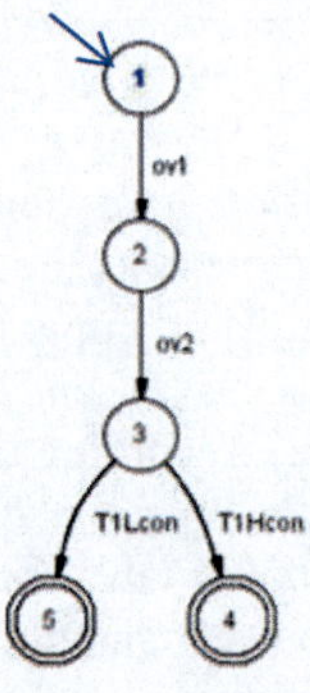

Fig. 8.18 Smallest diagnostic supervisor for Tr_{01} in liquid transfer system (Reprinted with permission from Kang and Chang 2014. Copyright 2014 Elsevier)

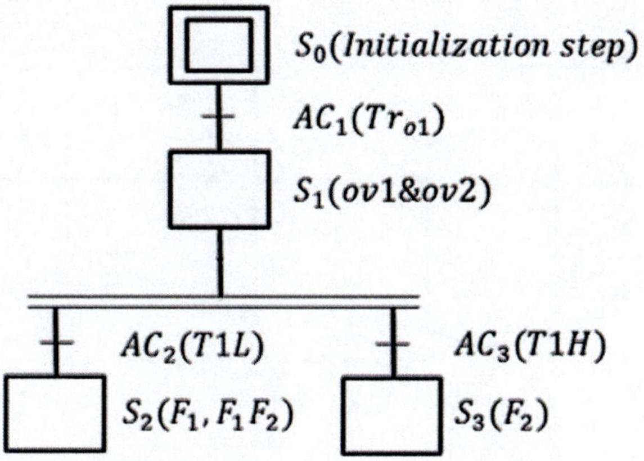

Fig. 8.19 Test plan for Tr_{01} in liquid transfer system (Reprinted with permission from Kang and Chang 2014. Copyright 2014 Elsevier)

8.5 Case Studies

To demonstrate the effectiveness of the proposed approach, the application results of two case studies are presented below:

8.5.1 A Three-Tank Buffer System

Let us consider the P&ID in Fig. 8.20 and its normal operating procedure (see Fig. 8.21). Notice that V-2 is a 3-way valve and the others are 2-way valves. The fluid in tank T-1 is directed to tank T-2 if V-2 is placed at the "+" position, while transported to T-3 if switched to the "-" position. All three tanks are equipped with level sensors. The one on T-1 is designed to detect three distinct states reflecting the low, intermediate, and high liquid levels, i.e., T1Lcon, T1Mcon, and T1Hcon, respectively, while that on T-2 (or T-3) is used to monitor only the states at low and high levels, i.e., T2Lcon (or T3Lcon) and T2Hcon (or T3Hcon). It is assumed that, initially, the liquid levels in all tanks are low, valves V-1 and V-3 are closed and V-2 is at the "-" position. In this example, let us consider the following seven failures:

i. F_1 (v1s_c): V-1 fails at the closed position;
ii. F_2 (v1s_o): V-1 fails at the open position;
iii. F_3 (v2M−): V-2 is mistakenly switched to the "-" position;
iv. F_4 (v2M+): V-2 is mistakenly switched to the " + " position;
v. F_5 (v3s_c): V-3 fails at the closed position;
vi. F_6 (v3s_o): V-3 fails at the open position;
vii. F_7 (T2leak): a leak develops in tank T-2.

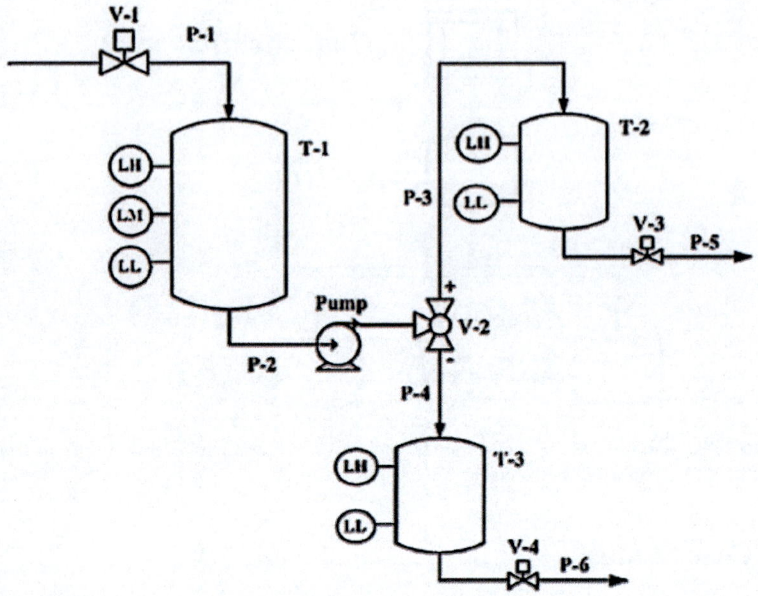

Fig. 8.20 P&ID of three-tank buffer system (Reprinted with permission from Kang and Chang 2014. Copyright 2014 Elsevier)

The aforementioned model construction procedure has been followed to build the diagnoser, in which a total of 12 observable event traces can be identified and, for the sake of brevity, only the undiagnosable ones are given in Fig. 8.22. Notice that the transition label "i cycles" denotes the event sequence in i complete normal cycles and $i = 0, 1, 2, \cdots$. The proposed synthesis has been applied to the three traces in Fig. 8.22 and the resulting test plans are summarized below:

- **Trace 7** (Tr_{07})**:** No effective test plan can be found in this case. This is due to the fact that F_1 (v1s_c) is certain to occur. After observing the readings T1Lcon, T2Lcon, and T3Lcon, there are really no ways to secure more fluid (by opening V-1) so as to render a change in the liquid level in T-2 for testing if failure F_6 (v3s_o) exists.
- **Trace 8** (Tr_{08})**:** After observing this trace in full, it is certain that failure v2M−(F_3) occurred at the time when the actions in step S_2 (see Fig. 8.21) were being executed during the current cycle. On the other hand, v3s$_o$ (F_6) can be neither confirmed nor rejected and, if present, the failure should develop after step S_4 (in the previous cycle) and before the most recent S_1. The diagnostic test at this point (see Fig. 8.23) calls for three consecutive actions, i.e., (1) switching off the pump (poff), (2) switching V-2 to the "+" position (tv2+), and (3) switching on the pump (pon). If the sensor readings are T1Lcon, T2Hcon, and T3Hcon, then V-3 should still be normal. If the sensor readings are T1Lcon, T2Lcon, and

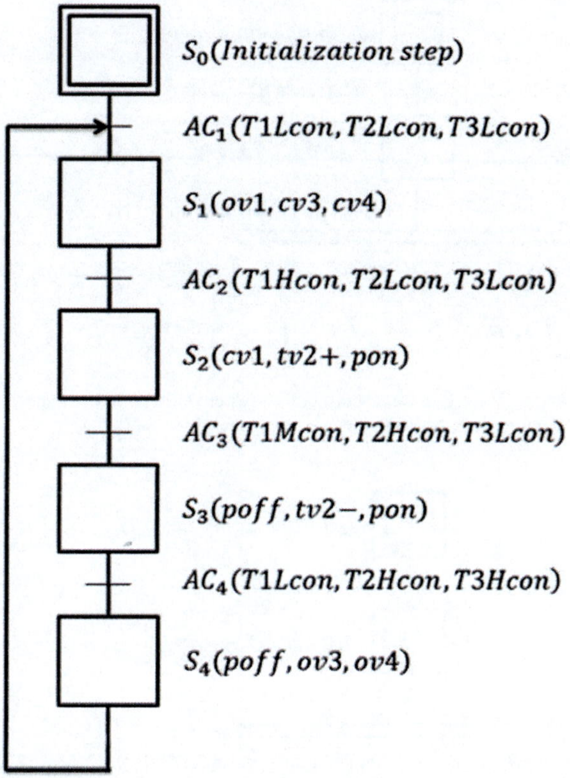

Fig. 8.21 Normal operating procedure of three-tank buffer system (Reprinted with permission from Kang and Chang 2014. Copyright 2014 Elsevier)

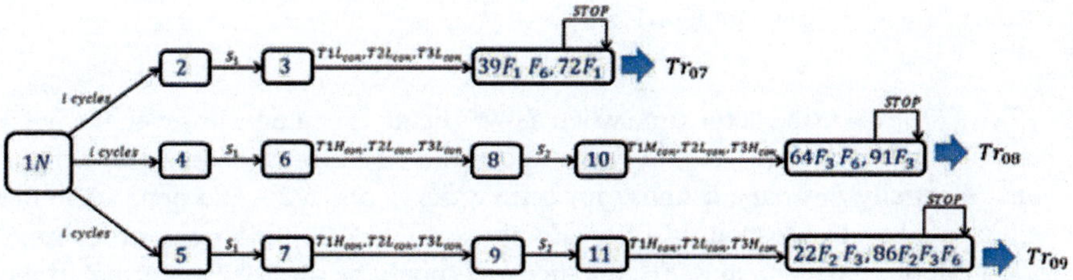

Fig. 8.22 Observable undiagnosable traces in the diagnoser of three-tank buffer system (Reprinted with permission from Kang and Chang 2014. Copyright 2014 Elsevier)

T3Hcon, then V-3 must have already failed at the open position, i.e., v3s_o (F_6) is an existing failure.

- **Trace 9** (Tr_{09})**:** Two events in the current cycle can be confirmed with this observable trace, i.e., (1) v1s_o (F_2) occurred at a time after S_1 and before S_2, and (2)

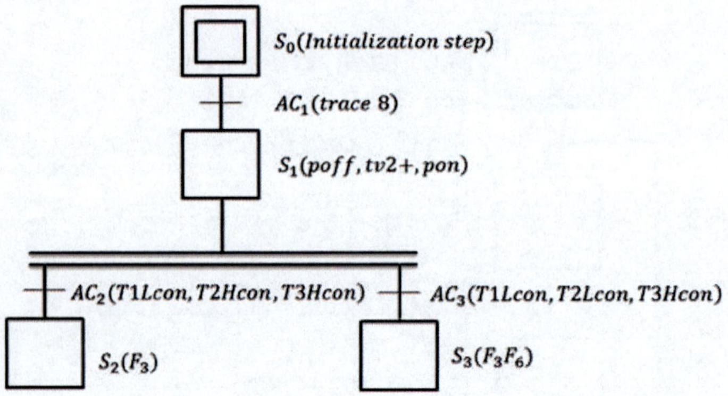

Fig. 8.23 Test plan for Trace 8 in three-tank buffer system (Reprinted with permission from Kang and Chang 2014. Copyright 2014 Elsevier)

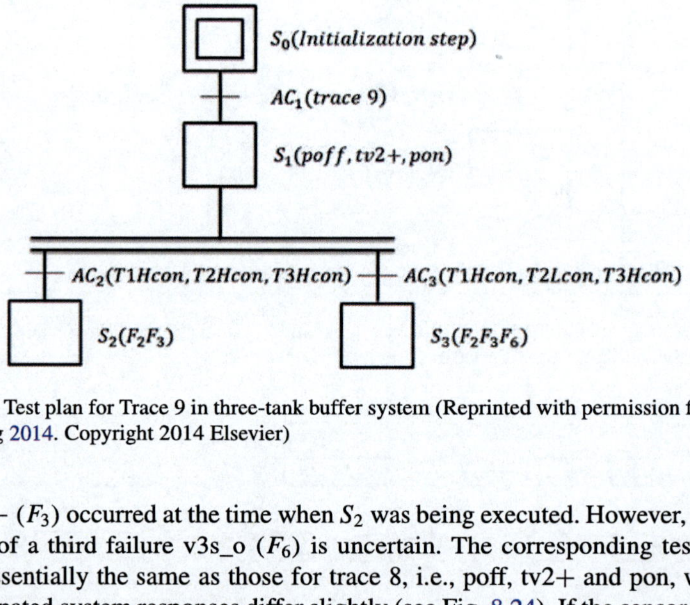

Fig. 8.24 Test plan for Trace 9 in three-tank buffer system (Reprinted with permission from Kang and Chang 2014. Copyright 2014 Elsevier)

v2M− (F_3) occurred at the time when S_2 was being executed. However, the presence of a third failure v3s_o (F_6) is uncertain. The corresponding test actions are essentially the same as those for trace 8, i.e., poff, tv2+ and pon, while the anticipated system responses differ slightly (see Fig. 8.24). If the sensor readings are T1Hcon, T2Hcon, and T3Hcon, then V-3 should be regarded as normal. If the sensor readings are T1Hcon, T2Lcon, and T3Hcon,, then V-3 must have already failed at the open position, i.e., v3s_o (F_6), in the given system.

8.5.2 A Beer Filtration Plant

The PFD of a beer filtration plant is shown in Fig. 8.25 (Lai et al., 2007; Chung

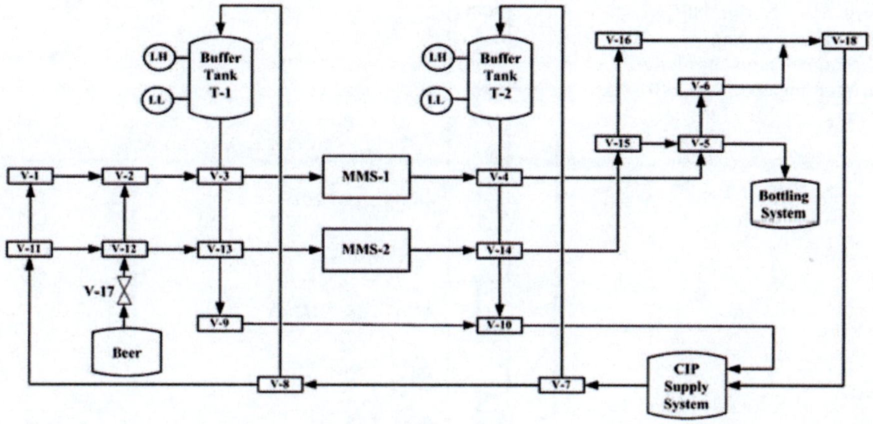

Fig. 8.25 Process flow diagram of beer filtration plant (Reprinted with permission from Kang and Chang 2014. Copyright 2014 Elsevier)

and Lai, 2008). This system consists of two multi-micro system filters (MMS-1 and MMS-2), two buffer tanks (T-1 and T-2), a supply and collection system for the cleanser (CIP), 17 double-disk piston valves (V-1 to V-16 and V-18) and a gate valve (V-17). Notice that each of the former valves can be switched to either ON or OFF position. When a valve is ON, the fluids entering the valve from vertical and horizontal pipelines are supposed to be mixed and then flow out from all outlet pipelines, whereas the fluids in vertical and horizontal pipelines flow separately when this valve is at the OFF position. There are four basic tasks to be performed in this plant, i.e., filling, filtration, bottling, and cleaning. The purpose of filling is to transport fresh beer from a source tank to the buffer tank T-1 by opening either (1) V-2 and V-3 or (2) V-12 and V-13. In the filtration operation, beer is transferred from tank T-1 to T-2 via filter MMS-1 or MMS-2. Valves V-3 and V-4 should be both switched to the ON positions in the former case, while V-13 and V-14 must be ON in the latter. Clearly, the filtered beer in T-2 should be moved to the bottling station either by opening V-4 and V-5 or by opening V-14 and V-15. Finally, the tasks of cleaning processing units can also be considered as four different material-transfer operations and they are listed below:

- Switch on V-8 and V-9 to clean T-1;
- Switch on V-7 and V-10 to clean T-2;
- Switch on V-1, V-6, and V-18 to clean MMS-1;
- Switch on V-11, V-16, and V-18 to clean MMS-2.

The normal operation steps and their activation conditions can be found in Fig. 8.26. Notice that, to enhance production efficiency, it is considered a good practice to clean equipment concurrently with one or more beer processing step. The initial beer level in each buffer tank is low and all seventeen double-piston disk valves

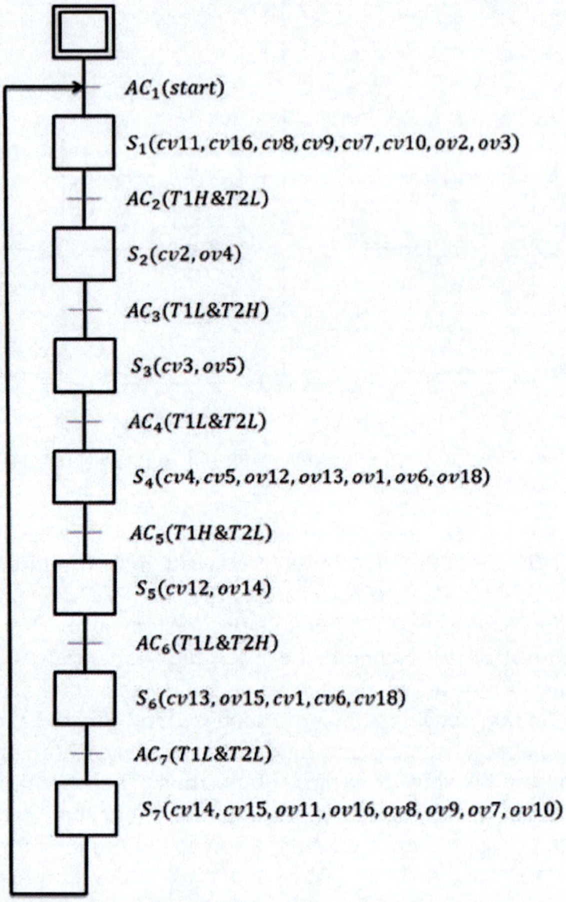

Fig. 8.26 Normal SFC of beer filtration plant (Reprinted with permission from Kang and Chang 2014. Copyright 2014 Elsevier)

(V-1 - V-16, V-18) are at the OFF positions when a cycle starts, while the gate valve (V-17) is always kept open during normal operation.

In this example, a total of five independent failures are considered:

i. F_1 (v2s_c): V-2 fails at the OFF (or CLOSE) position;
ii. F_2 (v2s_o): V-2 fails at the ON (or OPEN) position;
iii. F_3 (v6s_c): V-6 fails at the OFF (or CLOSE) position;
iv. F_4 (v6s_o): V-6 fails at the ON (or OPEN) position;
v. F_5 (T1leak): a leak develops in T-1.

The same model construction procedure has been followed to synthesize the diagnoser and its four undiagnosable traces are given in Fig. 8.27. Note that the transition label "*n* cycles" denotes the event sequence in more than one complete normal cycle, i.e., $n = 1, 2, 3, \cdots$. Notice that Trace 1 and Trace 2 can only be observed in the first cycle, while Trace 3 and Trace 4 should end in any of the later cycles. Note also that

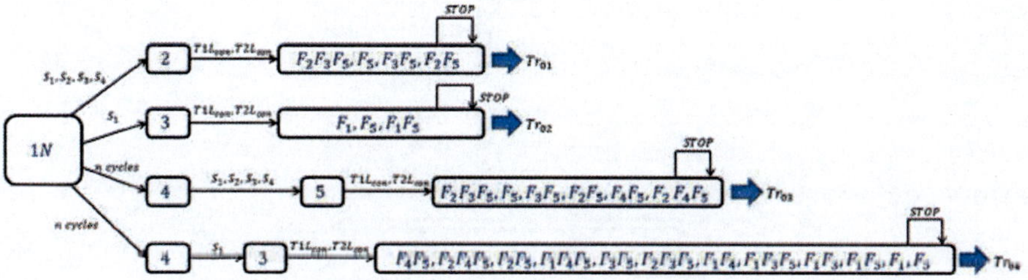

Fig. 8.27 Observable undiagnosable traces in the diagnoser of beer filtration plant (Reprinted with permission from Kang and Chang 2014. Copyright 2014 Elsevier)

Trace 1 is basically a substring of Trace 3 and, thus, the fault origins in the former case are also included in the latter. Finally, notice that the same conclusions can also be drawn from Trace 2 and Trace 4.

Since the given system is assembled with a large number of components (18 valves and 2 tanks), it is very difficult to evaluate all possible process configurations for the purpose of synthesizing a suitable emergency response procedure. To limit the search space and to avoid wasting beer, it is assumed that all diagnostic tests can be facilitated solely with a cleanser and, thus, V-17 should be closed in every plan. Notice also that the test effects must be observed with the level sensors on T-1 and T-2. Although only two level readings on each tank are needed for implementing the normal operating procedure (see Fig. 8.26), a third is incorporated in the component models for T-1 and T-2 respectively, i.e., T1Mcon is added to characterize the abnormal condition that an intermediate level in T-1 has continued for some time, while T2Mcon is added for the corresponding scenario in T-2.

The proposed test-plan synthesis method has been applied to all four traces in diagnoser and the resulting SFCs can be found in Figs. 8.28, 8.29, 8.30 and 8.31, respectively. For the sake of brevity, only the first and third are explained below in detail:

- **SFC for the test plan of Trace 1** (Fig. 8.28):

(A) Although the presence of T1leak (F_5) and the absence of v2s_c (F_1) and v6s_o (F_4) are verified by observing Trace 1 (AC_1), the remaining failures, i.e., v2s_o (F_2) and/or v6s_c (F_3), can be neither confirmed nor rejected. Knowing that only V-1, V-6, V-12, V-13, V-17, and V-18 are ON or open at this point in normal operation, one can apply S_1 (cv17 and ov4) to fill T-2 with cleanser via V-1 and V-4. It should also be noted that, without failure v6s_c, this flow is split into two at V-4 and one of them returns to the collection system via V-6 and V-18.

(B) If the level sensor on T-2 detects T2Mcon (AC_2) after completing S_1, then v6s_c (F_3) can be ruled out but the status of v2s_o (F_2) is still uncertain. The subsequent test step S_2 calls for cv1 and then ov10 to disconnect the inlet flow and also allow the cleanser in T-2 to be drained into the collection system via V-4, V-14 and V-10. As soon as T-2 is emptied or T2Lcon (AC_4) can be

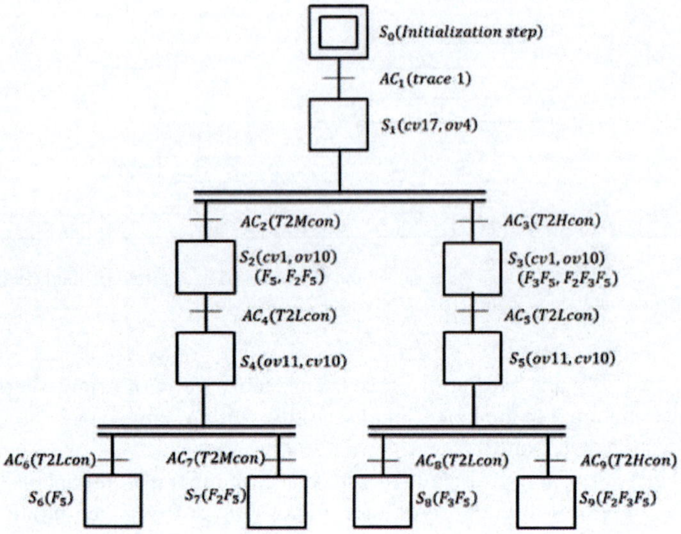

Fig. 8.28 Test plan for Trace 1 in beer filtration plant (Reprinted with permission from Kang and Chang 2014. Copyright 2014 Elsevier)

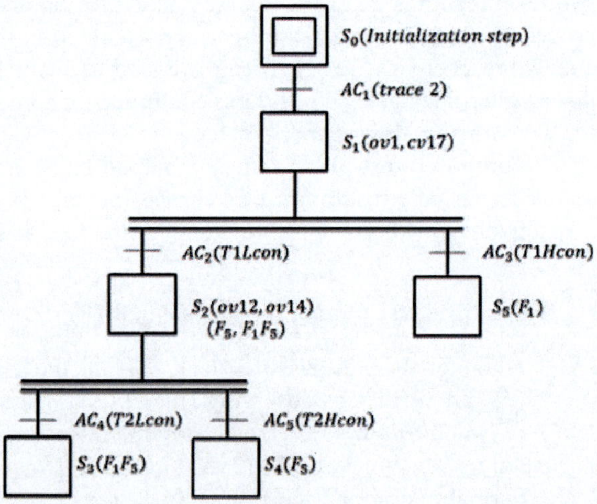

Fig. 8.29 Test plan for Trace 2 in beer filtration plant (Reprinted with permission from Kang and Chang 2014. Copyright 2014 Elsevier)

observed, the next step S_4 (cv10 and ov11) should be performed to fill T-2 via V-11, V-12, V-2, and V-4.

(C) If the condition T2Hcon (AC_3) is revealed with the level sensor on T-2 after executing all actions in S_1, then the presence of v6s_c (F_3) can be verified but the status of v2s_o (F_2) is still uncertain. Note that the subsequent event sequence for confirming/rejecting failure F_2 (see the events in S_3, AC_5, and

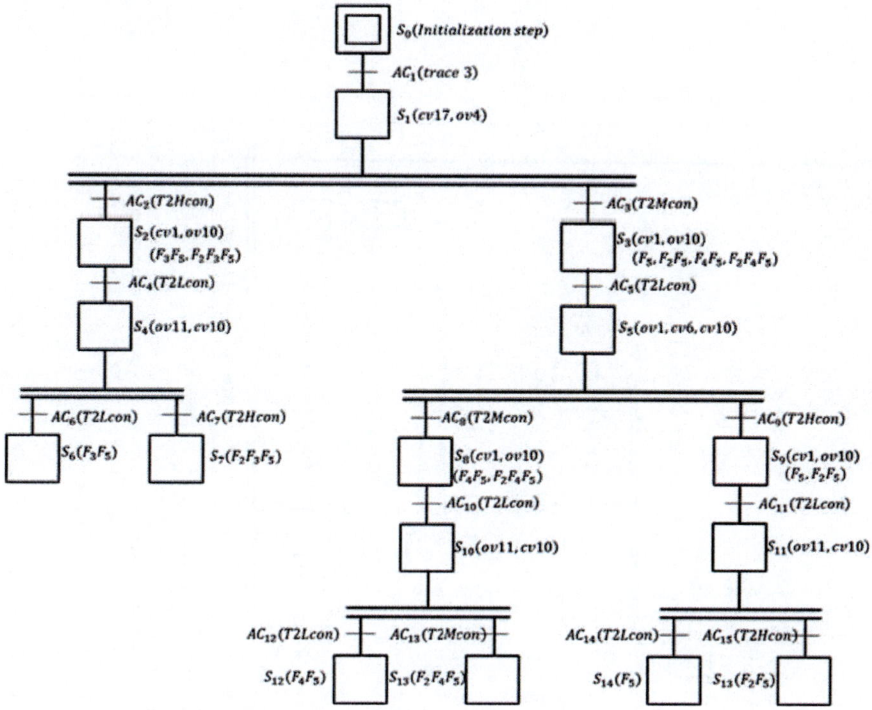

Fig. 8.30 Test plan for Trace 3 in beer filtration plant (Reprinted with permission from Kang and Chang 2014. Copyright 2014 Elsevier)

S_5) is essentially the same as that described in (B), i.e., cv1, ov10, T2Lcon, cv10, and ov11 in S_2, AC_4, and S_4.

(D) There should be two alternative outcomes after implementing S_4. If the sensor reading on T-2 is T2Lcon (AC_6), then v2s_o (F_2) can be disregarded and the corresponding fault origin is simply T1leak (F_5). If T2Mcon (AC_7) is observed, then F_2 can be confirmed and the fault origin is F_2F_5 (v2s_o & T1leak).

(E) The two scenarios resulting from S_5 are similar to those described in (D). If the sensor reading on T-2 is T2Lcon (AC_8), then v2s_o (F_2) can be ruled out and the corresponding fault origin is F_3F_5. If the reading shows T2Hcon (AC_9), then F_2 can be confirmed and the corresponding fault origin should consist of three coexisting failures, i.e., $F_2F_3F_5$ (v2s_o & $v6s\_c$&T1leak).

- **SFC for the test plan of Trace 3** (Fig. 8.30):

(A) Although the presence of T1leak (F_5) and the absence of v2s_c (F_1) are confirmed by observing trace 3 (AC_1), one still cannot be certain (a) whether V-2 is normal or fails at the ON position, i.e., v2s_o (F_2), and (b) whether V-6 is normal or fails at the ON or OFF positions, i.e., $v6s\_c$ (F_3) or $v6s\_o$ (F_4).

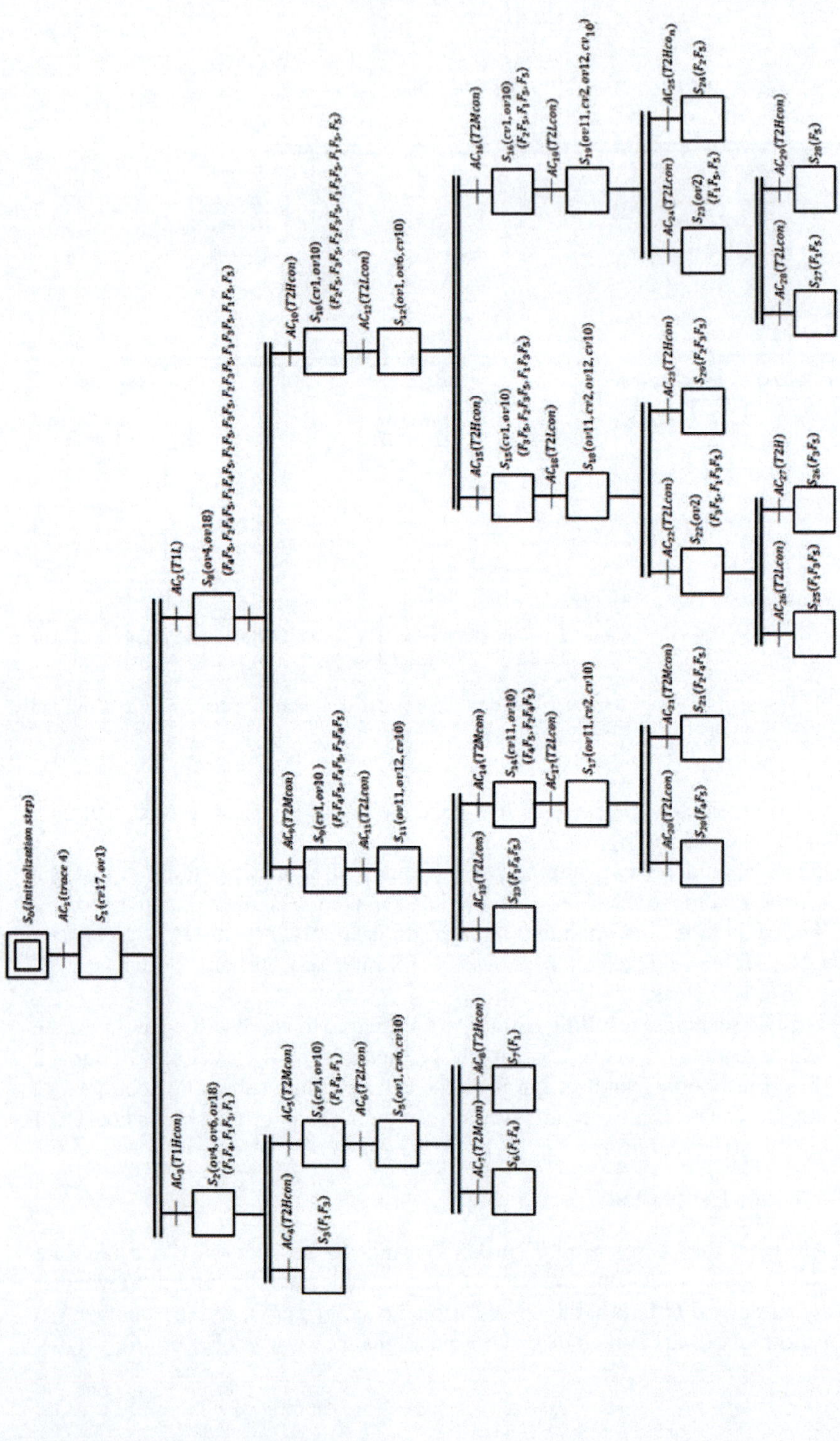

Fig. 8.31 Test plan for Trace 4 in beer filtration plant (Reprinted with permission from Kang and Chang 2014. Copyright 2014 Elsevier)

Knowing that only V-1, V-6, V-12, V-13, V-17, and V-18 are ON or open at this juncture in normal operation, one can try to implement S_1 (cv17 & ov4) to fill T-2 with cleanser via V-1 and V-4. It should also be noted that, without failure $v6s\_c$, this flow is split into two at V-4 and one of them should return to the collection system via V-6 and V-18.

(B) If the sensor reading on T-2 indicates T2Hcon (AC_2) after completing S_1, then $v6s\_c$ (F_3) can be confirmed but it is still uncertain if v2s_o (F_2) is present. The following step S_2 calls for cv1 and then ov10 to disconnect the inlet flow of T-2 and also allow the cleanser to be drained into the collection system via V-4, V-14, and V-10. After T-2 is emptied or T2Lcon (AC_4) is observed, S_4 (cv10 & ov11) should be performed to fill T-2 via V-11, V-12, V-2, and V-4.

(C) If the level sensor on T-2 shows T2Mcon (AC_3) after executing S_1, then $v6s\_c$ (F_3) can be rejected. However, one still cannot determine (a) whether V-2 is normal or fails at the ON position, i.e., v2s_o (F_2), and (b) whether V-6 is normal or fails at the ON position, i.e., $v6s\_o$ (F_4). The next step S_3 is essentially the same as S_2, i.e., cv1 & ov10, which is adopted primarily for the purpose of draining T-2. As soon as the subsequent condition T2Lcon (AC_4) is detected with the level sensor on T-2, step S_5 (ov1 & cv6 & cv10) should be performed to fill T-2 via V-1 and V-4. Note also that, if the failure $v6s\_o$ (F_4) is present, an additional flow may be branched out at V-4 to the collection system via V-6 and V-18.

(D) Only two alternative outcomes can be expected after implementing S_4. If the sensor reading on T-2 is T2Lcon (AC_6), then v2s_o (F_2) can be disregarded and the corresponding fault origin is $F_3 F_5$ (v6s_c & T1leak). On the other hand, if T2Hcon (AC_7) is observed, then F_2 can be confirmed and the fault origin is $F_2 F_3 F_5$ (v2s_o & $v6s\_c$ & T1leak).

(E) If the level sensor on T-2 detects T2Mcon (AC_8) after completing S_5, then $v6s\_o$ (F_4) can be confirmed but it is still uncertain if v2s_o (F_2) is present. The next test step S_8 again calls for cv1 and then ov10 to empty T-2 and transfer its content to the collection system via V-4, V-14, and V-10. After observing the subsequent condition T2Lcon (AC_{10}), S_{10} (cv10 & ov11) should be performed to fill T-2 via V-11, V-12, V-2, and V-4.

(F) If condition T2Hcon (AC_9) can be observed after executing S_5, then the presence of v6s_o (F_4) can be rejected but the status of v2s_o (F_2) is still uncertain. The required event sequence for confirming/rejecting F_2 (see S_9, AC_{11}, and S_{11}) is essentially the same as that described in (E), i.e., cv1, ov10, T2Lcon, cv10, and ov11 in S_8, AC_{10}, and S_{10}.

(G) There should be two possible scenarios after implementing S_{10}. If the sensor reading on T-2 is T2Lcon (AC_{12}), then v2s_o (F_2) can be rejected and the corresponding fault origin is F_4F_5 (v6s_o & T1leak). However, if T2Mcon (AC_{13}) is observed, then F_2 should be included and the corresponding fault origin is only $F_2F_4F_5$ (v2s$_o$ & v6s_o & T1leak).

(H) Only two possible outcomes can be produced by implementing S_{11}. If the sensor reading is T2Lcon (AC_{14}), then v2s_o (F_2) should be rejected and the corresponding fault origin is simply F_5 (T1leak). On the other hand, if T2Hcon (AC_{15}) is observed, then F_2 should be included and the corresponding fault origin is F_2F_5 (v2s_o & T1leak).

8.6 Concluding Remarks

A standardized methodology is presented in this chapter to systematically construct untimed automata for modeling sequential material- and/or-energy-transfer operations in a chemical plant, and to produce the corresponding diagnoser accordingly. A generic synthesis procedure has also been developed for creating the test plans of all undiagnosable traces in a diagnoser. It should be noted that, although the feasibility of the proposed approach has been successfully verified with several examples, additional model features may be incorporated to improve its effectiveness in a more realistic environment. In particular, the timed automata may be utilized to model operation times of the test steps so as to further enhance diagnostic resolution by making use of the online clocks.

References

Åkesson K, Fabian M, Malik R (2006) SUPREMICA—an integrated environment for verification, synthesis and simulation of discrete event systems. In: IEEE Proceedings of the 8th international workshop on discrete event systems, pp 384–385

Baroni P, Lamperti G, Pogliano P, Zanella M (1999) Diagnosis of large active systems. Artif Intell 110(1):135–183

Baroni P, Lamperti G, Pogliano P, Zanella M (2000) Diagnosis of a class of distributed discrete-event systems. IEEE Transactions on Systems Man and Cybernetics Part A-Systems and Humans 30(6):731–752

Benveniste A, Fabre E, Haar S, Jard C (2003) Diagnosis of asynchronous discrete-event systems: A net unfolding approach. IEEE Trans Autom Control 48(5):714–727

Cassandras CG, Lafortune S (2008) Introduction to Discrete Event Systems, 2nd edn. Springer Science + Business Media, LLC, New York, NY, USA

Chen YC, Yeh ML, Hong CL, Chang CT (2010) Petri-net based approach to configure online fault diagnosis systems for batch processes. Ind Eng Chem Res 49(9):4249–4268

Chung SL, Lai YH (2008) Process control of brewery plants. Journal of the Chinese Institute of Engineers 31(1):127–140

Debouk R, Lafortune S, Teneketzis D (2000) Coordinated decentralized protocols for failure diagnosis of discrete event systems. Discrete Event Dynamic System Theory and Applications 10(1–2):33–86

Kang A, Chang CT (2014) Automata generated test plans for fault diagnosis in sequential material- and energy-transfer operations. Chem Eng Sci 113:101–115

Kourti T, Macgregor JF (1995) Process analysis, monitoring and diagnosis, using multivariate projection methods. Chemometrics and Intelligent Laboratory Systems 28(1):3–21

Kourti T, Nomikos P, Macgregor JF (1995) Analysis monitoring and fault-diagnosis of batch processes using multiblock and multiway PLS. J Process Control 5(4):277–284

Lai JW, Chang CT, Hwang SH (2007) Petri-net based binary integer programs for automatic synthesis of batch operating procedures. Ind Eng Chem Res 46(9):2797–2813

Lee JM, Yoo CK, Lee IB (2004) Fault detection of batch processes using multiway kernel principal component analysis. Comput Chem Eng 28(9):1837–1847

Nomikos P, MacGregor JF (1994) Monitoring batch processes using multiway principal component analysis. AIChE J 40(8):1361–1375

Nomikos P, MacGregor JF (1995) Multivariate SPC charts for monitoring batch processes. Technometrics 37(1):41–59

Pierri F, Paviglianiti G, Caccavale F, Mattei M (2008) Observer-based sensor fault detection and isolation for chemical batch reactors. Eng Appl Artif Intell 21(8):1204–1206

Qiu WB, Kumar R (2006) Decentralized failure diagnosis of discreteevent system. IEEE Transactions on Systems Man and CyberneticsPart A-Systems and Humans 36(3):384–395

Ramadge PJ, Wonham WM (1987) Supervisory control of a class of discrete event processes. SIAM Journal on Control and Optimization 25:206–230

Ramadge PJ, Wonham WM (1989) The control of discrete event systems. Proc IEEE 77:81–98

Ruiz D, Nougues JM, Calderon Z, Espuna A, Puigjaner L (2001) Neural network based framework for fault diagnosis in batch chemical plants. Comput Chem Eng 24(2–7):777–784

Ruiz D, Canton J, Nougues JM, Espuna A, Puigjaner L (2001) On-line fault diagnosis system support for reactive scheduling in multipurpose batch chemical plants. Comput Chem Eng 25(4–6):829–837

Sampath M, Sengupta R, Lafortune S, Sinnamohideen K, Teneketzis D (1995) Diagnosability of discrete-event systems. IEEE Trans Autom Control 40(9):1555–1575

Sampath M, Sengupta R, Lafortune S, Sinnamohideen K, Teneketzis DC (1996) Failure diagnosis using discrete-event models. IEEE Trans Control Syst Technol 4(2):105–124

Sampath M, Lafortune S, Teneketzis D (1998) Active diagnosis of discrete-event systems. IEEE Trans Autom Control 43(7):908–929

Undey C, Ertunc S, Cinar A (2003) Online batch fed-batch process performance monitoring, quality prediction, and variable contribution analysis for diagnosis. Ind Eng Chem Res 42(20):4645–4658

Venkatasubramanian V, Rengaswamy R, Kavuri SN (2003) A review of process fault detection and diagnosis, part ii: Qualitative model and search strategies. Comput Chem Eng 27(3):313–326

Venkatasubramanian V, Rengaswamy R, Yin K, Kavuri SN (2003) A review of process fault detection and diagnosis, part i: Quantitative model based methods. Comput Chem Eng 27(3):293–311

Venkatasubranmanian V, Rengaswamy R, Kavuri SN, Yin K (2003) A review of process fault detection and diagnosis, part iii: Process history based methods. Comput Chem Eng 27(3):327–346

Yeh ML, Chang CT (2011) An automaton-based approach to evaluate and improve online diagnosis schemes for multi-failure scenarios in batch chemical processes. Chem Eng Res Des 89:2652–2666

Yeh ML, Chang CT (2012) An automata based method for online synthesis of emergency response procedures in batch processes. Comput Chem Eng 38:151–170

Yeh ML, Chang CT (2012) An automata-based approach to synthesis untimed operating procedures in batch chemical processes. Korean Journal of Chemical Engineers 29(5):583–594

Zad SH, Kwong RH, Wonham WM (2003) Fault diagnosis in discrete-event systems: Framework and model reduction. IEEE Trans Autom Control 48(7):1199–1204

Chapter 9
Synthesis of Diagnostic Tests Based on Timed Automata

A large number of high value-added chemical products, such as the specialty chemicals, foods, semiconductors, and pharmaceuticals, etc., are often manufactured in complex but flexible batch or sequential processes. Hardware failures are random but inevitable events over the lifespan of any such plant. If the root causes of a failure-induced event sequence cannot be correctly identified in time, the final consequences may be catastrophic. Generally speaking, the overall performance of a diagnostic system can be improved by capturing more online data. To this end, the obvious design strategy is to install additional sensors. However, since new hardware inevitably requires extra spending and, also, the related issues have already been discussed extensively in the literature, there are incentives to develop an alternative means for enhancing diagnostic resolution without capital investment. Yeh and Chang (2011) proposed to implement online test procedures for such a purpose, while Kang and Chang (2014) later developed an effective procedure-synthesis method to conjecture the diagnostic tests according to untimed automata.

Several studies have already been performed to address various issues concerning fault diagnosis in batch and sequential processes. Nomikos and MacGregor (1994, 1995) utilized the multi-way principal component analysis for batch process monitoring, which has later been extended for the purpose of online diagnosis (Kourti and Macgregor 1995; Kourti et al. 1995; Undey et al. 2003; Lee et al. 2004). Other fault identification tools, such as the artificial immune systems, artificial neural networks, and knowledge-based expert systems (Dai and Zhao 2011; Ghosh and Srinivasan 2011; Tan et al. 2012; Zhao 2014), have also been suggested for diagnosing the batch plants.

In order to expand the scope of diagnosis in realistic applications, Chen et al. (2010) developed several Petri net-based algorithms to configure fault identification systems for moderately complex plants. Since the event sequences (or traces) in multi-failure scenarios cannot be conveniently generated with the Petri-net models, their approach was limited to the single-failure incidents. On the other hand, this shortcoming can in general be avoided with the untimed automata (Sampath et al. 1995, 1996, 1998; Baroni et al. 1999, 2000; Debouk et al. 2000; Benveniste et al.

C.-T. Chang et al., *Process Plant Operating Procedures*, Advances in Industrial Control,
https://doi.org/10.1007/978-3-030-70978-5_9

2003; Zad et al. 2003; Qiu and Kumar 2006; Yeh and Chang 2011). A so-called "diagnoser" can be constructed accordingly to predict all observable fault propagation event sequences (or "traces") and to pinpoint the corresponding fault origins. In a later study, Gascard and Simeu-Abazi (2013) improved this approach by using the timed automata to build diagnosers for the dynamic discrete-event systems.

Since the root cause(s) of a trace in the diagnoser may or may not be unique, it is desirable to further enhance the diagnostic resolution with nonconventional means. As mentioned previously, Kang and Chang (2014) have developed a systematic method to generate the test plans for upgrading a given diagnoser without capital investment (see Chap. 8). However, due to the lack of time-tracking mechanisms in their models, the failure-induced behaviors cannot always be characterized adequately. To overcome this drawback, it is obviously reasonable to make use of the timed automata for the purpose of generating more comprehensive test plans. Notice that such models have already been utilized to address other closely-related issues. For examples, they were used to verify if any given procedure conforms to the design specifications (Lohmann et al. 2006; Kim and Moon 2009, 2011; Lahtinen et al. 2012), and Li et al. (2014) also proposed a systematic approach to synthesize controller actions for periodic operations.

To facilitate clear illustration of the proposed approach, the general procedure for test-plan synthesis is first summarized below.

1. All embedded components in the given process are first modeled with the timed automata.
2. All possible fault propagation scenarios and their observable event traces (OETs) are next enumerated exhaustively.
3. The optimal test plan for every OET is then established by generating the supervisory controller to achieve a higher degree of diagnostic resolution.

The resulting test plans can then be implemented online after observing any of the OETs online during actual operation.

9.1 Model-Building Approach

The basic approach described in the previous chapter is also adopted here to build the time automata. For clarity, this model construction method is illustrated here with a simple example. Specifically, let us revisit the fictitious liquid transfer system in Fig. 8.1 but replace the sequential function chart (SFC) in Fig. 8.2 with the one in Fig. 9.1. For illustration convenience, let us omit the process configuration level and classify the components here into a simplified hierarchy of 4 different levels: (1) the programmable logic controller (PLC); (2) the actuators, i.e., the three-way valves (V-1 and V-3) and the two-way valves (V-2 and V-4); (3) the processing units, i.e., the buffer tank, and (4) the online sensor(s). If a three-way valve is closed, the port connecting to the horizontal pipeline in Fig. 8.1, i.e., pipe P-2 in the case of V-1 or pipe P-3 in the case of V-3, is assumed to be blocked. Otherwise, its inlet flow(s)

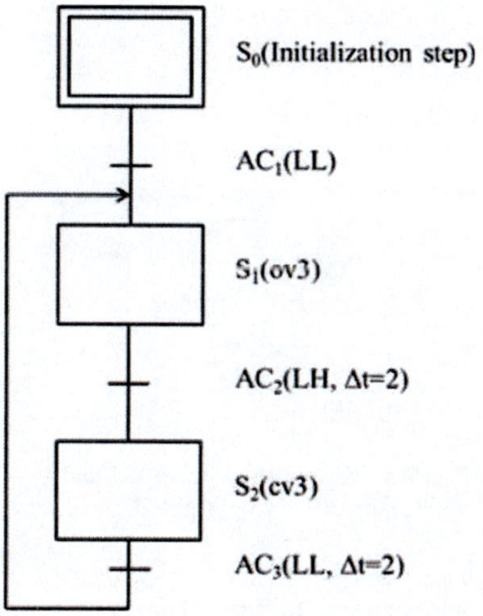

Fig. 9.1 Normal SFC of liquid transfer operation in Example 9.1 (Reprinted with permission from Hsieh and Chang 2016. Copyright 2016 Elsevier)

should be directed to every outlet pipeline. It is assumed that all valves except V-4 are closed initially. For illustration clarity, the above system is refferred to as Example 9.1 in this chapter. Thus, it is clear from the SFC in Fig. 9.1 that, during the normal operation, the buffer tank is filled in two time units ($\Delta t = 2$) with liquid via pipelines P-1, P-3 and P-4 after opening V-3 and then drained via P-5 by gravity in another two time units ($\Delta t = 2$). For the sake of brevity, only 4 fault origins are considered in this example and they are denoted, respectively, by

- f_{1A} (or *Tank_LEAK*), i.e., a large leak develops in tank,
- f_{1B} (or *Tank_leak*), i.e., a minor leak develops in tank,
- f_2 (or *V3SC*), i.e., V-3 fails at the "close" position, and
- f_3 (or *V3SO*), i.e., V-3 fails at the "open" position.

As mentioned before, the plant model can in general be obtained by first building automata to model all components in the given process and then integrating them via the parallel decomposition operation (Cassandras and Lafortune 2008). The free software UPPAAL (Behrmann et al. 2006) can be used for model building and verification. Note also that the controller and the remaining components are characterized differently. Let us first outline the construction principles for the latter components, i.e., the valves, the tank, and the sensors. Specifically, a timed automaton should be used to represent a finite set of all identifiable normal and abnormal states of the hardware item under consideration and also the specific events facilitating the state transitions. The prerequisite conditions of an event can be imposed with the so-called "guard" of the corresponding transition in UPPAAL, while the updated integer and clock variables after transition may also be specified as attributes (Behrmann et al. 2006). Finally, it is assumed that the failures and the resulting abnormal states included in these component models can be identified in advance with a hazard

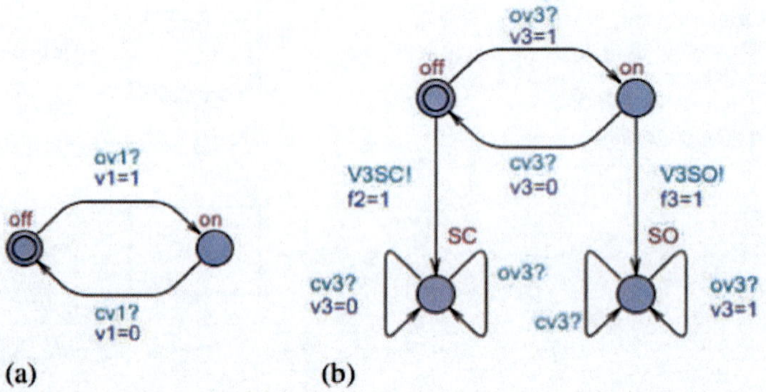

Fig. 9.2 Valve models for **a** V-1 and **b** V-3 in Example 9.1 (Reprinted with permission from Hsieh and Chang 2016. Copyright 2016 Elsevier)

assessment method. The corresponding models in the present example are briefly described as follows:

- **Valve models**: Since only the failures of V-3, i.e., *V3SC!* and *V3SO!*, are considered in the present example, V-1, V-2 and V-4 can be modeled with automata of the same structure and, therefore, only the component models of V-1 and V-3 are presented in Fig. 9.2a and b, respectively. The place *off* in the former model is used to represent the closed position of V-1, while *on* the opposite. The "receiver" events are labeled with question marks, i.e., *ov1?* and *cv1?*, indicating that these events must occur in other components at some prior instances. The former triggers the *off*-to-*on* state transition and then resets the binary variable *v1* to 1, while the latter activates the *on*-to-*off* transition and resets *v1* to 0. The normal behavior of V-3 described in Fig. 9.2b is essentially the same as that in Fig. 9.2a, while the sender events *V3SO!* and *V3SC!* activate the transitions to the failed states *SO* and *SC*, respectively. Note that the exclamation mark (*!*) is used here to specify an initiator or "sender" event that takes place in the present component as long as all prerequisite conditions can be satisfied. Notice also that a system deadlock, i.e., a state that no further event can be executed, is usually formed after reaching any of the failed states.
- **Tank model**: Only the elapsed times of state transitions in the tank model are assumed to be nonzero and the corresponding automaton can be found in Fig. 9.3. Note that the places *LL* and *LH* are used to, respectively, represent the low and high liquid levels under normal conditions, while *LL_leak* and *LH_leak* denote the corresponding liquid levels after a minor leak develops and *LL_LEAK* is the low level eventually reached after a large leak. For the sake of illustration brevity, let us consider only the attributes associated with the *LL*-to-*LH* transition. Three prerequisite conditions (guards), characterized with four binary variables (*v1*–*v4*) and a clock variable x, must be satisfied before triggering this transition. The binary variables *v1*–*v4* are used to represent the corresponding valve states, i.e., 1 denotes open and 0 otherwise, and the clock variable x records time needed

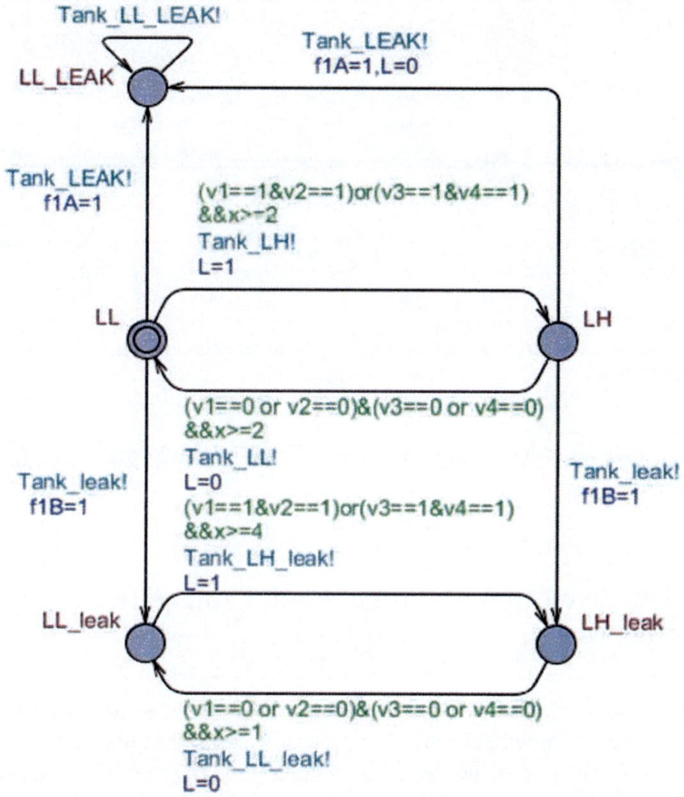

Fig. 9.3 Tank model in Example 9.1 (Reprinted with permission from Hsieh and Chang 2016. Copyright 2016 Elsevier)

to complete the corresponding state transition. The second attribute *Tank_LH!* denotes the corresponding state-transition event itself and, as mentioned before, the exclamation mark (*!*) is used to specify that it is an initiator or "sender" event. The last attribute is the reset condition, i.e., the liquid level is reset to high ($L = 1$) after completing the transition.

- **Sensor model**: Since sensor failures are not considered in this example, the online measurements always accurately reflect the tank states and, thus, the sensor model is omitted for the sake of brevity. It should be emphasized that this practice does not result in a loss in generality since the sensor models can always be built with the same principles described previously.

The controller model can be constructed on the basis of the given SFC, the failures and also the failure-induced events (see Fig. 9.4). The controller actions in this model, i.e., open V-3 (*ov3!*) and close V-3 (*cv3!*), should naturally be viewed as senders, while the observable state-transition events (i.e., *Tank_LH?* and *Tank_LL?*), the unobservable failures (i.e., *V3SC?*, *V3SO?*, *Tank_leak?* and *Tank_LEAK?*), and the

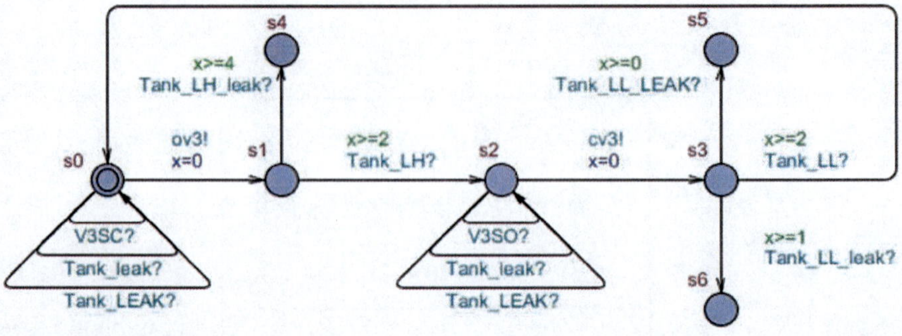

Fig. 9.4 Controller model in Example 9.1 (Reprinted with permission from Hsieh and Chang 2016. Copyright 2016 Elsevier)

failure-induced events (i.e., *Tank_LL_leak?*, *Tank_LH_leak?* and *Tank_LL_LEAK?*) are receivers.

9.2 Exhaustive Enumeration of Fault Propagation Scenarios

According to Clarke et al. (1986), "model checking" is essentially an algorithmic procedure for verifying whether a given system is compliant with the target specifications. A well-tested software verifier can be applied to determine if a set of timed automata conform to the desired system properties. In cases when there is any specification violation, the verifier can provide a counter scenario, from which the user should be able to find error(s) and then modify the models accordingly. The model checking tool provided in UPPAAL (Behrmann et al. 2006) is utilized for enumerating all fault propagation paths described in this chapter. Figure 9.5 summarizes the required reasoning procedure for logic deduction.

As mentioned before, failures may occur randomly at any time over the lifespan of a batch plant. Since these events are usually not directly perceptible with sensors, the fault origins can only be diagnosed with other available information. Under the assumption that the sensor measurements, the actuator signals and the clock readings can be obtained online, all OETs in the present example can be identified according to the proposed reasoning procedure and summarized in Fig. 9.6. The rectangles in this figure are used to specify the implied system states, which may be either normal (*N*) or under the influence of one or more failure, while the arrows are transitions triggered by the corresponding observable events. Note also that every abnormal event is marked with a double quote. Let us consider these traces one-by-one:

- **Trace 1**: The first transition on this trace is used to represent the event sequence that may be experienced in i (where, $i = 0, 1, 2, \ldots$) completed normal cycles. The subsequent action is the first step of SFC, i.e., *ov3*, which should normally

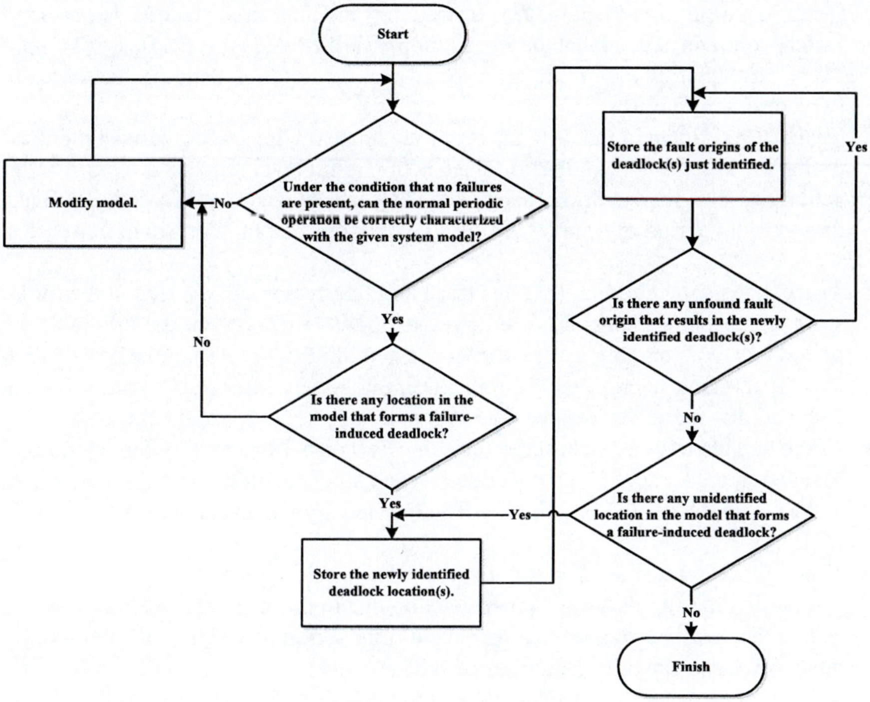

Fig. 9.5 Deductive reasoning procedure for identifying fault propagation paths (Reprinted with permission from Hsieh and Chang 2016. Copyright 2016 Elsevier)

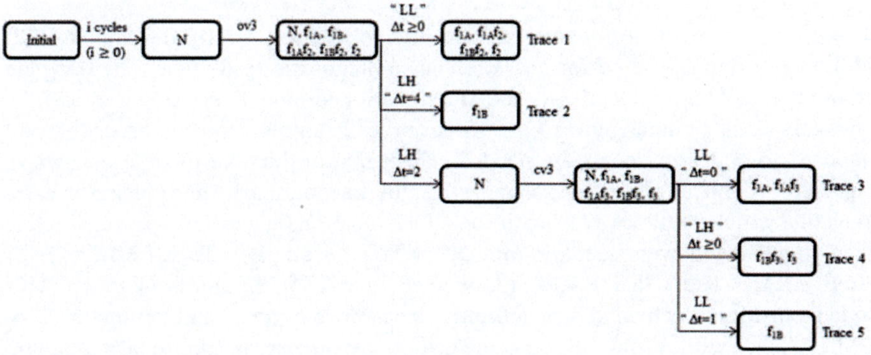

Fig. 9.6 Observable event traces in Example 9.1 (Reprinted with permission from Hsieh and Chang 2016. Copyright 2016 Elsevier)

result in a high liquid level (*LH*). However, if the abnormal state *LL* is observed instead, one can deduce that there are four possibilities: (1) f_{1A}; (2) $f_{1A}f_2$; (3) $f_{1B}f_2$; (4) f_2. Note that, in either 2nd or 3rd scenario, the notation denotes that there are two coexistent failures.

- **Trace 2**: As shown in Fig. 9.6, the event sequence on this trace is almost identical to that on Trace 1. The only difference is the liquid level reached after controller action *ov3*, i.e., the resulting target state (*LH*) is achieved in this case in a longer-than-normal period of time ($\Delta t = 4$). The only implied fault origin in this case is f_{1B}.
- **Trace 3**: As shown in Fig. 9.6, this trace and first two traces overlap in the initial stage. After completing *ov3*, the target state *LH* is reached in the allotted time period ($\Delta t = 2$) and, thus, the subsequent step in SFC, i.e., *cv3*, must be executed next. If the tank is emptied (*LL*) almost immediately afterwards, then it can be deduced that there are two possible root causes, i.e., (1) f_{1A} and (2) $f_{1A}f_3$.
- **Trace 4**: This trace is essentially the same as Trace 3 except the final symptom. The anticipated state *LL* can never be reached since the liquid level is maintained at the original level *LH* indefinitely. The implied fault origins are (1) $f_{1B}f_3$ and (2) f_3.
- **Trace 5**: This trace is also the same as the previous two except for the final symptom. After applying the control action *cv3* in this case, if the target state *LL* can be reached in a shorter-than-expected time period ($\Delta t = 1$), this abnormally quick response can only be attributed to f_{1B}.

9.3 Construction of Test Plans

If two or more fault origins are implied after observing a fully developed OET, additional tests may be performed to enhance diagnostic resolution. The test plan of an OET can be produced with the aforementioned model checking tool and the synthesis steps summarized in Fig. 9.7. In order to implement this procedure, all standard component models are needed except that of the controller. As shown in Fig. 9.8, the controller model in Example 9.1 must be modified by introducing an extra transition which points away from the deadlock location (at which there are no active events) reached in every scenario implied by Trace 1, i.e., f_{1A} (*Tank_LEAK?*), $f_{1A}f_2$ (*Tank_LEAK?* and *V3SC?*), $f_{1B}f_2$ (*Tank_leak?* and *V3SC?*), and f_2 (*V3SC?*). This added transition is activated by a fictitious receiver event (*test?*) and terminated at an artificial place without outputs. To guide test-plan synthesis, an additional automaton should also be constructed and subsequently modified repeatedly according to the following two procedures:

- Procedure A:
  - If $n_{stage} = 0$, connect artificial places *s0*, *s1* and *s2* in series and, then, connect *s2* to *s3a*, *s3b* and *s3c* etc., in parallel. The guards on transition *s0* $\rightarrow$ *s1* should be all fault origins implied by the given OET, while its triggering event

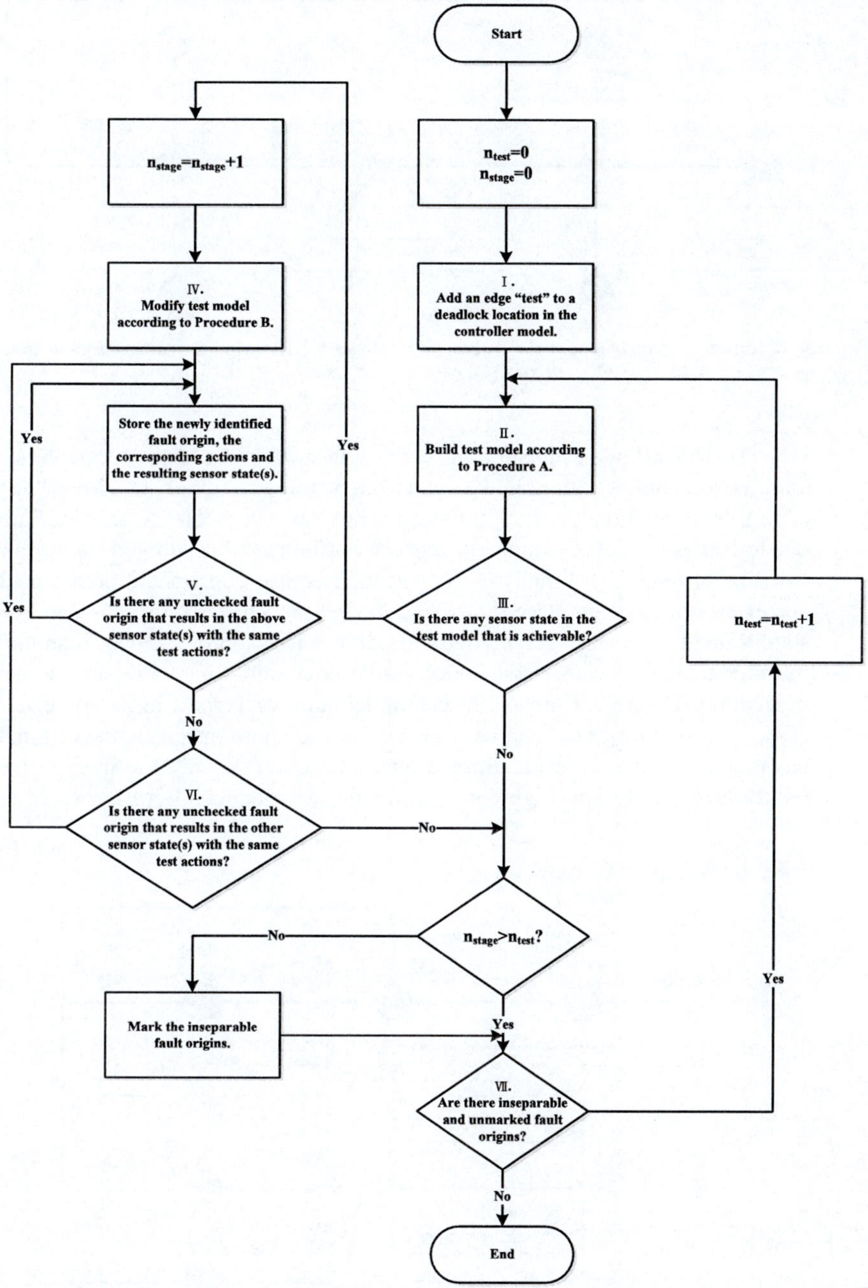

Fig. 9.7 Test-plan synthesis procedure (Reprinted with permission from Hsieh and Chang 2016. Copyright 2016 Elsevier)

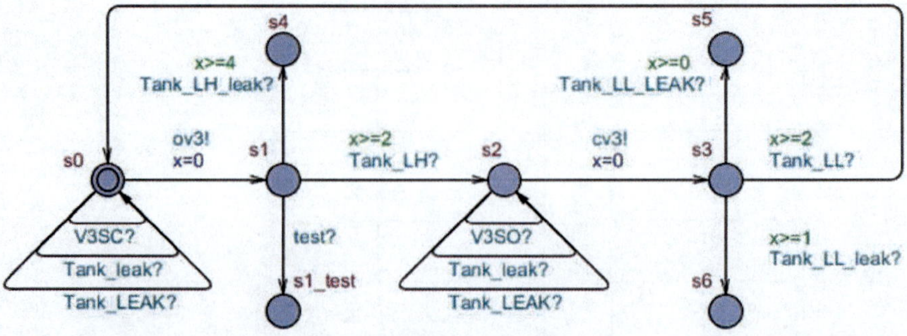

Fig. 9.8 Modified controller model for Trace 1 in Example 9.1 (Reprinted with permission from Hsieh and Chang 2016. Copyright 2016 Elsevier)

is *test1!*. To facilitate evaluation of all feasible steps in the test, multiple loops are then assembled with *s1* and *s2* and each is associated with an allowed test action. Note also that, although these two places can in fact be merged into one to form self-recycle loops, the current configuration is adopted simply for the sake of legibility. Finally, a distinct loop is also constructed between *s2* and every downstream place (i.e., *s3a*, *s3b*, and *s3c*, etc.) to represent a unique state-transition event observed online with a sensor and/or a clock. It should be noted that all aforementioned loops are used mainly for creating multi-step procedures. Figure 9.9 shows the test model built for Trace 1 in Example 9.1.

– If $n_{stage} > 0$, remove the guards for the confirmed fault origins in the original test model and then insert additional places between *s0* and *s1* to incorporate the confirmed test steps. Figure 9.10 shows the test model built for Trace 1 after

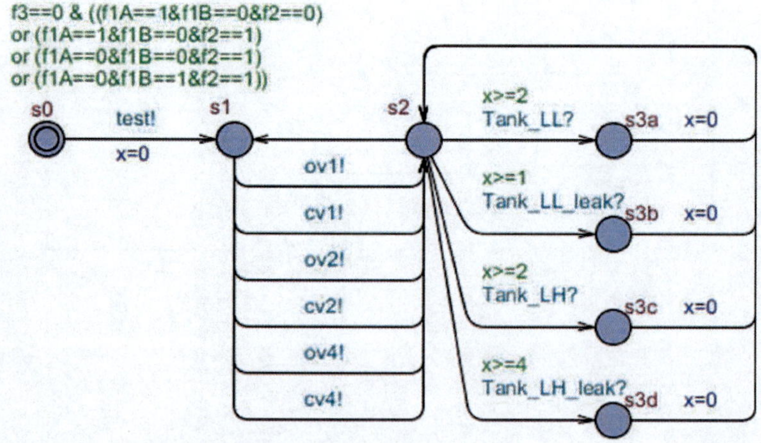

Fig. 9.9 Test model generated for Trace 1 with Procedure A in Example 9.1 ($n_{stage} = 0$) (Reprinted with permission from Hsieh and Chang 2016. Copyright 2016 Elsevier)

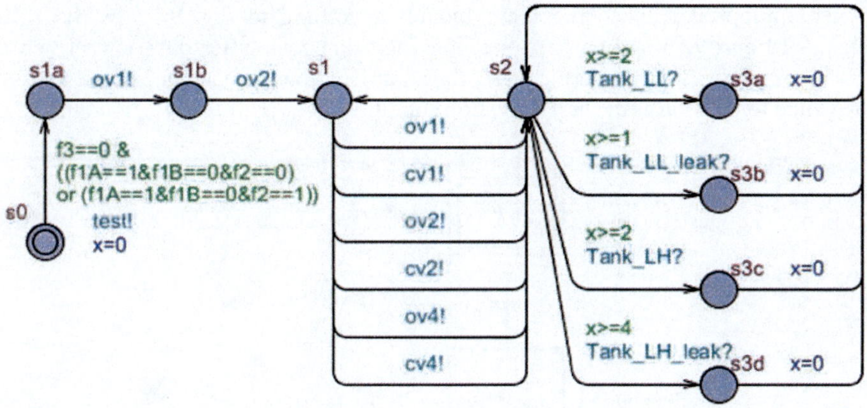

Fig. 9.10 Test model generated for Trace 1 with Procedure A in Example 9.1 (n_{stage} = 1) (Reprinted with permission from Hsieh and Chang 2016. Copyright 2016 Elsevier)

implementing the test steps *ov1* and *ov2* in Example 9.1. Note that, since these actions can be applied to produce unique responses for the 3rd and 4th implied fault origins, i.e., $f_{1B}f_2$, and f_2, respectively, only the first two are imposed as guards on the transition *s0* → *s1a*.

- Procedure B:
  Remove all loops between *s1* and *s2* in the test model established according to Procedure A and a specific n_{stage}, and then insert additional places between them to incorporate the confirmed test steps. Figure 9.11 shows the modified model built for Trace 1 in Example 9.1 after identifying the test steps *ov1* and *ov2* in the initial stage (n_{stage} = 1).

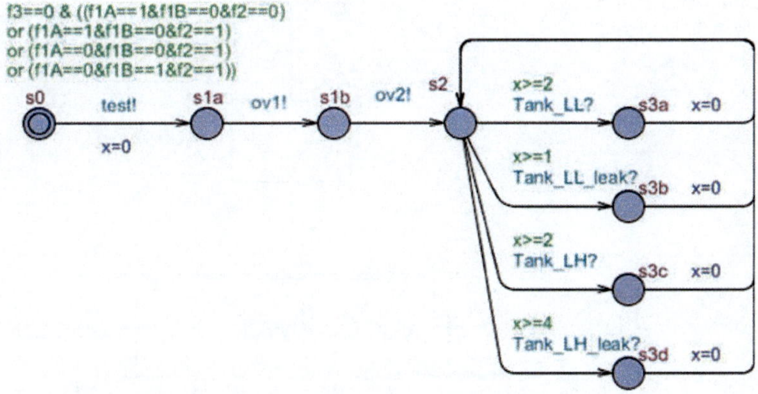

Fig. 9.11 Test model modified with Procedure B for Trace 1 in Example 9.1 (n_{stage} = 1) (Reprinted with permission from Hsieh and Chang 2016. Copyright 2016 Elsevier)

By appropriately checking these models according to Fig. 9.7, the SFCs in Figs. 9.12 and 9.13 can be generated for fault diagnosis after observing Trace 1 and 4, respectively. In the former case, the control actions *ov1* and *ov2* in the step S1 are called for as soon as the first activation condition, i.e., AC_1 (Trace 1), is confirmed.

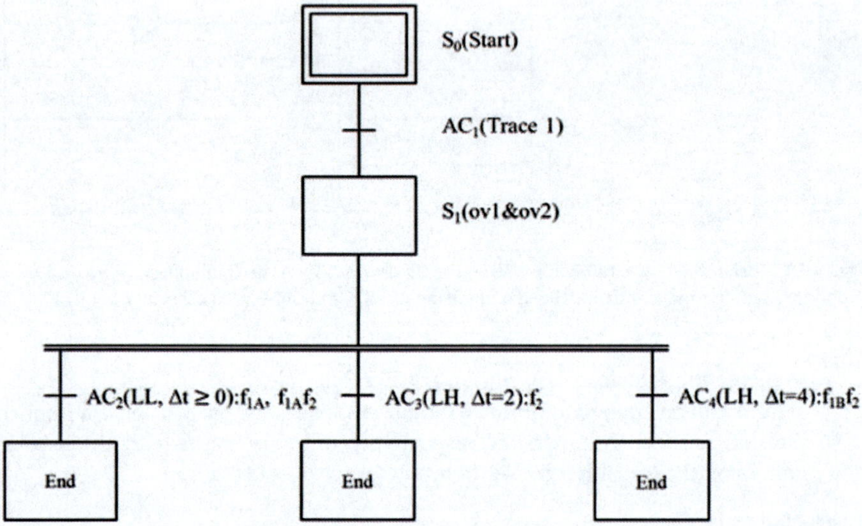

Fig. 9.12 Optimal test procedure for Trace 1 in Example 9.1 (Reprinted with permission from Hsieh and Chang 2016. Copyright 2016 Elsevier)

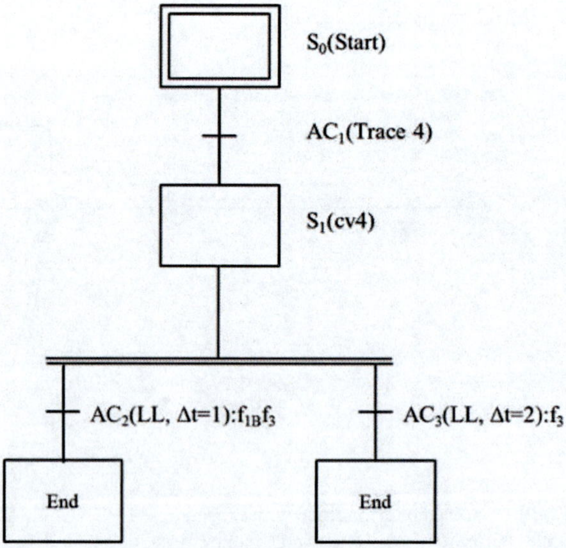

Fig. 9.13 Optimal test procedure for Trace 4 in Example 9.1 (Reprinted with permission from Hsieh and Chang 2016. Copyright 2016 Elsevier)

There may be three resulting scenarios depending on the sensor and clock readings obtained after the above test actions, i.e., (1) AC_2 (*LL*, $\Delta t \geq 0$), (2) AC_3 (*LH*, $\Delta t = 2$), and (3) AC_4 (*LH*, $\Delta t = 4$), and the corresponding fault origins should be: (1) f_{1A} or $f_{1A}f_2$; (2) f_2; (3) $f_{1B}f_2$. On the other hand, the first activation condition in the latter trace, i.e., AC_1 (Trace 4), prompts the test action in S_1 (*cv4*). There may be two possibilities, i.e., (1) AC_2 (*LL*, $\Delta t = 1$), which implies that the fault origin is $f_{1B}f_3$, and (2) AC_3 (*LL*, $\Delta t = 2$), which implies f_3.

Note that the test plans for the other three OETs are not presented here. There are obviously no needs to perform tests for Trace 2 or 5 since there is only one possible cause in either case, while none can be identified with the proposed procedure for Trace 3. Finally, if the above results are compared with those presented in the previous chapter, it is clear that the present test plans are superior since they are capable of differentiating different degrees of tank leaks. This is due to the inherent feature of time automata that allows proper representation of the elapsed time associated with every state-transition process.

9.4 Case Studies

To verify the effectiveness of the test-plan synthesis approach presented in this chapter, a series of extensive case studies have been carried out and two of them are summarized below.

Example 9.2 *(A Three-Tank Buffer System)*

Let us also revisit the P&ID shown in Fig. 8.20 in the previous chapter and replace the corresponding normal operating procedure specified in Fig. 8.21 with the one given in Fig. 9.14. For the sake of illustration clarity, let us repeat the process description for this example in the sequel. Notice first that V-2 is a 3-way valve and V-1, V-3 and V-4 are traditional gate valves. The fluid in tank T-1 is directed to tank T-2 if V-2 is placed at the + position and pump is switched on, while transported to T-3 if switched to the −position. All three tanks are equipped with level sensors. The one on T-1 is designed to detect three distinct states reflecting the low, intermediate and high liquid levels, i.e., *T1L*, *T1M*, and *T1H*, respectively, while that on T-2 (or T-3) is used to monitor only the states at low and high levels, i.e., *T2L* (or *T3L*), and *T2H* (or *T3H*). It is assumed that, initially, the liquid levels in all tanks are low, valves V-1, V-3 and V-4 are closed and V-2 is at the −position. In the present example, let us again consider the following seven failures:

i. f_1 (*V1SC*), i.e., V-1 fails at the closed position,
ii. f_2 (*V1SO*), i.e., V-1 fails at the open position,
iii. f_3 (*V2M_p*), i.e., V-2 is mistakenly switched to the + position,
iv. f_4 (*V2M_m*), i.e., V-2 is mistakenly switched to the −position,
v. f_5 (*V3SC*), i.e., V-3 fails at the closed position,
vi. f_6 (*V3SO*), i.e., V-3 fails at the open position, and

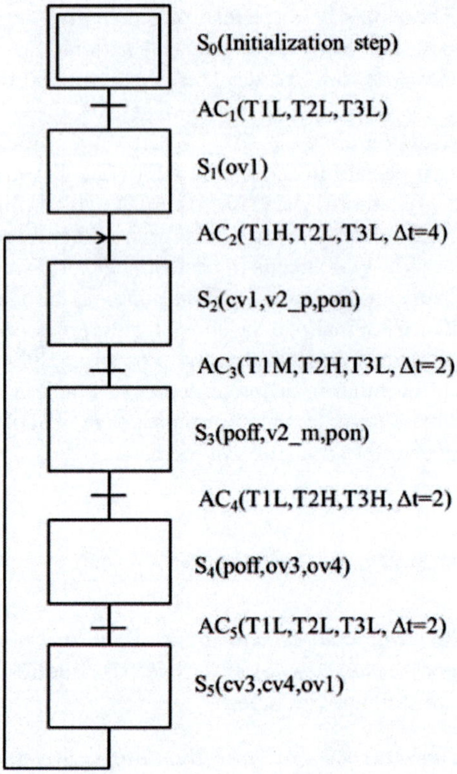

Fig. 9.14 Normal SFC of the three-tank buffer system in Example 9.2 (Reprinted with permission from Hsieh and Chang 2016. Copyright 2016 Elsevier)

vii. f_7 (*T2_leak*), i.e., a leak develops in tank T-2.

The modeling approach presented in this chapter and the reasoning procedure provided in Fig. 9.5 can be followed to identify the diagnosable OETs in Fig. 9.15a and the undiagnosable ones in Fig. 9.15b. For the sake of brevity, only the latter scenarios are discussed in the sequel:

- **Trace 9**: This trace develops before the initial cycle ($i = 0$) can be completed. Four consecutive event sets are observed in the precedence order given below
  (1) V-1 is opened (*ov1*) immediately after operation starts.
  (2) Liquid level in T-1 reaches *T1H* in 4 time units ($\Delta t = 4$), while those in T-2 and T-3 are maintained at *T2L* and *T3L* respectively.
  (3) V-1 is closed (*cv1*), V-2 is switched to the + position (*v2_p*) and then pump is switched on (*pon*).
  (4) Liquid level in T-1 reaches *T1M* in 2 time units ($\Delta t = 2$), while the abnormal states in T-2 and T-3, i.e., *T2L* and *T3H*, are observed at this time.

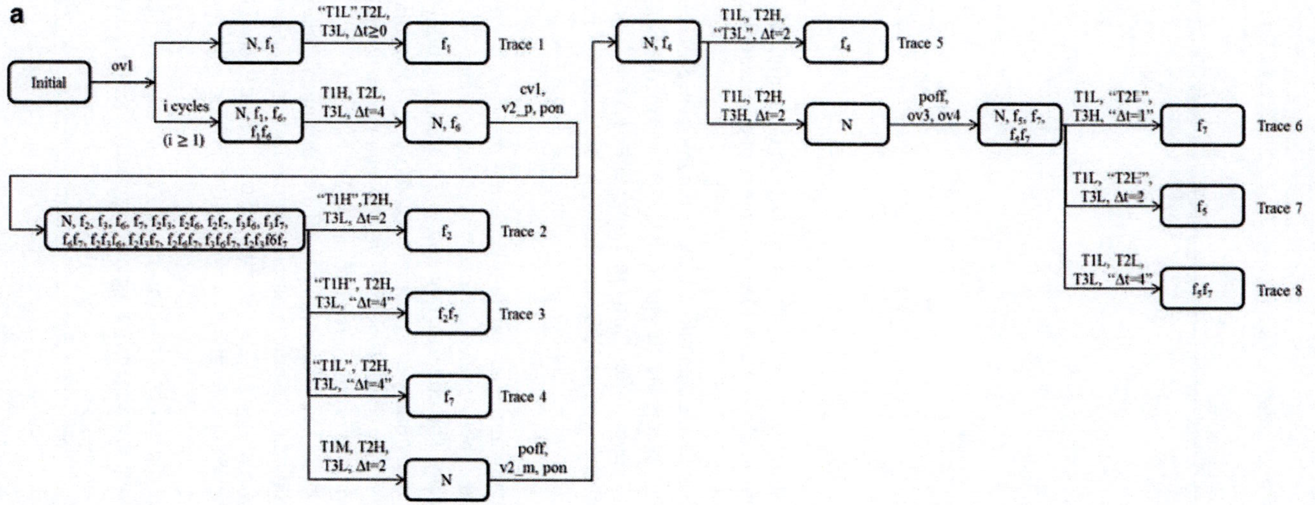

Fig. 9.15 **a** Diagnosable OETs in Example 9.2 (Reprinted with permission from Hsieh and hang 2016. Copyright 2016 Elsevier). **b** Undiagnosable OETs in Example 9.2 (Reprinted with permission from Hsieh and Chang 2016. Copyright 2016 Elsevier)

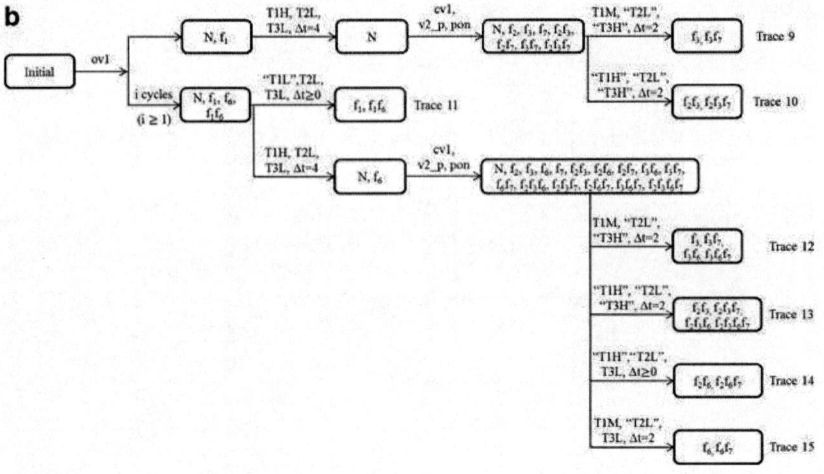

Fig. 9.15 (continued)

It can be observed that the system behaves normally in the initial stage until when the symptoms *T2L* and *T3H* show up. The possible root causes in this case should be (1) f_3 and (2) f_3f_7.

- **Trace 10**: This event sequence also takes place in the initial cycle ($i = 0$). The first three groups of normal events are the same as those on Trace 9, while abnormal conditions in all tanks appear afterwards in 2 time units ($\Delta t = 2$), i.e., the liquid level in T-1 is kept unchanged at *T1H*, and those in T-2 and T-3 reach *T2L* and *T3H*, respectively. The implied fault origins in this case should be (1) f_2f_3 and (2) $f_2f_3f_7$.
- **Trace 11**: Notice that the initial action on this trace is *ov1* (i.e., open V-1) and, on the next transition, the label *i cycles* denotes the event sequence in one or more completed normal cycle. The abnormal symptom *T1L*, i.e., the liquid level in T-1 is low, is maintained indefinitely after completion of i ($i \geq 1$) normal cycles and, thus, the desired activation condition *T1H* in AC_2 can never be satisfied in this scenario. The implied fault origins are (1) f_1 and (2) f_1f_6.
- **Trace 12**: The initial event sequence of this trace, i.e., *ov1* and *i cycles*, is identical to that of Trace 11, while the remaining part is essentially the same as that of Trace 9 after the first control action *ov1*. When compared with Trace 9, one could observe that more fault origins can be implicated with this OET, i.e., (1) f_3, (2) f_3f_7, (3) f_3f_6, and (4) $f_3f_6f_7$.
- **Trace 13**: The initial event sequence of this trace, i.e., *ov1* and *i cycles*, is identical to that of Trace 11, while the remaining part is essentially the same as that of Trace 10 after the first control action *ov1*. The implied fault origins in this case, i.e., (1) f_2f_3 (2) $f_2f_3f_7$, (3) $f_2f_3f_6$, and (4) $f_2f_3f_6f_7$, are also more than those associated with Trace 10.

- **Trace 14**: The event sequence of this trace is essentially the same as those of Traces 12 and 13 except the final symptoms, i.e., the liquid levels in tanks T-1, T-2, and T-3 are always kept unchanged at *T1H*, *T2L*, and *T3L*, respectively. The first two tank states are abnormal since the previous control actions (*cv1*, *v2_p* and *pon*) have been applied to transfer material from T-1 to T-2. The corresponding fault origins could be (1) f_2f_6 and (2) $f_2f_6f_7$.
- **Trace 15**: The event sequence of this trace is also the same as those of Traces 12 and 13 except the final conditions, i.e., the liquid levels in tanks T-1, T-2, and T-3 become *T1M*, *T2L*, and *T3L*, respectively, after 2 time units ($\Delta t = 2$). The liquid level in T-2 is not expected in SFC and this abnormality may be attributed to (1) f_6 or (2) f_6f_7.

It was found that, after applying the synthesis procedure in Fig. 9.7, the root causes implied by Traces 11, 14, and 15 cannot be further distinguished via diagnostic tests. On the other hand, note that the fault origins associated with Trace 9 actually form a subset of those corresponding to Trace 12 and, similarly, every fault origin implied by Trace 10 is also by Trace 13. Thus, only the test plans of Traces 12 and 13 are presented in Figs. 9.16 and 9.17, respectively, and these plans are also summarized in the sequel:

- **Plan 2.Tr12**: As shown in Fig. 9.16, the test action *v2_p* (i.e., switch V-2 to the + position) in step S1 is applied as soon as AC_1 (Trace 12) is observed online. There may be three resulting scenarios: (1) AC_2 (*T1L*, *T2H*, *T3H*, $\Delta t = 2$), and a single-failure fault origin f_3 can be confirmed; (2) AC_3 (*T1L*, *T2H*, *T3H*, $\Delta t = 4$), and the two-failure fault origin f_3f_7 can be confirmed; (3) AC_4 (*T1L*, *T2L*, *T3H*, $\Delta t = 2$), and two multi-failure fault origins can be implicated, i.e., f_3f_6 and $f_3f_6f_7$.

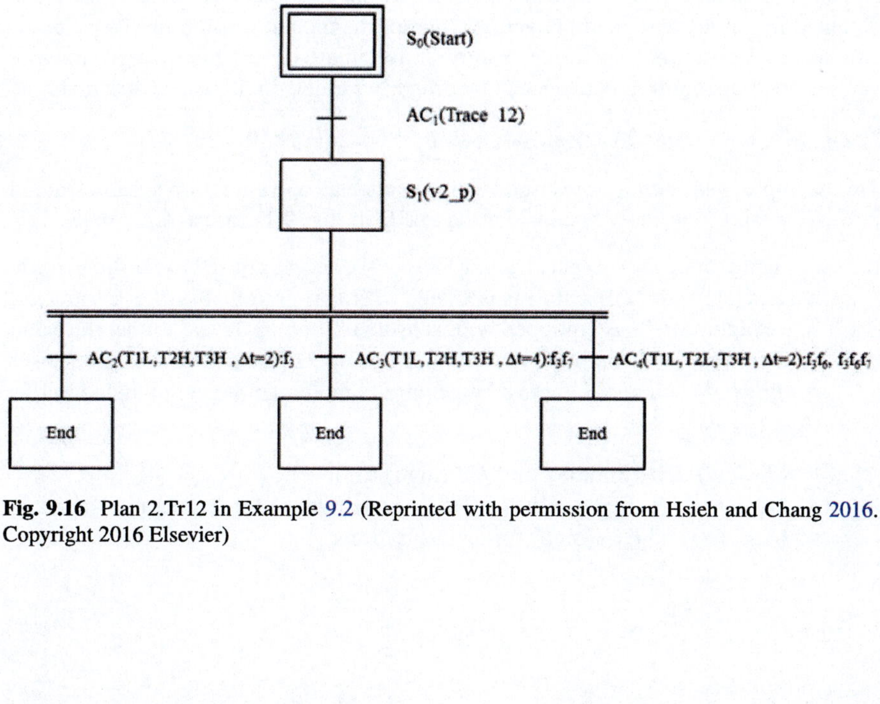

Fig. 9.16 Plan 2.Tr12 in Example 9.2 (Reprinted with permission from Hsieh and Chang 2016. Copyright 2016 Elsevier)

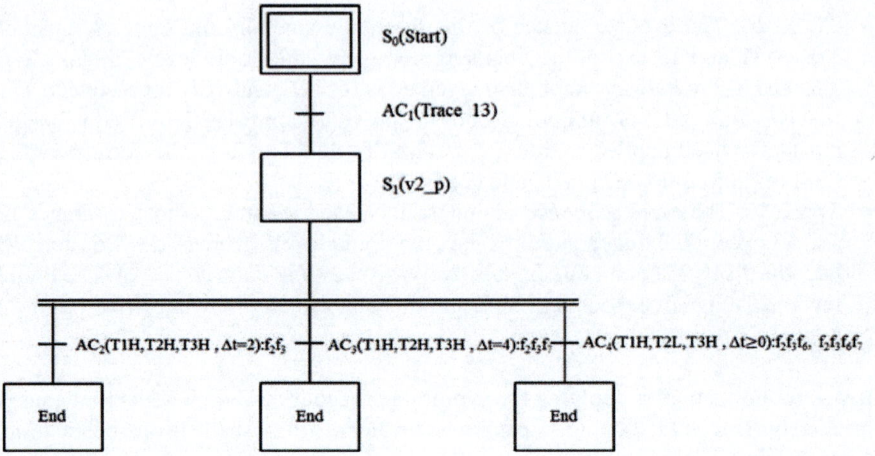

Fig. 9.17 Plan 2.Tr13 in Example 9.2 (Reprinted with permission from Hsieh and Chang 2016. Copyright 2016 Elsevier)

- **Plan 2.Tr13**: As shown in Fig. 9.17, the test action *v2_p* (i.e., switch V-2 to the + position) in step S1 is needed as soon as AC_1 (Trace 13) is observed online. This action may result in three possible outcomes: (1) AC_2 (*T1H*, *T2H*, *T3H*, $\Delta t = 2$), and a two-failure fault origin f_2f_3 can be confirmed; (2) AC_3 (*T1H*, *T2H*, *T3H*, $\Delta t = 4$), and the three-failure fault origin $f_2f_3f_7$ can be confirmed; (3) AC_4 (*T1H*, *T2L*, *T3H*, $\Delta t \geq 0$), and two multi-failure fault origins can be implicated, i.e., $f_2f_3f_6$ and $f_2f_3f_6f_7$.

Again, if compared with the results reported in the previous chapter, the test plans obtained in the present example achieve higher diagnostic resolution. Specifically, tank leaks of different magnitudes can be differentiated since the elapsed times of various level-changing processes can be properly modeled with timed automata.

Example 9.3 (*A Batch Evaporation System*)

This example is essentially a revised version of the batch evaporation system studied in Bauer et al. (2004). Let us consider the P&ID in Fig. 9.18 and assume that.

(a) All actuators, i.e., the gate valves (V1 –V4), the pump (P1) and the electric heaters (H1 and H2), can be manipulated with a programmable logic controller;
(b) The evaporator T1 is equipped with sensors to monitor level, temperature and concentration. To facilitate normal and test operations, several target sensor readings of each variable must be acquired online and they are labeled in this example as

  - *LL* (low), *LM* (middle), and *LH* (high) for levels,
  - *TL* (low), *TH* (high), and *THH* (higher than high) for temperatures, and
  - *QL* (low) and *QH* (high) for concentrations.

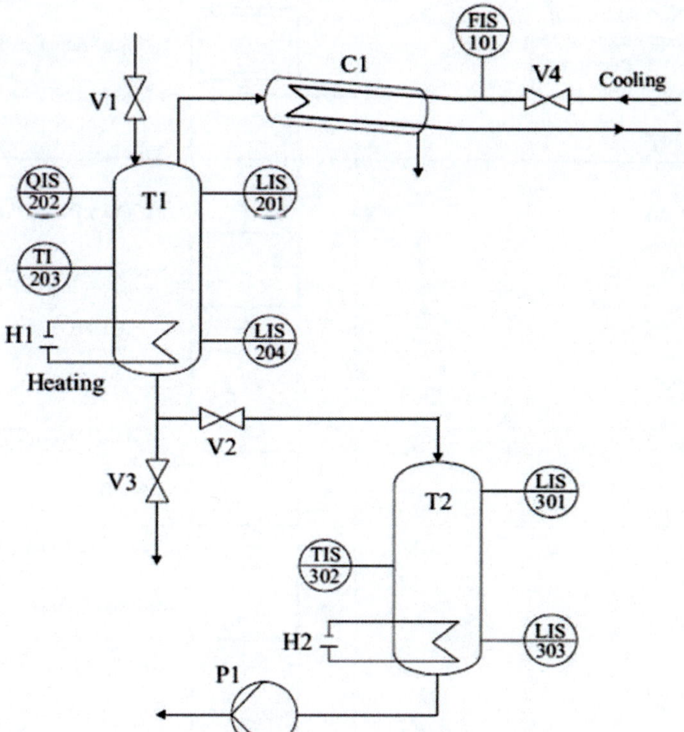

Fig. 9.18 P&ID of the batch evaporation system in Example 9.3 (Reprinted with permission from Hsieh and Chang 2016. Copyright 2016 Elsevier)

(c) The condenser C1 is equipped with a flow sensor to determine if the flow rate of cooling medium reaches *FL* (low or zero) and *FH* (high).

(d) The buffer vessel T2 is also equipped with level and temperature sensors. The level targets is also denoted as *LL* (low) and *LH* (high), while the temperature targets *TL* (low) and *TH* (high).

The initial system state is set according to the additional assumptions that, before the evaporation operation begins, all actuators are switched off, evaporator T1 and buffer tank T2 are both empty, and cooling in C1 is not provided. Based on these initial conditions, the normal operating procedure adopted in the present example can be specified with the SFC in Fig. 9.19. To facilitate discussions, let us consider the following seven hardware failures:

i. f_1 (*V1SC*), i.e., V1 fails at the closed position;
ii. f_2 (*V1SO*), i.e., V1 fails at the open position;
iii. f_3 (*ov4_M*), i.e., controller fails to open V4;
iv. f_4 (*H1_failure*), i.e., heater H1 fails;
v. f_5 (*T2_leak*), i.e., a leak develops in T2;

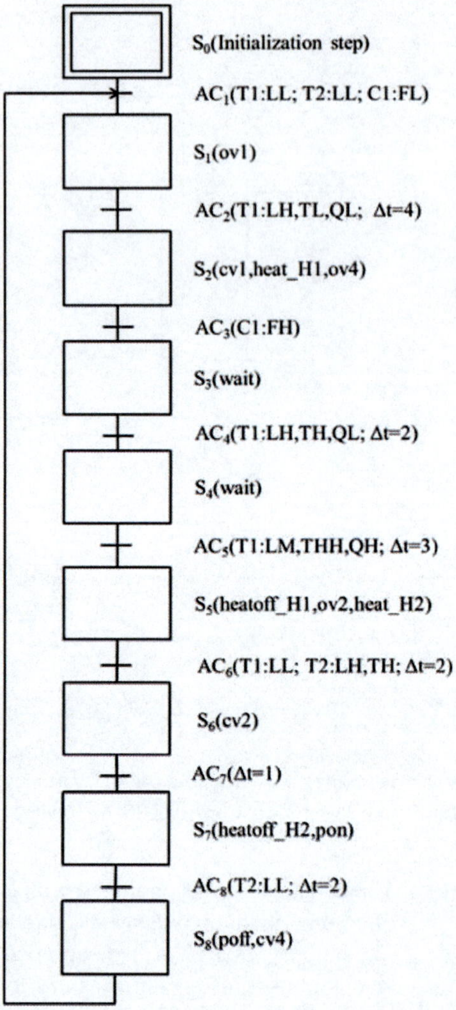

Fig. 9.19 Normal SFC of the batch evaporation system in Example 9.3 (Reprinted with permission from Hsieh and Chang 2016. Copyright 2016 Elsevier)

vi. f_6 (*P1_failure*), i.e., pump P1 fails;
vii. f_7 (*QIS_failure*), i.e., concentration analyzer on T1 fails.

The aforementioned modeling approach and the reasoning procedure in Fig. 9.5 have again been applied to produce the OETs in Fig. 9.20. For the sake of brevity, only the undiagnosable scenarios are described below in detail:

- **Trace 1**: After experiencing the initial state mentioned above and also the event sequence in i ($i \geq 0$) normal cycles, the control action *ov1* (i.e., opening valve V1) must be executed next. According to the SFC given in Fig. 9.19, the level sensors on T1 and T2 should reach targets *LH* and *LL*, respectively, in four time units ($\Delta t = 4$) and the corresponding reading of the flow sensor on C1 must be

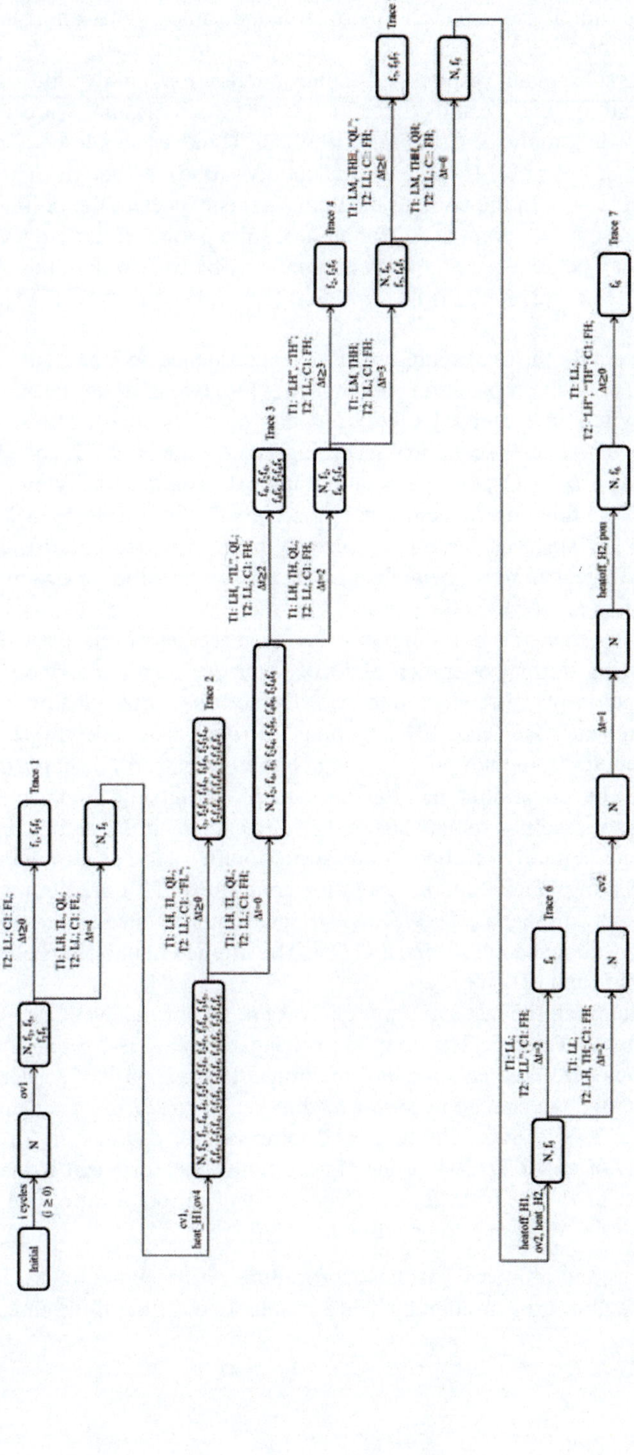

Fig. 9.20 Observable event traces in Example 9.3 (Reprinted with permission from Hsieh and Chang 2016. Copyright 2016 Elsevier)

FL. However, the level reading of T1 at this time remains at *LL*, which may be attributed to f_1 or f_1f_5.

- **Trace 2**: The initial sequence on this trace is the same as that on Trace 1, i.e., *i cycles* and then *ov1*. Instead of the final symptom observed in the previous scenario, the subsequent responses of control action *ov1* after 4 time units, i.e., the sensor readings *LH*, *TL* and *QL* for T1, *LL* for T2 and *FL* for C1, are all normal events. According to the SFC in Fig. 9.19, the next moves should be to close V1 (*cv1*), switch on heater H1 (*heat_H1*), and open V4 (*ov4*). Although an immediate change to target *FH* in the cooling medium flow is expected, the online reading of flow sensor on C1 is kept at *FL* for an indefinite period of time in the present scenario. It can be deduced that this outcome may be caused by 8 possible fault origins, i.e., (1) f_3, (2) f_2f_3, (3) f_3f_4, (4) f_3f_5, (5) $f_2f_3f_4$, (6) $f_2f_3f_5$, (7) $f_3f_4f_5$, and (8) $f_2f_3f_4f_5$.
- **Trace 3**: Except for the final symptom, the event sequence on Trace 2 also appears on Trace 3. The fault propagation path branches into two after the actions to close V1 (*cv1*), switch on heater H1 (*heat_H1*), and open V4 (*ov4*). The subsequent observations in the latter scenario concerning Trace 3, i.e., *LH*, *TL* and *QL* in T1, *LL* in T2, and *FH* in C1, can be acquired almost instantaneously and they are all normal. Note that, after 2 additional time units, the temperature in T1 should be raised to *TH* since heater H1 has already been switched on. However, this temperature target can never be realized on Trace 3 due to the following reasons: (1) f_4; (2) f_2f_4; (3) f_4f_5; (4) $f_2f_4f_5$.
- **Trace 4**: Other than the last symptom, the event sequence on Trace 3 can be observed during normal operation and this sequence also appears on Trace 4. This propagation path branches into two distinct ones after closing V1 (*cv1*), switching on heater H1 (*heat_H1*), opening V4 (*ov4*), and observing *FH* in C1. On the present trace, the normal operating conditions can still be confirmed after 2 time units. The temperature in T1 is raised to *TH* at this time as expected, while the other sensor readings remain stable, i.e., *LH* and *QL* in T1 and *LL* in T2. An abnormal state is finally reached in this scenario after another 3 time units, i.e., the sensor readings show that the operating conditions of T1 are not responding to the heat input and stay at *LH*, *TH*, and *QL* indefinitely. Note that the anticipated targets for T1 should be *LM*, *THH*, and *QH*. The implied fault origins of this trace should be (1) f_2 and (2) f_2f_5.
- **Trace 5**: Other than the last symptom on Trace 4, its normal event sequence can also be found on Trace 5. The common propagation path is branched after the control actions *cv1*, *heat_H1*, and *ov4*, the immediate response of sensor reading on C1 to *FH*, and the change in sensor reading on T1 to *TH* in 2 time units later. After another 3 time units, the level and temperature readings on T1 are still normal, i.e., *LM* and *THH*, while the concentration measurement is maintained at abnormally low value (*QL*) for a sufficiently long period of time. The possible root cause for this case should be either f_7 or f_5f_7.

After applying the proposed synthesis procedure to the above traces, it can be found that the fault origins implied by Trace 1 cannot be further differentiated with diagnostic tests. Thus, only the test plans of the remaining four traces are provided in Figs. 9.21, 9.22, 9.23 and 9.24 and, also, the activation conditions of these SFCs contain only the sensor readings that are affected by the test actions. A brief summary is presented in the sequel:

- **Plan 3.Tr2**: As shown in Fig. 9.21, the test actions *heatoff_H1* (i.e., switch off heater H1) and *ov2* (i.e., open V2) in step S_1 are executed as soon as AC_1 (Trace 2) is observed online. Four resulting conditions may appear (1) AC_2 (T1: *LM*; T2: *LH*; $\Delta t = 2$); (2) AC_3 (T1: *LH*; T2: *LH*; $\Delta t = 2$); (3) AC_4 (T1: *LM*; T2: *LL*; $\Delta t = 2$); (4) AC_5 (T1: *LH*; T2: *LH*; $\Delta t = 4$). The subsequent diagnostic tests in the second stage are outlined below
  - **AC_2**—The online test results in this case may be attributed to f_3 (i.e., controller fails to open V4) or f_3f_4 (i.e., controller fails to open V4 and also heater H1 fails). If the test actions suggested in step S_2 (i.e., *ov4*, *cv2*, and *heat_H1*) are performed, there may be two possible outcomes, i.e., AC_6 (T1: *TH*; C1: *FH*; $\Delta t = 1$) and AC_7 (T1: *TL*; C1: *FH*; $\Delta t \geq 0$). A single fault origin can then be diagnosed upon reaching each of these conditions. AC_6 indicates that the root cause is f_3, while AC_7 confirms f_3f_4.
  - **AC_3**—The online test results in this activation condition may be caused by f_2f_3 (i.e., V1 sticks at the open position and controller fails to open V4) or $f_2f_3f_4$ (i.e., V1 sticks at the open position, controller fails to open V4 and also heater H1 fails). If the test actions listed in step S_3 (i.e., *ov4*, *cv2*, and *heat_H1*) are implemented, there may be two different responses, i.e., AC_8 (T1: *TH*; C1: *FH*; $\Delta t = 2$) and AC_9 (T1: *TL*; C1: *FH*; $\Delta t \geq 0$). AC_8 indicates that the root cause is f_2f_3, while AC_9 confirms $f_2f_3f_4$.
  - **AC_4**—The online test results in this scenario are resulted from f_3f_5 (i.e., controller fails to open V4 and T2 leaks) or $f_3f_4f_5$ (i.e., controller fails to open V4, heater H1 fails and T2 leaks). If the diagnostic tests specified in step S_4 (i.e., *ov4*, *cv2*, and *heat_H1*) are carried out, two sets of possible symptoms may be obtained, i.e., AC_{10} (T1: *TH*; C1: *FH*; $\Delta t = 1$) and AC_{11} (T1: *TL*; C1: *FH*; $\Delta t \geq 0$). The former suggests that the fault origin is f_3f_5, while the latter $f_3f_4f_5$.
  - **AC_5**—The online test results in this case imply that $f_2f_3f_5$ (i.e., V1 sticks at the open position, controller fails to open V4 and T2 leaks) or $f_2f_3f_4f_5$ (i.e., V1 sticks at the open position, controller fails to open V4, heater H1 fails and T2 leaks). If the test actions given in step S_5 (i.e., *ov4*, *cv2*, and *heat_H1*) are applied, the test results may be either AC_{12} (T1: *TH*; C1: *FH*; $\Delta t = 2$) or AC_{13} (T1: *TL*; C1: *FH*;$\Delta t \geq 0$). The former condition implies that the fault origin is $f_2f_3f_5$, while the latter $f_2f_3f_4f_5$.

- **Plan 3.Tr3**: As shown in Fig. 9.22, it is required to implement the test action *ov2* (i.e., open V2) in step S_1 after triggering AC_1 (Trace 3). There may be four subsequent scenarios: (1) AC_2 (T1: *LM*; T2: *LH*; $\Delta t = 2$); (2) AC_3 (T1: *LH*; T2: *LH*; $\Delta t = 2$); (3) AC_4 (T1: *LM*; T2: *LL*; $\Delta t = 2$); (4) AC_5 (T1: *LH*; T2: *LH*; $\Delta t = 4$). The corresponding fault origins are listed below

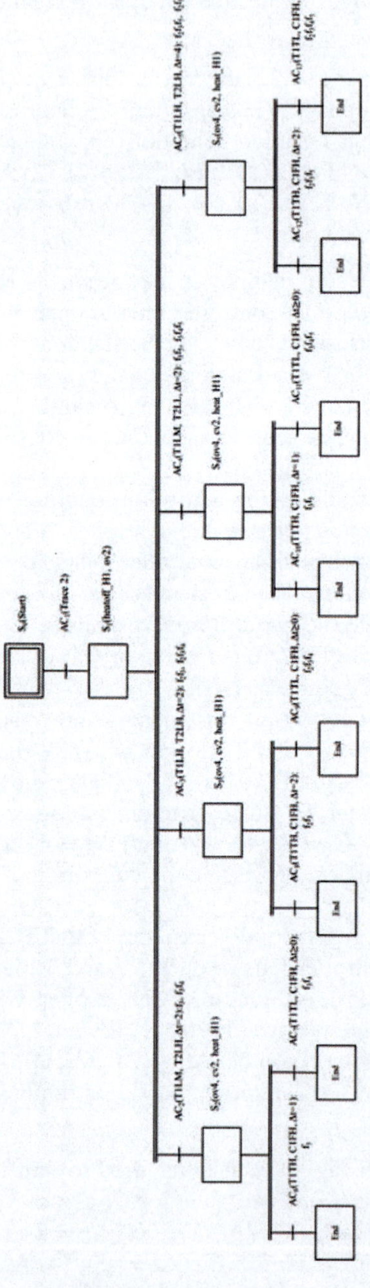

Fig. 9.21 Plan 3.Tr2 in Example 9.3 (Reprinted with permission from Hsieh and Chang 2016. Copyright 2016 Elsevier)

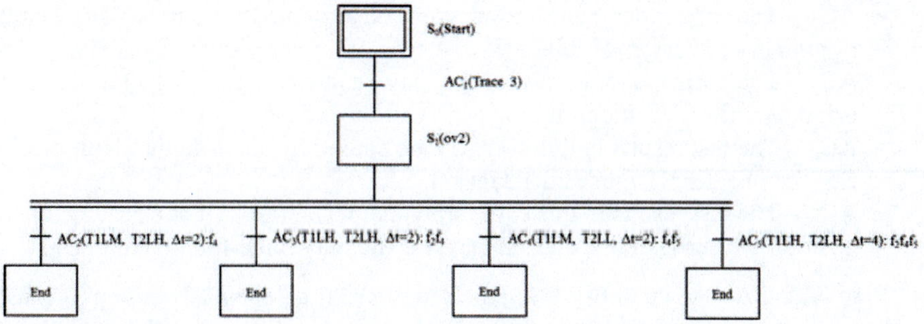

Fig. 9.22 Plan 3.Tr3 in Example 9.3 (Reprinted with permission from Hsieh and Chang 2016. Copyright 2016 Elsevier)

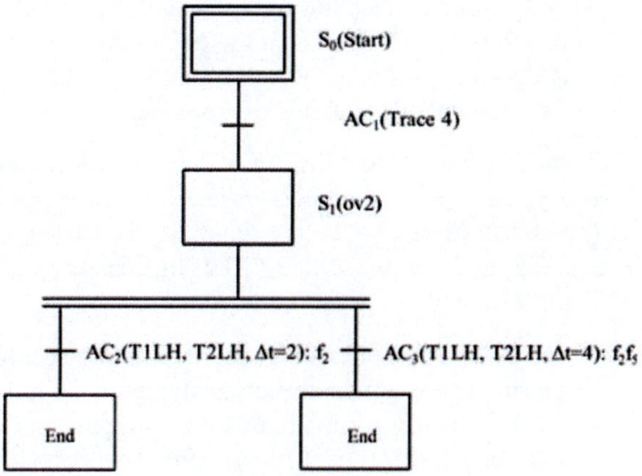

Fig. 9.23 Plan 3.Tr4 in Example 9.3 (Reprinted with permission from Hsieh and Chang 2016. Copyright 2016 Elsevier)

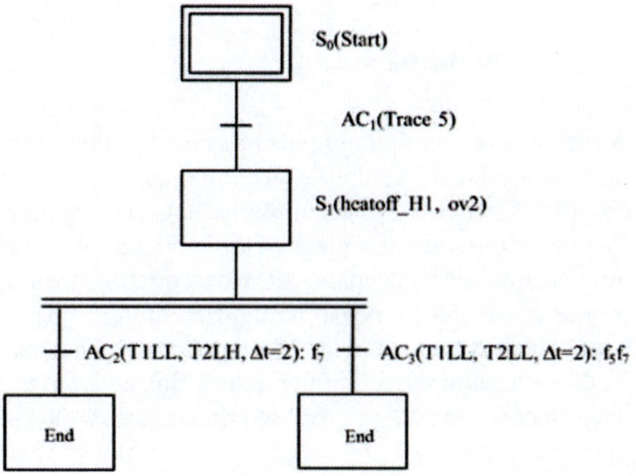

Fig. 9.24 Plan 3.Tr5 in Example 9.3 (Reprinted with permission from Hsieh and Chang 2016. Copyright 2016 Elsevier)

- **AC_2**—The corresponding test results may be attributed to a single-failure fault origin f_4 (i.e., heater H1 fails).
- **AC_3**—The corresponding test results may be attributed to a two-failure fault origin f_2f_4 (i.e., V1 sticks at the open position and also heater H1 fails).
- **AC_4**—The test results in this scenario are caused by a two-failure fault origin f_4f_5 (i.e., heater H1 fails and T2 leaks).
- **AC_5**—The test results in this case imply that $f_2f_4f_5$ (i.e., V1 sticks at the open position, heater H1 fails, and T2 leaks) is the only three-failure fault origin.

- **Plan 3.Tr4**: As shown in Fig. 9.23, the action *ov2* (i.e., open V2) in step S_1 must be applied after activating AC_1 (Trace 4). There may be two possible outcomes, i.e., (1) AC_2 (T1: *LH*; T2: *LH*;$\Delta t = 2$) and (2) AC_3 (T1: *LH*; T2: *LH*; $\Delta t = 4$). The corresponding diagnosis is summarized below:

  - **AC_2**—The corresponding test results may be attributed to a single-failure fault origin f_2 (i.e., V1 sticks at the open position).
  - **AC_3**—The test results in this case imply that f_2f_5 (i.e., V1 sticks at the open position and T2 leaks) is the root cause.

- **Plan 3.Tr5**: As shown in Fig. 9.24, the test actions *heatoff_H1* (i.e., switch off H1) and *ov2* (i.e., open V2) in step S_1 must be performed when observing AC_1 (Trace 5). There may be two possible responses, i.e., (1) AC_2 (T1: *LL*; T2: *LH*; $\Delta t = 2$) and (2) AC_3 (T1: *LL*; T2: *LL*; $\Delta t = 2$). The corresponding diagnosis is outlined below

  - **AC_2**—The corresponding test results may be attributed to a single-failure fault origin f_7 (i.e., concentration analyzer on T1 fails).
  - **AC_3**—The test results in this case suggest that the root cause is a two-failure fault origin f_5f_7 (i.e., T2 leaks and also concentration analyzer on T1 fails).

9.5 Concluding Remarks

A standardized methodology is presented in this chapter to systematically construct timed automata for modeling all components in any given plant, and to enumerate the observable event traces accordingly. A generic synthesis procedure is also provided for conjecturing the test plans of all undiagnosable traces. It should be noted that the proposed method is capable of differentiating various time delays caused by fault origins of the same type but with different intensities. This is a unique feature which has not been discussed in the previous chapter. The diagnosis results of extensive case studies are summarized at the end of this chapter for demonstrating the feasibility and effectiveness of the proposed procedure synthesis strategy.

References

Baroni P, Lamperti G, Pogliano P, Zanella M (1999) Diagnosis of large active systems. Artif Intell 110(1):135–183

Baroni P, Lamperti G, Pogliano P, Zanella M (2000) Diagnosis of a class of distributed discrete-event systems. IEEE Trans Syst Man Cybern Part A Syst Hum 30:731–752

Bauer N, Engell S, Huuck R, Lohmann S, Lukoschus B, Remelhe M, Stursberg O (2004) Verification of PLC programs given as sequential function charts. In: Integration of software specification techniques for applications in engineering. Springer, Berlin, pp 517–540

Behrmann G, David A, Larsen KG (2006) A tutorial on UPPAAL 4.0. Aalborg University, Denmark

Benveniste A, Fabre E, Haar S, Jard C (2003) Diagnosis of asynchronous discrete-event systems: a net unfolding approach. IEEE Trans Autom Control 48:714–727

Cassandras CG, Lafortune S (2008) Introduction to discrete event systems, 2nd edn. Springer Science + Business Media, LLC, New York, NY, USA

Chen YC, Yeh ML, Hong CL, Chang CT (2010) Petri-net based approach to configure online fault diagnosis systems for batch processes. Ind Eng Chem Res 49:4249–4268

Clarke EM, Emerson A, Sistla KL (1986) Automatic verification of finite-state concurrent systems using temporal logic specification. ACM Trans Programm Lang Syst 8:244–263

Dai Y, Zhao J (2011) Fault diagnosis of batch chemical processes using a dynamic time warping (DTW)-based artificial immune system. Ind Eng Chem Res 50:4534–4544

Debouk R, Lafortune S, Teneketzis D (2000) Coordinated decentralized protocols for failure diagnosis of discrete event systems. Discret Event Dyn Syst Theory Appl 10:33–86

Gascard E, Simeu-Abazi Z (2013) Modular modeling for the diagnostic of complex discrete-event systems. IEEE Trans Autom Sci Eng 10:1101–1123

Ghosh K, Srinivasan R (2011) Immune-system-inspired approach to process monitoring and fault diagnosis. Ind Eng Chem Res 50:1637–1651

Hsieh WC, Chang CT (2016) Timed-automata based method for synthesizing diagnostic tests in batch processes. Comp Chem Eng 84:12–27

Kang A, Chang CT (2014) Automata generated test plans for fault diagnosis in sequential material- and energy-transfer operations. Chem Eng Sci 113:101–115

Kim J, Moon I (2009) Automatic verification of control logics in safety instrumented system design for chemical process industry. J Loss Prevent Proc Ind 22:975–980

Kim J, Moon I (2011) Model checking for automatic verification of control logics in chemical process. Ind Eng Chem Res 50:905–915

Kourti T, Macgregor JF (1995) Process analysis, monitoring and diagnosis, using multivariate projection methods. Chemo Intell Lab Sys 28:3–21

Kourti T, Nomikos P, Macgregor JF (1995) Analysis monitoring and fault-diagnosis of batch processes using multiblock and multiway PLS. J Proc Cont 5:277–284

Lahtinen J, Valkonen J, Björkman K, Frits J, Niemelä I, Heljanko K (2012) Model checking of safety-critical software in the nuclear engineering domain. Rel Eng Syst Safe 105:104–113

Lee JM, Yoo CK, Lee IB (2004) Fault detection of batch processes using multiway kernel principal component analysis. Comp Chem Eng 28:1837–1847

Li JH, Chang CT, Jiang D (2014) Systematic generation of cyclic operating procedures based on timed automata. Chem Eng Res Des 92:139–155

Lohmann S, Stursberg O, Engell S (2006) Systematic design of logic controllers for processing plants starting from informal specifications. Comp Aid Chem Eng 21:1317–1322

Nomikos P, MacGregor JF (1994) Monitoring batch processes using multiway principal component analysis. AIChE J 40:1361–1375

Nomikos P, MacGregor JF (1995) Multivariate SPC charts for monitoring batch processes. Technometrics 37:41–59

Qiu WB, Kumar R (2006) Decentralized failure diagnosis of discrete event system. IEEE Trans Syst Man Cybern Part A Syst Hum 36:384–395

Sampath M, Lafortune S, Teneketzis D (1998) Active diagnosis of discrete-event systems. IEEE Trans Autom Control 43:908–929
Sampath M, Sengupta R, Lafortune S, Sinnamohideen K, Teneketzis DC (1995) Diagnosability of discrete-event systems. IEEE Trans Autom Control 40:1555–1575
Sampath M, Sengupta R, Lafortune S, Sinnamohideen K, Teneketzis DC (1996) Failure diagnosis using discrete-event models. IEEE Trans Control Syst Technol 4:105–124
Tan WL, Nor NM, Abu Bakar MZ, Ahmad Z, Sata SA (2012) Optimum parameters for fault detection and diagnosis system of batch reaction using multiple neural networks. J Loss Prevent Proc Ind 25:138–141
Undey C, Ertunc S, Cinar A (2003) Online batch fed-batch process performance monitoring, quality prediction, and variable contribution analysis for diagnosis. Ind Eng Chem Res 42:4645–4658
Yeh ML, Chang CT (2011) An automaton-based approach to evaluate and improve online diagnostic schemes for multi-failure scenarios in batch processes. Chem Eng Res Des 89:2652–2666
Zad SH, Kwong RH, Wonham WM (2003) Fault diagnosis in discrete-event systems: framework and model reduction. IEEE Trans Autom Control 48:1199–1204
Zhao C (2014) Quality-relevant fault diagnosis with concurrent phase partition and analysis of relative changes for multiphase batch processes. AIChE J 60:2048–2062

Chapter 10
Synthesis of Diagnostic Test Plans Based on Hybrid Automata

Unexpected faults and failures in a chemical plant may result in catastrophic consequences. The offline practices of hazard assessment can reduce the expected loss of accidents only to a certain extent, while online fault diagnosis is an alternative option for further improving plant safety. Numerous effective modeling methods have already been proposed to facilitate diagnosis according to the online measurement data, e.g., see Nomikos and MacGregor (1994, 1995), Chen and Jiang (2011), and Dai and Zhao (2011). In addition, Yeh and Chang (2011) showed that another viable approach is to apply diagnostic tests.

It should be noted from the outset that, in the literature, the automata have been utilized primarily for modeling the discrete-event systems (DESs) (Debouk et al. 2000; Benveniste et al. 2003; Zad et al. 2003; Qiu and Kumar 2006; Malik et al. 2011). Although Gascard and Simeu-Abazi (2013) and Gomes Cabral et al. (2015) developed fault diagnosers according to such models, these approaches precluded the use of sequential tests. Yeh and Chang (2011) proposed to follow a trial-and-error automata-based procedure to identify extra actions needed for enhancing diagnostic resolution, while Kang and Chang (2014) developed a systematic synthesis strategy to search for the test plans. Although two subsequent studies have also been carried out to address various implementation issues related to diagnostic tests (Hsieh and Chang 2016; Wang et al. 2017), all of them dealt with only simple material-handling processes and, thus, there are still incentives to improve these automata-based modeling methods for more realistic systems that involve coupled reactive and heat/mass transfer processes.

A hybrid modeling strategy is provided in this chapter to facilitate systematic synthesis of credible diagnostic tests. To this end, the extended finite automata (EFA) (Åkesson et al. 2006) are utilized as the primary building blocks. An extensive set of model configurations are presented in the sequel to incorporate both the simulation data generated by commercial software, e.g., Aspen Plus Dynamics®, and also the generic engineering knowledge into a system automaton. Since the generic engineering knowledge itself is not specific enough for stipulating a manageable

C.-T. Chang et al., *Process Plant Operating Procedures*, Advances in Industrial Control,
https://doi.org/10.1007/978-3-030-70978-5_10

number of event paths even in a moderately complex dynamic system, the simulation data are used to construct a reliable model that characterizes the normal operation succinctly. On the other hand, since not all failure-induced scenarios can be exhaustively simulated in advance, the knowledge-based models are utilized to predict the fault propagation behaviors during diagnostic tests. The feasibility and effectiveness of such a hybrid modeling approach are demonstrated in this chapter with two examples concerning the startup operations of a two-component flash drum and a three-component distillation column.

10.1 Illustrative Example

To facilitate clear illustration of the proposed modeling approach, let us first consider the flash startup operation defined by the process flow diagram (PFD) in Fig. 10.1 and the sequential function chart (SFC) in Fig. 10.2. In this system, there are a heater (HEATER), an inlet valve (Vin), two outlet valves (Vvap and Vliq), and four PID controllers (INPUT_FC, HEATER_TC, FLASH_PC and FLASH_LC) for controlling the feed flowrate, the feed temperature, and the vapor pressure and liquid level in the flash drum. It is assumed that, at steady state, the feed is a mixture of 30 wt% H_2O and 70 wt% methanol and its flowrate, temperature and pressure are kept at 1000 kmol/h, 20 °C and 1.1 bar, respectively. In addition, the initial conditions are set as follows:

- Vin, Vvap, and Vliq are all closed, while the heater is off;
- All controllers are on manual;
- Flash drum is empty and at room temperature.

Finally, let us limit the scope of our consideration here to only six failures, i.e., F1–F6, for simplicity. F1 represents the event that valve Vin sticks at close position (f_VinSC); F2 is a heater failure that stops its energy output (f_H_failed); F3 denotes the faulty condition of drum leaking (f_leaking); F4 is the valve failure that keeps

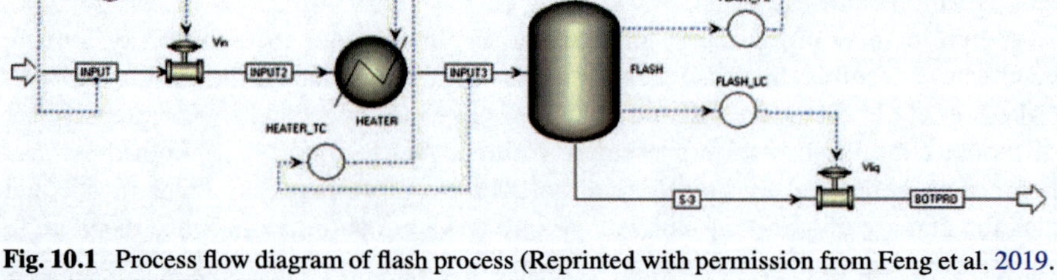

Fig. 10.1 Process flow diagram of flash process (Reprinted with permission from Feng et al. 2019. Copyright 2019 Elsevier)

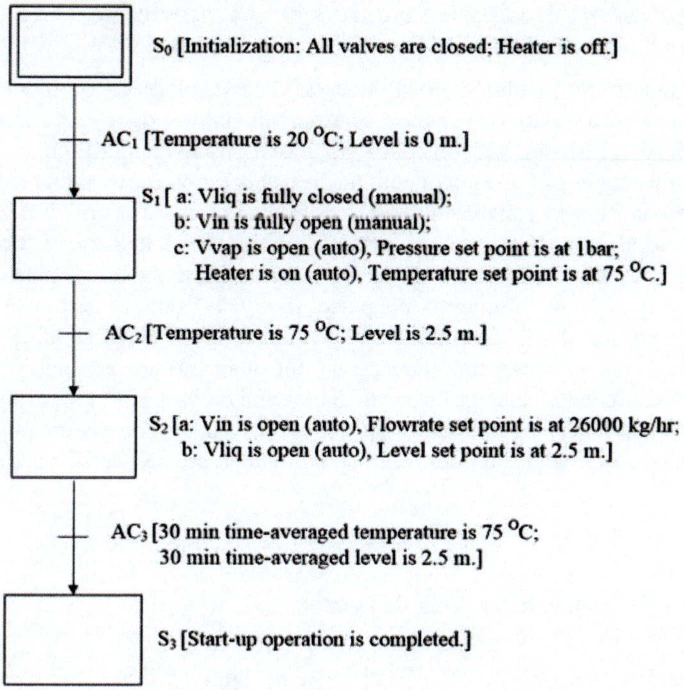

Fig. 10.2 Sequential function chart for startup operation of flash process (Reprinted with permission from Feng et al. 2019. Copyright 2019 Elsevier)

Vin stay at the open position (f_VinSO); F5 is the valve failure that keeps Vliq stay at the close position (f_VliqSC); F6 denotes the scenario that an erroneous set point of temperature controller HEATER_TC causing an excessively large heat flow.

As mentioned in Chap. 3, all sequential procedures can be executed according to a hierarchical structure (see Fig. 3.1). A total of five levels can always be identified, i.e., (1) programmable logic controller (PLC) or operator, (2) PID controller and actuator, (3) process configuration, (4) processing units, and (5) online sensors. The components in the aforementioned flash startup process can be classified as follows:

- Level 1: PLC;
- Level 2: INPUT_FC/Vin, HEATER_TC/Heater, FLASH_PC/Vvap, FLASH_LC/Vliq;
- Level 3: material input and output flows and energy input flow;
- Level 4: flash drum;
- Level 5: level, temperature and pressure sensors.

10.2 Automata Built with Engineering Knowledge

Each of the above-mentioned components can be modeled with an untimed automaton according to the generic engineering knowledge. Traditionally, this automaton is constructed with an ad hoc approach (Wang et al. 2017):

All normal states of the component under consideration must be first enumerated and, if there is a need to analyze the effects of its failure(s), all corresponding failed states should also be taken into account. Each normal or failed state is treated as a distinct place in the automaton, and the place representing the initial state should be marked with an arrow pointing to this place. Every realizable state-transition event should be introduced into the automaton by connecting the corresponding input and output states with a directed arc. If necessary, the guard(s) and variable(s) on this arc can also be included to specify its prerequisites and outcomes, respectively.

For the flash startup operation in Figs. 10.1 and 10.2, the component models built with the above approach and/or the corresponding compressed version are presented below:

10.2.1 PID Controller and Actuator

The lumped model of INPUT_FC/Vin can be built with the knowledge-based approach according to the assumptions that INPUT_FC is direct acting and Vin is air-to-open (A/O). By following the traditional modeling approach described above, one can construct the automaton for the latter shown previously in Chap. 6 (see Fig. 6.3 and the corresponding illustration in Sect. 6.3.1). Since each component state is characterized with at least two places in this automaton and also in automata described in Chaps. 3, 6, and 9, it is clear that the traditional approach is only effective for modeling systems with relatively few state variables. To facilitate easy construction and concise presentation of automata, the "compressed" versions are adopted in the present chapter. In particular, the components Vin and INPUT_FC are modeled alternatively with automata using significantly fewer places and transitions (see Figs. 6.16 and 6.17 together with the corresponding explanations in Appendix 6.1). Note that the other level-2 components, i.e., HEATER_TC/Heater, FLASH_PC/Vvap, and FLASH_LC/Vliq, can be modeled with the same approach.

10.2.2 Process Configuration

The so-called process configuration is represented with automata to describe the material and energy flows that connect major processing units. Since in the present example there is only a single unit, i.e., the flash drum itself, it is only necessary to model its inlet flow (Fin), its outlet vapor and liquid flows (Fvap and Fliq) and its energy input flow from HEATER (Feng). All four automata can be found in Fig. 6.18.

10.2.3 Processing Units

Without loss of generality, let us neglect the pressure effects and use only two state variables, i.e., liquid temperature and level, to model the flash drum. Let us further assume that, on the basis of prior operational experiences, the entire ranges of temperature and level variations may be discretized into 3 and 4 qualitative states, respectively. The former range is partitioned into $\left[25^{-}\,°\text{C}, 25^{+}\,°\text{C}\right)$, $\left[25^{+}\,°\text{C}, 75^{+}\,°\text{C}\right)$, and $\left[75^{+}\,°\text{C}, \infty\,°\text{C}\right)$ and each is assigned with an integer value, e.g., 0, 1, and 2, while the latter may be divided into $\left[0\,\text{m}, 0^{+}\,\text{m}\right)$, $\left[0^{+}\,\text{m}, 2.5^{-}\,\text{m}\right)$, $\left[2.5^{-}\,\text{m}, 2.5^{+}\,\text{m}\right)$, $\left[2.5^{+}\,\text{m}, 4.2^{-}\,\text{m}\right)$, and $\left[4.2^{+}\,\text{m}, \infty\,\text{m}\right)$ and labeled with 0, 1, 2, 3, and 4, respectively. Thus, the flash drum may be characterized with 15 ($= 3 \times 5$) possible states and at most 210 ($= 15 \times 14$) transitions. Since it is quite difficult to depict and visualize the corresponding automaton obtained with the traditional modeling approach, the compressed automata in Figs. 10.3 and 10.4 are used instead to respectively represent the dynamic behaviors of temperature and level in flash operation.

Ftemp_inc in Fig. 10.3 denotes the event causing an increase in temperature, while Ftemp_des is the event resulting in a temperature decrease. On the other hand, Ftemp_average is used to model the steady-state scenario. The corresponding causal relations are summarized as follows:

- Ftemp_inc: If the liquid temperature in flash drum is lower than that of the heating fluid in HEATER (i.e., PU_Ftemp $<$ A_H), then the former should be increased by one unit (PU_Ftemp+ $=$ 1);
- Ftemp_des: If the mixture temperature in flash drum is higher than that of the heating fluid in HEATER (i.e., PU_Ftemp $>$ A_H), then the former should be decreased by one unit (PU_Ftemp− $=$ 1);
- Ftemp_average: If in a specified period of time (1) the liquid temperature in flash drum always equals that of the heating fluid in HEATER (i.e., PU_Ftemp $==$

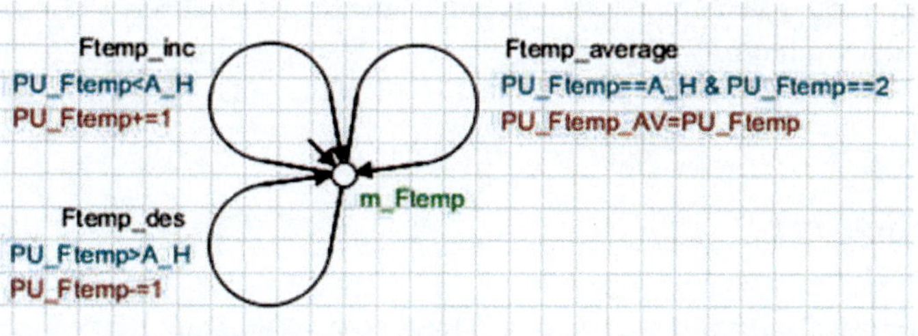

Fig. 10.3 Automaton model of temperature variation in flash drum (Reprinted with permission from Feng et al. 2019. Copyright 2019 Elsevier)

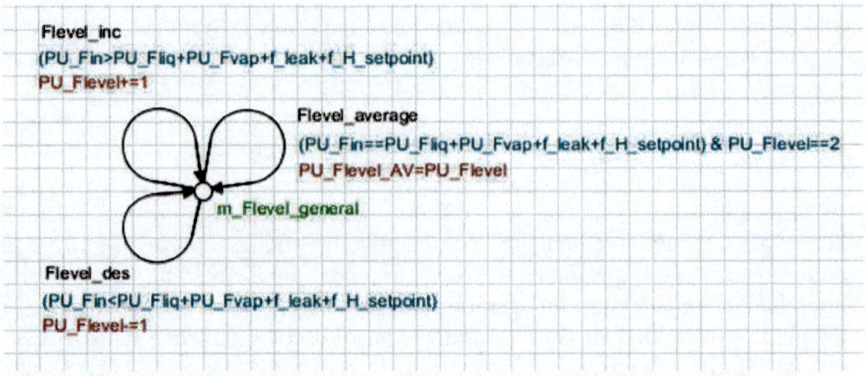

Fig. 10.4 Automaton model of level variation in flash drum (Reprinted with permission from Feng et al. 2019. Copyright 2019 Elsevier)

A_H) and also (2) this temperature stays at the designated steady-state value of 2 (PU_Ftemp == 2), then the time-averaged temperature should be at the same value of 2 (PU_Ftemp_AV = PU_Ftemp).

Similarly, Flevel_inc in Fig. 10.4 clearly denotes the event resulting in a level increase, while Flevel_des an event producing the opposite effect. The third event Flevel_average is used to characterize the steady-state operation in which material balance is satisfied. The sufficient conditions of these three events and their outcomes are specified in the corresponding guards and variable updates respectively. More specifically, these causal relations can be described as follows:

- Flevel_inc: If the inlet flow rate is greater than the sum of liquid and vapor product flow rates and leak rate (i.e., PU_Fin > PU_Fliq + PU_Fvap + f_leak), then the height of liquid level should be increased by one unit (PU_Flevel+ = 1);
- Flevel_des: If the inlet flow rate is less than the sum of liquid and vapor product flow rates and leak rate (i.e., PU_Fin < PU_Fliq + PU_Fvap + f_leak), then the height of liquid level should be decreased by one unit (PU_Flevel− = 1);
- Flevel_average: If in a specified period of time (1) the inlet flow rate always equals the sum of liquid and vapor product flow rates and leak rate (i.e., PU_Fin == PU_Fliq + PU_Fvap + f_leak) and also (2) the liquid level stays at the designated steady-state value of 2 (PU_Flevel == 2), then the time-averaged liquid level should be at the same value of 2 (PU_Flevel_AV = PU_Flevel).

10.2.4 Sensors

Each sensor can also be modeled with a compressed automaton. Every measurement-taking action is represented with a self-recycle loop as shown in Fig. 10.5 for the temperature sensor in the flash startup example. As mentioned previously, the entire

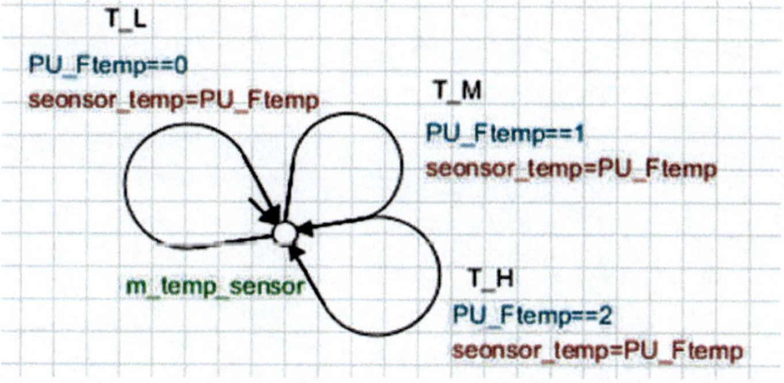

Fig. 10.5 Automaton model for temperature sensor in flash drum (Reprinted with permission from Feng et al. 2019. Copyright 2019 Elsevier)

range of temperature variation is discretized into 3 intervals. A distinct self-recycle loop is introduced into the sensor model to represent each of these measurements taking actions. To simplify the discussion, sensor failures are excluded in the present example and it is assumed that the sensor measurements always truthfully reflect the corresponding state variables of flash drum. Note that other sensor models can be built with the same approach.

10.2.5 PLC or Operator

The PLC (or operator) model can be constructed according to the given SFC. The operation steps in SFC should be represented with the corresponding events in an automaton, while the activation conditions of these steps are specified in the event guards. The actuator states reached after executing a particular group of simultaneous steps in SFC should be explicitly stipulated in the variable updates of the corresponding transition. The precedence order of activation conditions and operation steps in the given SFC should be exactly the same as that of guards and events in the automaton.

By following the above principles, one should be able to translate the SFC in Fig. 10.2 into the automaton in Fig. 10.6. It can be observed from Fig. 10.2 that a time period of at least 30 min is required to confirm activation condition AC_3 after completing the operation steps in S_2. Since the elapsed time cannot be expressed explicitly in an untimed automaton, the wait action required in S_3 is treated as a fictitious step and it is reflected with the self-recycle loops in Figs. 10.3 and 10.4, i.e., Ftemp_average and Flevel_average. These two events should take place after implementing S_2 when s_flow = 2 and s_level = 1, and the level and temperature updates, i.e., PU_Flevel_AV == 2 and PU_Ftemp_AV == 2, can be regarded as the results of online measurements and calculations needed for confirmation of steady

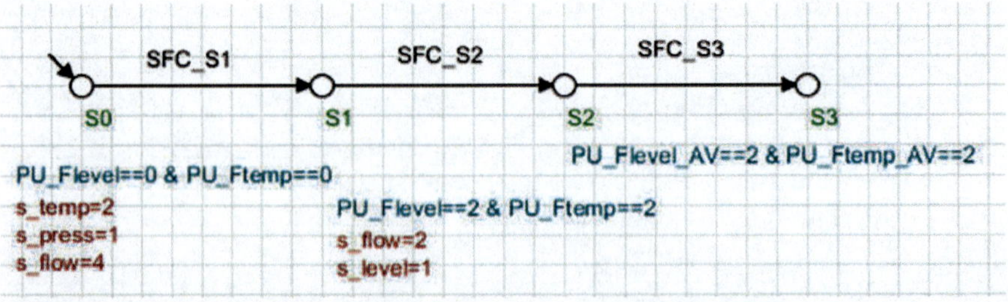

Fig. 10.6 Automaton model of PLC/operator in flash startup example (Reprinted with permission from Feng et al. 2019. Copyright 2019 Elsevier)

state and, consequently, they are used as the guards of SFC_S3 for terminating the startup operation.

10.2.6 System Hierarchy

To impose the causal relationships implied by the generic system hierarchy in Fig. 3.1, it is necessary to build an automaton accordingly. Figure 10.7 shows this model for the flash startup operation. Notice that there is a one-to-one correspondence between the five places in this automaton, i.e., layer1 to layer5, and the five hierarchical levels in Fig. 3.1, and that the events allowed in each level are specified on the self-recycle loop at the corresponding place. All such events are listed and discussed below.

- On layer1, events SPC_S1, SPC_S2, and SPC_S3 represent the operation steps S_1, S_2, and S_3 in SFC (see Figs. 10.2 and 10.6);

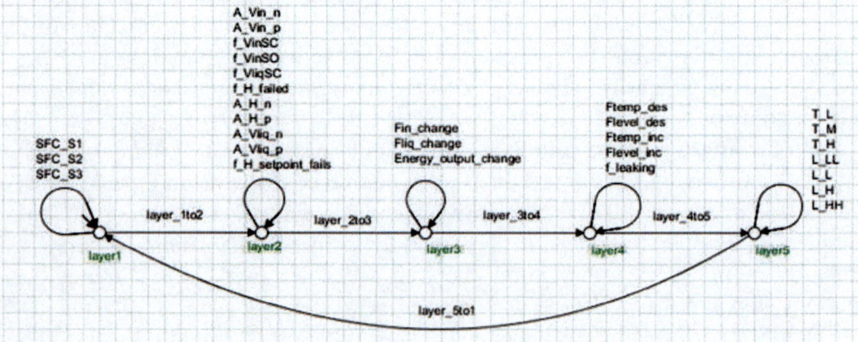

Fig. 10.7 Automaton representing system hierarchy (Reprinted with permission from Feng et al. 2019. Copyright 2019 Elsevier)

- On layer2, there are two types of events. The normal actuator movements, i.e., A_Vin_n, A_Vin_p, A_Vliq_n, A_Vliq_p, A_H_n, and A_H_p, are classified as events of the first type, while the abrupt failures of actuator and/or PID controller, i.e., f_Vin_SC (F1), f_Vin_SO (F4), f_Vliq_SC (F5), f_H_failed (F2), and f_H_setpoint_fails (F6), are type-2 events.
- On layer3, the system configuration is manipulated by varying the material and energy flows among processing units. For the flash drum, the inlet and outlet material flows are adjusted with two corresponding events, i.e., Fin_change and Fliq_change, and the heat flow from the heater to flash drum is controlled via Energy_output_change.
- On layer4, the allowed changes in state variables of the processing unit(s), i.e., level and temperature variations in the flash drum, are characterized with corresponding events om self-recycle loops, i.e., Flevel_des, Flevel_inc, $\mathrm{Ftemp}_{\mathrm{des}}$, and Ftemp_inc. As mentioned before, the vessel failure f_leaking (F3) should also be included as an event in level 4.
- On layer5, the allowed events can be regarded as the measurement-taking actions, i.e., L_LL, L_L, L_H, L_HH, T_L, $\mathrm{T_M}$, and T_H.

10.2.7 Path Explosion

The system model can be synthesized by integrating all automata developed in the previous six subsections via the parallel composition operation described in Chap. 3. Although several equally effective tools are available, the free software SUPREMICA (Åkesson et al. 2006) is utilized here to synchronize all component events. Since so far only the generic engineering knowledge is utilized to build the component models, an overwhelmingly large number of paths (strings) may be extracted from the integrated automaton even when the system dynamics is moderately complex.

To illustrate this drawback more clearly, let us consider the flash startup procedure under normal operating conditions only. After removing all failures and failed states from the aforementioned component models, the parallel composition operation can be carried out to produce an extremely complicated path network. This undesirable phenomenon of path explosion can be attributed to the fact that, since the dynamic behavior of any MIMO system cannot be adequately described with the untimed automata developed on the basis of generic engineering knowledge only, all unspecified feasible paths during operation are exhaustively enumerated. For example, as indicated in the SFC in Fig. 10.2, the level and temperature in flash drum are supposed to be raised to 2.5 m and 75 °C respectively (i.e., $\mathrm{AC_2}$) after implementing the operation steps in $\mathrm{S_1}$. However, since the precedence order of level and temperature increases is not given, all possible paths ending at these targets are generated according to the fundamental definition of parallel composition. Thus, it is obvious that some of the paths generated by SUPREMICA may not be real and, moreover, the flawed operating procedures that contain deadlocks and/or live-locks (Cassandras and Lafortune 2008) may even be present in this unrealistic path network.

10.3 Automata Built with Simulation and/or Historical Data

From the above discussions, it is clear that the knowledge-based models must be further constrained to facilitate effective identification of feasible paths for diagnostic tests. To this end, additional automata are constructed according to simulation and/or historical data for better representations of the dynamic behaviors of the fourth-level units under *normal* operating conditions. This modeling practice can be justified on the basis of the following observations:

- The undesired phenomenon of path explosion can be mainly attributed to the complex dynamic behaviors of processing units during normal operations;
- Every failure-induced abnormal trace is always emanated from a particular normal state and there can be a large number of paths leading to this normal state;
- Simulation and historical data of fault propagation scenarios are rarely available, while such data are abundant for the normal processes.

To facilitate clear illustration of this data-based model building strategy, let us again consider the flash startup operation described in Figs. 10.1 and 10.2. Figures 10.8, 10.9, 10.10 and 10.11 show the corresponding simulation data generated with Aspen Plus Dynamics, and these continuous data must be first discretized into several intervals. The discretization schemes of the state variables of the processing unit, i.e., the liquid level and temperature of flash drum, have already been given

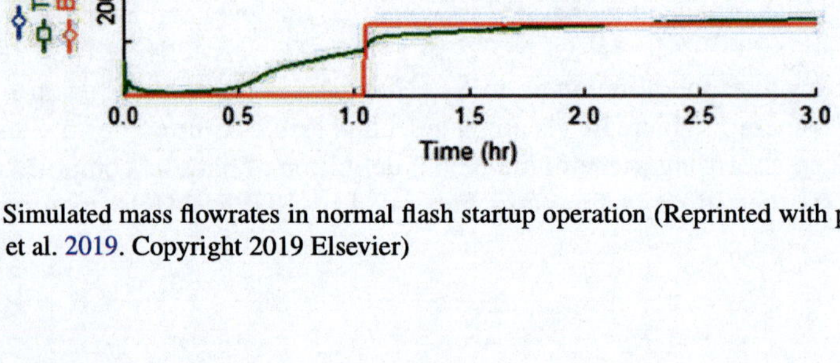

Fig. 10.8 Simulated mass flowrates in normal flash startup operation (Reprinted with permission from Feng et al. 2019. Copyright 2019 Elsevier)

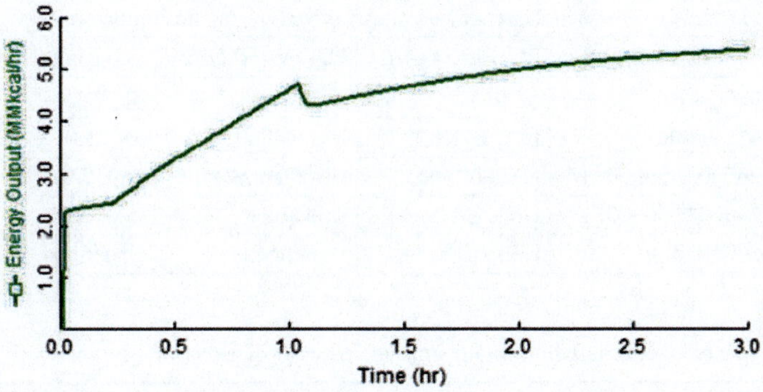

Fig. 10.9 Simulated energy flowrate in normal flash startup operation (Reprinted with permission from Feng et al. 2019. Copyright 2019 Elsevier)

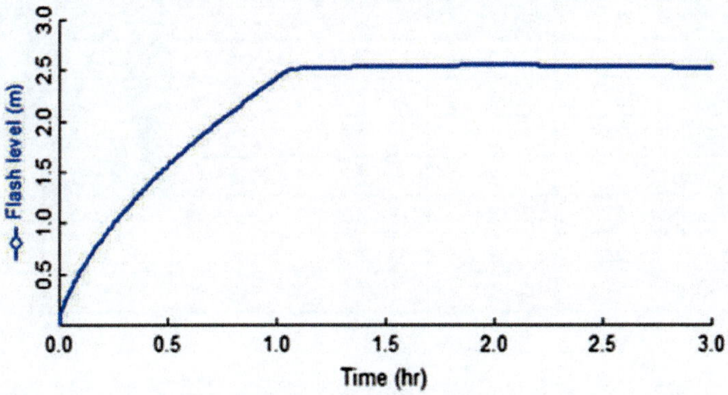

Fig. 10.10 Simulated liquid level in normal flash startup operation (Reprinted with permission from Feng et al. 2019. Copyright 2019 Elsevier)

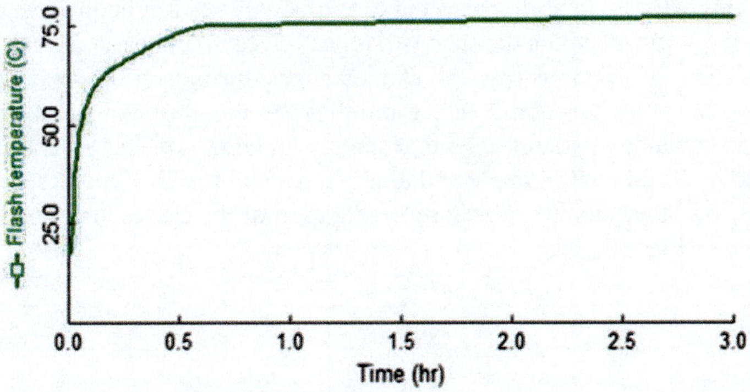

Fig. 10.11 Simulated liquid temperature in normal flash startup operation (Reprinted with permission from Feng et al. 2019. Copyright 2019 Elsevier)

Table 10.1 Discretized mass and energy flowrates for configuring flash process

Value	PU_Fin (kg/h)	PU_Fliq (kg/h)	PU_Fvap (kg/h)	PU_Feng (MMkcal/h)
0	$[0, 0^+)$	$[0, 0^+)$	$[0, 0^+)$	$[0, 0^+)$
1	$[0^+, 26000^-)$	$[0^+, 13000^-)$	$[0^+, 13000^-)$	$[0^+, 3.0)$
2	$[26000^-, 26000^+)$	$[13000^-, \infty)$	$[13000^-, \infty)$	$[3.0, 5.0)$
3	$[26000^+, 50000)$	Undefined	Undefined	$[5.0, \infty)$
4	$[50000, \infty)$	Undefined	Undefined	Undefined

Table 10.2 Discretized simulation data in normal startup operation of flash process

Time	PU_Ftemp	PU_Flevel	PU_Feng	PU_Fin	PU_Fliq	PU_Fvap
0	0	0	0	0	0	0
0.01	0	0	2	4	0	0
0.03	1	0	2	4	0	0
0.09	1	1	2	4	0	0
0.55	1	1	2	4	0	0
0.65	2	1	2	4	0	0
0.89	2	1	2	4	0	1
1.05	2	2	2	2	1	1
2.05	2	2	2	2	1	1
3	2	2	2	2	1	1

previously in Sect. 10.2.3, while those of the material and energy input and/or output flows, can be found in Table 10.1. Based on these discretization schemes, all simulation data in Figs. 10.8, 10.9, 10.10 and 10.11 can be converted to their qualitative values in Table 10.2. An abridged version of this data set can be produced next by removing every row in which the state variables, i.e., temperature and level, are identical to those in the previous row. All state changes during startup operation can be easily extracted from this abridged set and they are incorporated in the automaton model as a sequence of state-transition events. A total of six (6) consecutive events were found in the present example and their guards and variable updates are listed in Table 10.3. An alternative graphical representation of the corresponding automaton is presented in Fig. 10.12.

Table 10.3 State-transition events in normal startup operation of flash process

	Guard						Variable	
	PU_Ftemp	PU_Flevel	PU_Feng	PU_Fin	PU_Fliq	PU_Fvap	db_Ftemp	db_Flevel
1	0	0	0	0	0	0	0	0
2	0	0	2	4	0	0	1	0
3	1	0	2	4	0	0	0	1
4	1	1	2	4	0	0	1	0
5	2	1	2	4	0	1	0	1
6	2	2	2	2	1	1	0	0

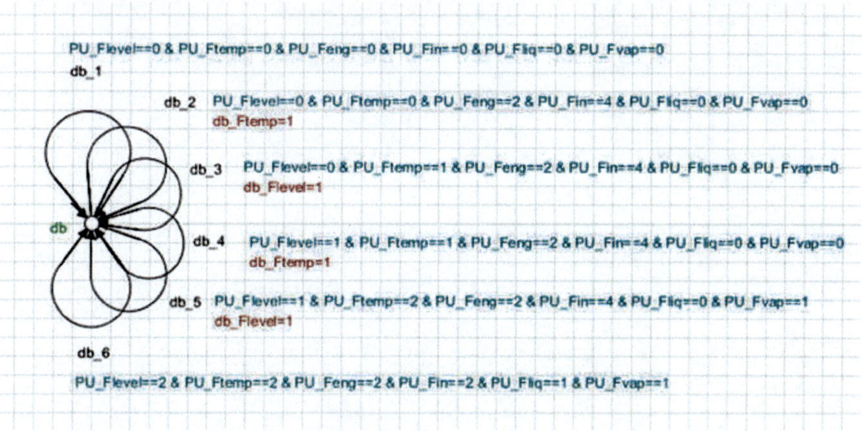

Fig. 10.12 Graphical representation of automaton specified in Table 10.3 (Reprinted with permission from Feng et al. 2019. Copyright 2019 Elsevier)

10.4 Hybrid Models

- Since simulation and historical data of failure-induced scenarios are often not available in practical applications, it becomes necessary to predict, at least qualitatively, the fault propagation behaviors based on the engineering knowledge in these situations. Thus, for the purpose of characterizing the normal and abnormal operation modes of every processing unit, the data-based model and its knowledge-based counterpart must be both incorporated into a hybrid automaton. In the flash startup example, the switch actions from the data-based models of the normal operations to the knowledge-based models after the inception of failure(s) can be triggered with the automata shown in Figs. 10.13 and 10.14. The implied switching mechanisms are summarized below:

  – The state m_Ftemp_db in Fig. 10.13 refers to the data-based temperature prediction model, while m_Ftemp_general denotes the knowledge-based counterpart. Since the system is assumed to be normal initially, the former is adopted

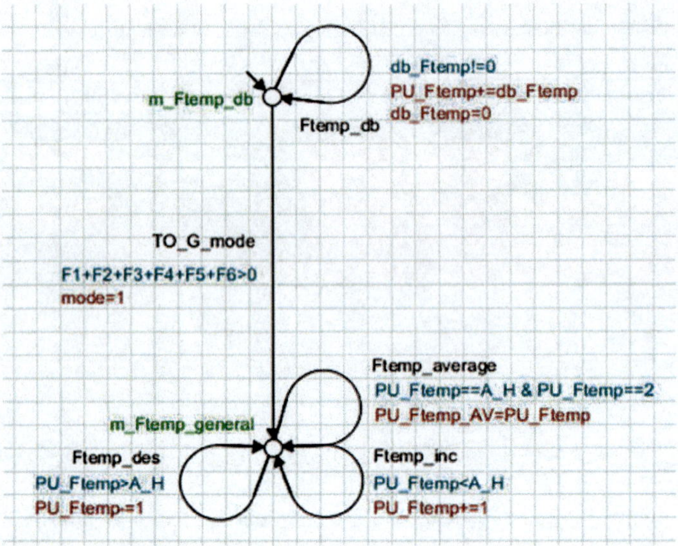

Fig. 10.13 Switching mechanism for predicting temperature variation (Reprinted with permission from Feng et al. 2019. Copyright 2019 Elsevier)

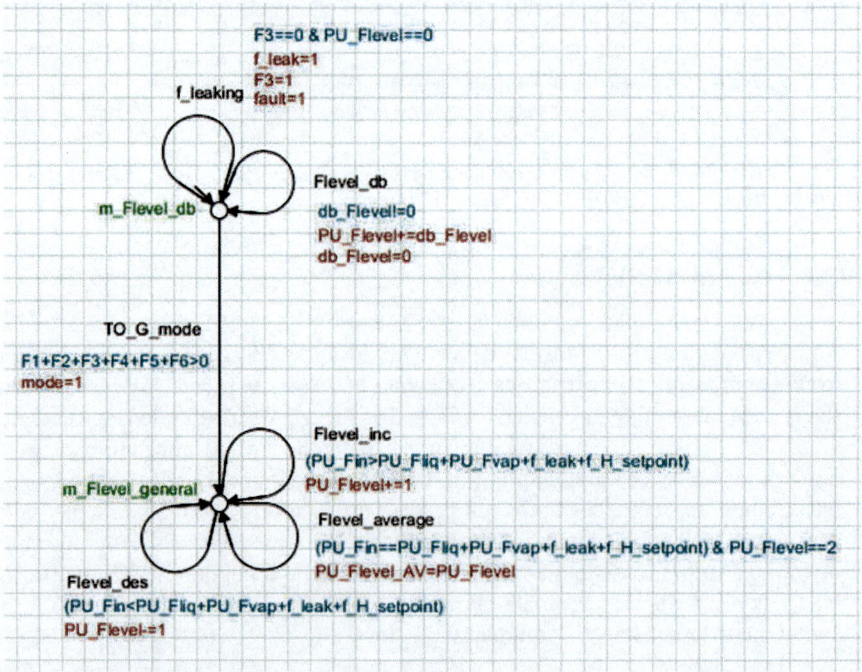

Fig. 10.14 Switching mechanism for predicting level variation (Reprinted with permission from Feng et al. 2019. Copyright 2019 Elsevier)

to determine the temperature variation during the initial stage. Notice that the guard of event Ftemp_db can be satisfied after finishing the transitions specified in the second and fourth rows of Table 10.3 (or the corresponding events db_2 and db_4 in Fig. 10.12). However, if one or more failure occurs, i.e., $F1+F2+F3+F4+F5+F6 > 0$, at an instance between any two consecutive transitions, the switch action To_G_mode can be triggered to start utilizing the knowledge-based model for predicting temperature (which has already been presented previously in Fig. 10.3).
– The state m_Flevel_db in Fig. 10.14 refers to the data-based level prediction approach, while m_Flevel_general denotes the knowledge-based counterpart. Since the initial system state is normal, the former should be adopted to determine the level variation initially. Notice that the guard of event Flevel_db is satisfied after completing the third and fifth transitions in Table 10.3 (or the corresponding events db_3 and db_5 in Fig. 10.12). Notice that, since F3 (i.e., a leak develops in the flash drum) is a failure of the processing unit under consideration, it is modelled as a self-loop on m_Flevel_db. If at an instance between any two consecutive transitions one or more failure occurs, i.e., $F1 + F2 + F3 + F4 + F5 + F6 > 0$, the switching action To_G_mode can be activated to predict level variation according to the knowledge-based model and this model is already given in Fig. 10.4.

Finally, in order to integrate the above hybrid models into the system model for generating realistic event paths, it is necessary to incorporate the additional events utilized in data-based models into the event list on layer4 of the system hierarchy. For the flash startup example, these extra events, i.e., db_1–db_6, Ftemp_db and Flevel_db, should be included on the self-recycle loop of layer4 in Fig. 10.7.

10.5 Observable Event Traces

Although a hardware item may fail at any instance during routine operation, this failure is usually unobservable. Fault diagnosis can be performed to identify the root cause(s) of abnormal system state based on available current and past online data. It is assumed here that the observable events are limited to those associated with actuator actions and sensor measurements. On the basis of this assumption, the parallel composition can be applied to integrate all automata mentioned above so as to synthesize a diagnoser in which all observable event traces (OETs) are embedded.

Let us again use the same flash startup process as an example for illustration convenience. All aforementioned automata can be synchronized to produce the diagnoser in Fig. 10.15. Notice that the normal system states are denoted by rectangles inscribed with label "N" and they are connected by directed arcs marked with appropriate activation conditions (AC_1, AC_2, and AC_3) and operation steps (S_1 and S_2) in SFC. If executing a particular step does not yield the anticipated activation condition, then the corresponding abnormal state, which is also denoted by a rectangle, should be

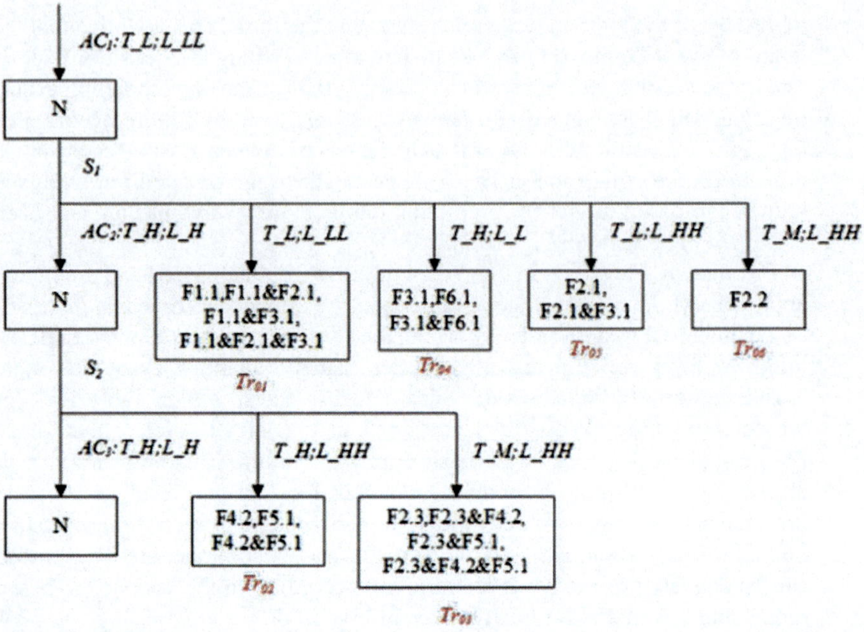

Fig. 10.15 Diagnoser in flash startup example (Reprinted with permission from Feng et al. 2019. Copyright 2019 Elsevier)

placed on an additional branch emanating from the step under consideration to highlight the strayed system condition. Notice that more than one abnormal state may appear and they can usually be identified according to the online measurements. It can be observed from Fig. 10.15 that four OETs emerge after S_1 and two after S_2. For each OET in this diagnoser, the abnormal measurements are underlined and the corresponding fault origins are listed in the rectangle denoting the end state. Note that the fault origins of an OET are separated by commas and, if there are multiple failures in a fault origin, then they are connected by the symbol &. All multi-origin OETs in Fig. 10.15 are clearly not diagnosable, while Tr_{06} is the only exception.

Note that all six failures considered in the flash startup example have been expressed previously in Sect. 10.1 according to a unified standard format, i.e., Fi ($i = 1, 2, \ldots, 6$). In order to further specify the occurrence time of every failure, it is necessary to attach an additional index to this notation. In particular, a failure in diagnoser is represented in the form of Fi.j and the additional index j ($= 1, 2, 3$) denotes the time period between the two instances when the consecutive activation conditions AC_{j-1} and AC_j are satisfied. Notice also that, for any given failure, not all periods are considered in the present example for illustration conciseness. Following is a list of failure events included in the diagnoser in Fig. 10.15:

- Since F1 and F5 are both valve failures and all valves are closed initially, their effects can be detected when the operator starts to open Vin and Vliq, respectively.

Thus, it is only necessary to analyze the scenarios in which F1 takes place before AC_1 and F5 before AC_2. As a result, only F1.1 and F5.1 are included in diagnoser.
- A heater failure (F2) may be revealed at any time when the heater is on and, thus, the possible events should be F2.1, F2.2, and F2.3.
- A leak in the flash drum (F3) may develop at any time during the startup operation. However, only F3.1 is considered here to simplify the subsequent analysis.
- Since F4 (Vin sticks at the open position) can happen after it is opened manually in S_1 and this failure is detectable after the inlet flow controller is switched to the auto mode in S_2, it is therefore only necessary to consider F4.2.
- Since F6 is due to an incorrect controller setting, it can be revealed after HEATER_TC is switched to the auto mode at step S_1. Consequently, F6.1 should be the only possible failure event in this situation.

10.6 Test Plan Synthesis

If more than one fault origin is implicated by an OET, a test plan may be devised to enhance diagnostic resolution by carrying out a dedicated operating procedure or test plan. To this end, a series of structurally identical automata should be built to drive the given system to produce a unique set of sensor measurements for each fault origin. Figure 10.16 shows the generalized auxiliary automaton for generating a diagnostic step. The embedded events can be interpreted as follows:

- The failure_maybe_events are OET-implicated failures which may or may not be present;
- The failure_shouldbe_events denote the OET-confirmed failures;
- The actuators_events include all possible actuator adjustments;
- The sensors_events include all possible measurement-taking actions;
- The diagnostic_test represents a single step in the test plan. Its guards should be an exhaustive list of all possible combinations of sensor measurements except that associated with the operating conditions just before implementing the present step. In other words, each combination in this list can be viewed as a unique reachable system state.

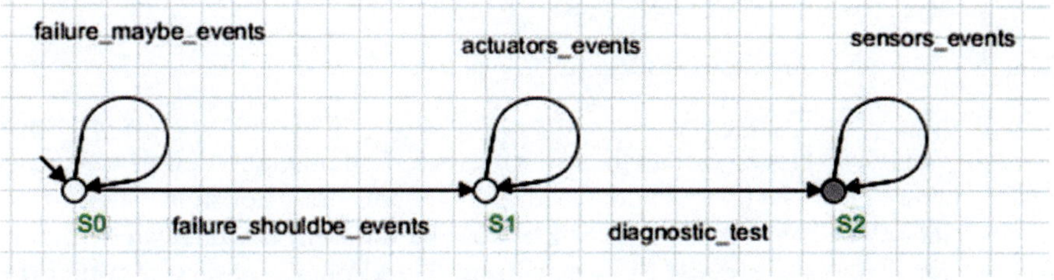

Fig. 10.16 Auxiliary automaton for conjecturing a single diagnostic step (Reprinted with permission from Feng et al. 2019. Copyright 2019 Elsevier)

From Fig. 10.16, it is clear that multiple actuator actions may be required in a single diagnostic step. It should also be noted that more than one step may be needed in a complete test plan. As suggested previously in Chap. 9, the total number of diagnostic steps (L) should be bounded from above and below, i.e., $m \leq L \leq M$, and these bounds for a given OET can be determined with the following formulas:

$$M = F - 1 \tag{10.1}$$

$$m = \left\lceil \frac{\ln F}{\ln R} \right\rceil \tag{10.2}$$

where F is the number of fault origins; R denotes the total number of online sensors; $\lceil \cdot \rceil$ denotes the ceiling operator. Once obtaining the upper and lower bounds, it is practically feasible to synthesize a multi-step test plan by trial and error. Since the generalized search procedure has already been detailed in the previous chapter, a specific example is given here to illustrate the synthesis procedure. Let us revisit the diagnoser of the flash startup process in Fig. 10.15 and consider only trace Tr_{02} to provide a concise explanation of the test-plan synthesis strategy. The required tasks are explained below:

i. Discard the PLC/operator model in level 1 (see Fig. 10.6).
ii. Identify the final component states on Tr_{02}, and use them as the initial conditions of the corresponding test procedure. If a component is always normal in this scenario, its state achieved after implementing step S_2 can be extracted from the given SFC. On the other hand, since the fault origins of Tr_{02} may be F4 (f_VinSO), F5 (f_VliqSC) or F4&$F5$ (f_VinSO and f_VliqSC), the failed states of Vin and Vliq cannot be confirmed without the diagnostic tests. Therefore, the component states of Vin and Vliq before the diagnostic tests should be set respectively as follows: A_Vin = 4 (i.e., the inlet valve is normal and fully open) and A_Vliq = 0 (i.e., the outlet valve of the liquid output is normal and fully closed), and a list of all corresponding component states is given below.

    - level 2: A_Vin = 4, A_Vliq = 0, A_H = 2;
    - level 3: PU_Fin = 4, PU_Fliq = 0, PU_Fvap = 1, PU_Feng = 2;
    - level 4: PU_Flevel = 3, PU_Ftemp = 2;
    - level 5: sensor_level = 3, sensor_temp = 2.

iii. Modify the component models for test-plan synthesis.

    - Since the SFC of a test plan is supposed to be synthesized with this proposed diagnostic procedure and therefore the level-1 component should be unavailable a priori, let us start with the level-2 components. The original automaton model of Vin (see Fig. 6.16) can now be tailored to a dedicated version for Tr_{02} (see Fig. 10.17). In addition, based on the observation that the implicated failures of Tr_{02} do not include F1 (f_VinSC), the corresponding places and arcs can be removed from the original model of Vin. On the

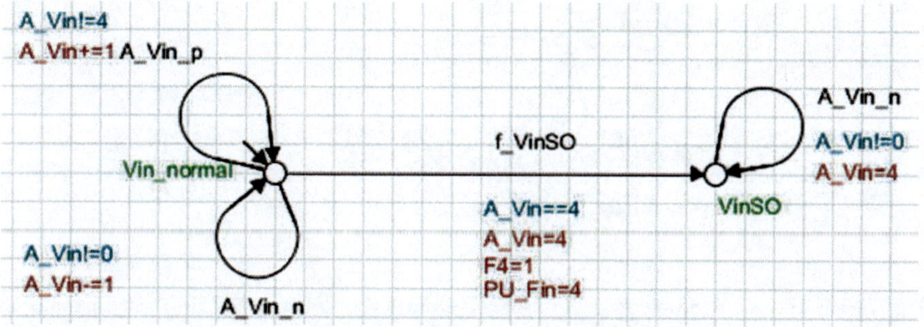

Fig. 10.17 The modified automaton model of Vin for Tr_{02} (Reprinted with permission from Feng et al. 2019. Copyright 2019 Elsevier)

other hand, since the presence (or absence) of failure F4 (f_VinSO) cannot be categorically confirmed or rejected after observing Tr_{02}, it is necessary to set the initial condition of valve Vin to be that just before this failure, i.e., A_Vin = 4, so as to incorporate all possibilities in the test plan. Finally, the other level-2 component models, i.e., those for HEATER_TC/Heater, FLASH_PC/Vvap, and FLASH_LC/Vliq, can be modified with the same method and, for the sake of brevity, they are omitted here.

- Since none of the assumed failures in the present example, i.e., F1 – F6, are included in the level-3 automata, no modifications (except the initial conditions specified in step 2) are needed to model the process configuration.
- The level-4 automata in Figs. 10.18 and 10.19 are the modified versions of Figs. 10.13 and 10.14 for modeling the temperature and level variations of flash drum, respectively, in the diagnostic tests for OET Tr_{02}. It is determined in step 2 that, although the drum temperature is still normal (PU_Ftemp = 2) before implementing the test plan, the liquid level is abnormally high (PU_Flevel = 3). Since failure F3 is not included in the fault origins of Tr_{02}, the self-looping event f_leaking should not be included in Fig. 10.19. Finally, because of the fact that the data-based models are

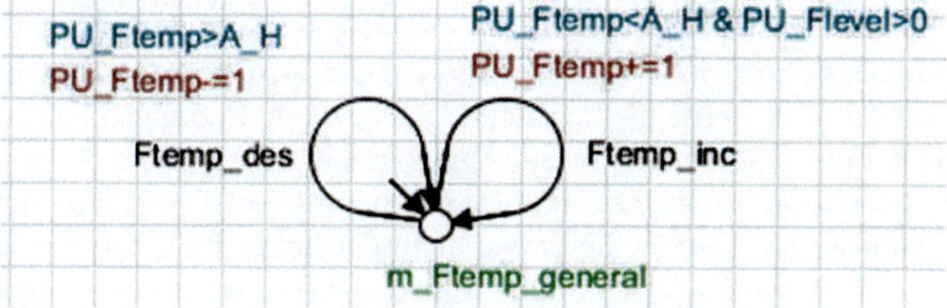

Fig. 10.18 The modified automaton model of drum temperature variation for Tr_{02} (Reprinted with permission from Feng et al. 2019. Copyright 2019 Elsevier)

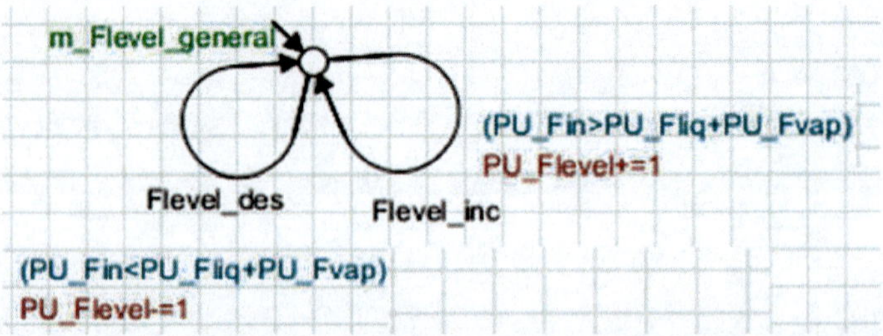

Fig. 10.19 The modified automaton model of drum level variation for Tr_{02} (Reprinted with permission from Feng et al. 2019. Copyright 2019 Elsevier)

constructed according to simulation results of the *normal* operations only, they should also be removed from the hybrid models.

- Since the assumed failures in the present example, i.e., F1 − F6, are not concerned with the level-5 components, no modifications (except the initial conditions specified in step 2) are needed to model the online sensors.

iv. Construct the auxiliary automata and perform parallel compositions iteratively to maximize the total number of distinct traces and each trace must end at a unique set of online measurements.

- Since the number of fault origins of Tr_{02} is three, i.e., F = 3, the upper bound (M) and lower bound (m) of the step number of the diagnostic tests (L) can be determined according to Eqs. (10.1) and (10.2), i.e., M = 2 and m = 1. Therefore, at most two auxiliary automata are needed.
- Figure 10.20a shows the auxiliary automaton used for generating the first diagnostic step. The self-looping events attached on node S0 are failures that may or may not be present, i.e., F4 (f_VinSO) and F5 (f_VliqSC), while those on S1 should be all feasible actuator actions. Finally, the self-looping events on the marked state S2 should include all possible sensor readings. The guard-less event between S0 and S1 (void) indicates that no incidences are required to take place, while either of the two guards of the next event between S1 and S2 (diagnostic_test), i.e., PU_Ftemp! = 2 or PUFlevel! = 3, is adopted primarily to prohibit reaching the starting conditions of the test process. By applying the parallel composition on all aforementioned modified automata and the auxiliary automaton given in Fig. 10.20a, it is possible to isolate only one fault origin, i.e., F5 (f_VliqSC), on a unique trace that represents this first diagnostic step.
- Figure 10.20b shows the auxiliary automaton used for generating the next diagnostic step. Since the two indistinguishable fault origins before the second step are F4 (f_VinSO) and F4 & $F5$ (f_VinSO and f_VliqSC), this automaton can be constructed simply by moving f_VliqSO from the self-looping arc on S0 in Fig. 10.20a and replacing event void with this failure. This practice implies that failure F4 is treated as a certain event.

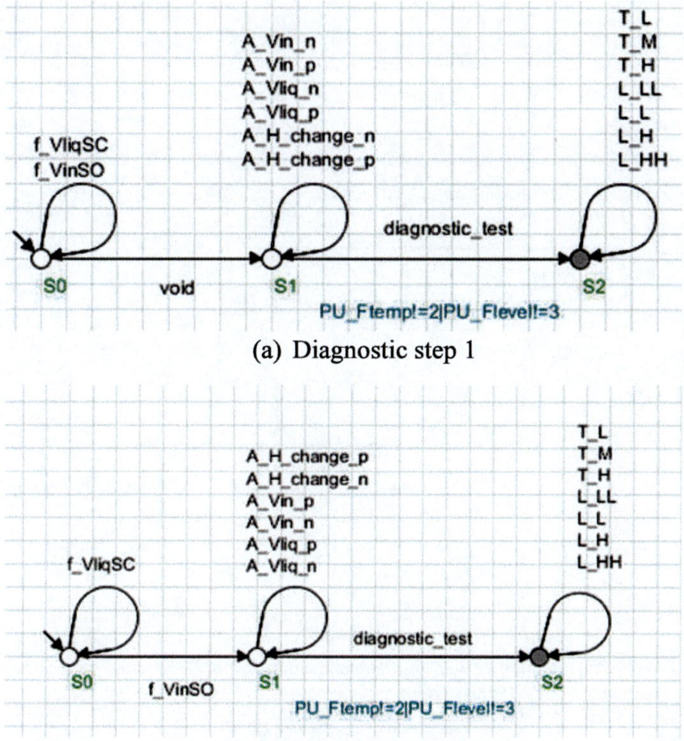

(a) Diagnostic step 1

(b) Diagnostic step 2

Fig. 10.20 Auxiliary automata for diagnostic tests of Tr_{02} (Reprinted with permission from Feng et al. 2019. Copyright 2019 Elsevier)

v. Summarize the OETs generated in the previous step with a SFC. This test plan for Tr_{02} can be found in Fig. 10.21.

10.7 Validation of Test Plans

10.7.1 Trace $Tr_{02.1}$

This OET and the subsequent diagnostic tests specified in Fig. 10.21 were simulated with Aspen Plus Dynamics by introducing fault origin F5 (f_VliqSC) at 0 h. Let us consider only the simulated level variation in this scenario in Fig. 10.22. A vertically upward arrow is inserted into this figure at 1.55 h to denote the time of fault detection and the entire horizon can then be divided into two separate periods accordingly.

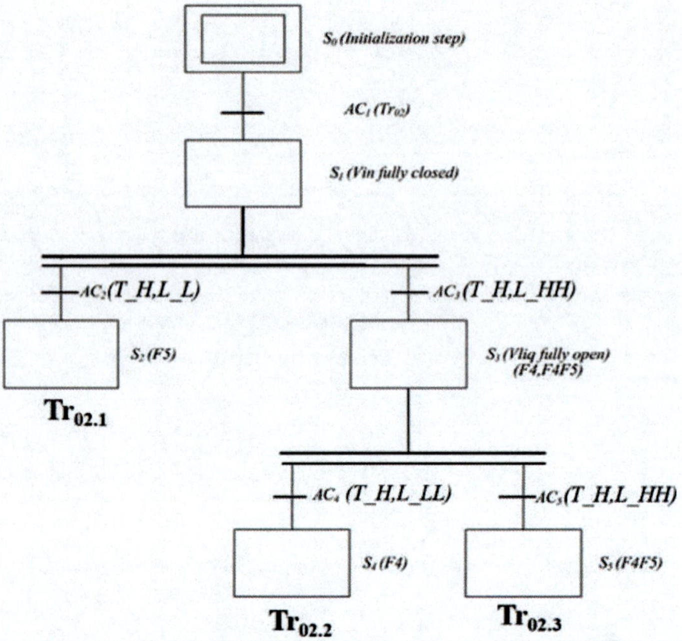

Fig. 10.21 Diagnostic test plan of Tr_{02} in flash startup example (Reprinted with permission from Feng et al. 2019. Copyright 2019 Elsevier)

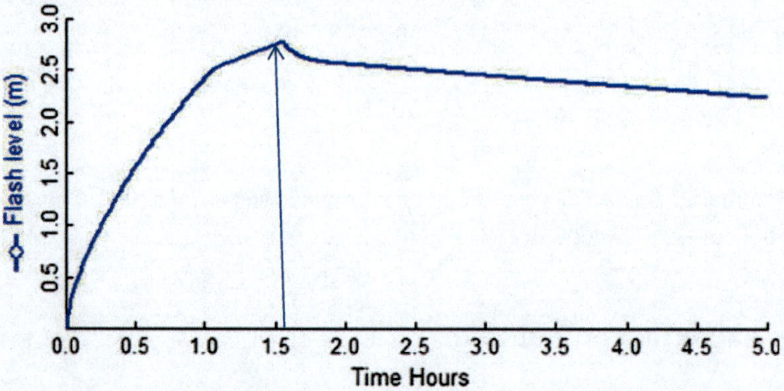

Fig. 10.22 Simulated time profile of level variation for $Tr_{02.1}$ (Reprinted with permission from Feng et al. 2019. Copyright 2019 Elsevier)

The SFC in Fig. 10.2 was executed in the first period. Notice that, after observing condition AC_2 (i.e., level reaching 2.5 m) at time 1.5 h and then applying step S_2, i.e., switching the level and flow controllers to AUTO and adjusting their set points to 2.5 m and 260000 kg/h, respectively, the level-rising trend still continued until 1.55 h due to the presence of failure F5(f_VliqSC). Since condition AC_3 in the normal operating procedure could not be satisfied in this case, the first diagnostic action shown in Fig. 10.21, i.e., manually closing the inlet valve Vin, was performed next. Since the heater was still on at this time, the liquid level in the flash drum was gradually lowered due to vaporization. This symptom, i.e., AC_2 in Fig. 10.21, confirmed the corresponding fault origin F5.

10.7.2 Trace $Tr_{02.2}$

Figure 10.23 shows the simulated level variation resulting from fault origin F4 (f_VinSO) and the subsequent tests. Since F4 is supposed to take place after implementing S_1 and before reaching AC_2, it was introduced in the simulation run at the earliest possible time near 0 h. The SFC in Fig. 10.2 was also followed in the period before the fault detection time at 1.55 h. After observing AC_2 (i.e., level reaching 2.5 m) at time 1.5 h and then executing S_2, i.e., switching the level and flow controllers to AUTO and adjusting their set points to 2.5 m and 260000 kg/h, respectively, the subsequent level-rising trend was still evident due to the presence of F4(f_VinSO), i.e., Vin failed at the open position. Since it was confirmed that condition AC_3 in the normal operating procedure could not be met after half an hour at 1.55 h, the first diagnostic action, i.e., manually closing inlet valve Vin, was taken at once. However, when compared with the level-decreasing trend produced by the same action in the case of $Tr_{02.1}$, it can be found that the system responded differently after 1.55 h in the present scenario. According to AC_3 in Fig. 10.21, the next online measurement to be

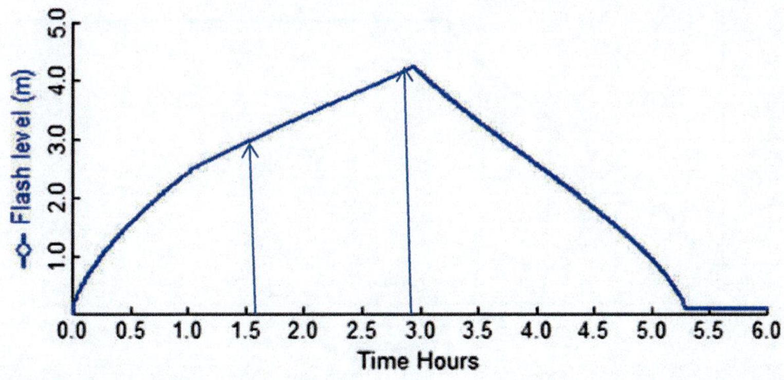

Fig. 10.23 Simulated time profile of level variation for $Tr_{02.2}$ (Reprinted with permission from Feng et al. 2019. Copyright 2019 Elsevier)

confirmed is L_HH, i.e., the level is extremely high and, more specifically, L_HH is set to be 4.2 m or higher in this example. As shown by the second arrow in Fig. 10.23, this chosen criterion was realized at 2.95 h. Notice that the next diagnostic action in test plan, i.e., S_3 in Fig. 10.21, is to manually open the outlet valve Vliq so as to produce the maximum possible liquid output flow. Since the heater was still on at this time, the combined flowrate of vapor and liquid outputs was larger than the input flow rate despite the fact that the inlet valve Vin was stuck at the open position. As a result, the liquid level in flash drum dropped to the next designated level L_LL in AC_4 of the test plan (see Fig. 10.21) at 5.25 h and, in the present example, L_LL was chosen to be 0 m. In this scenario, observing AC_3 and then AC_4 confirmed the corresponding fault origin F4.

10.7.3 Trace $Tr_{02.3}$

Figure 10.24 shows the combined effects of F4 (f_VinSO) and F5 (f_VliqSC) and the subsequent diagnostic tests on the transient behavior of liquid level in flash startup operation. Failure F5 was again introduced in the simulation run at time 0 h, while F4 at a slightly later time. As in the previous two scenarios, the SFC in Fig. 10.2 was followed in the period before 1.55 h. After observing AC_2 (i.e., when the liquid level was raised to 2.5 m) at time 1.5 h and then executing S_2 in the normal operating procedure, i.e., switching the level and flow controllers to AUTO and adjusting their set points to 2.5 m and 260000 kg/h, respectively, the fast level-rising rate was still maintained due to both F4, i.e., Vin stuck at the open position, and F5, i.e., Vliq stuck at the closed position. Therefore, after confirming at 1.55 h that AC_3 in the normal operating procedure could not be satisfied, the diagnostic step S_1 in the test plan, i.e., manually closing the inlet valve Vin, was again carried out at this point. According to condition AC_3 in Fig. 10.21, the next online observation in this scenario

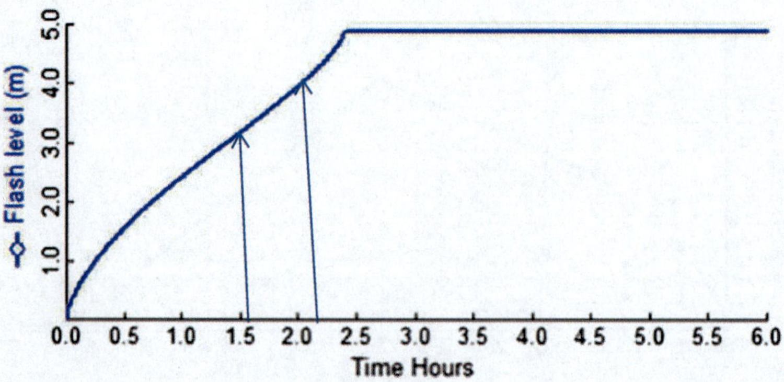

Fig. 10.24 Simulated time profile of level variation for $Tr_{02.2}$ (Reprinted with permission from Feng et al. 2019. Copyright 2019 Elsevier)

should be L_HH, i.e., the level reaching 4.2 m or higher. As shown by the second arrow in Fig. 10.24, this condition is realized at around 2.15 h and the subsequent action in test plan, i.e., S_3 in Fig. 10.21, was implemented at this time. However, due to coexistence of both F4 (f_VinSO) and F5 (f_VliqSC), the fast level-rising rate was unaffected by S_3 and eventually the liquid filled the entire drum. Thus, in this scenario, observing AC_3 and then AC_5 indicated coexistence of F4 and F5.

10.8 Additional Case Studies

10.8.1 Process Description

Let us next consider a *modified* version of the distillation startup process in a demonstrative example (ColumnStratup) published by Aspen Plus Dynamics (Al-Malah 2017). This startup operation can be characterized with the PFD in Fig. 10.25 and the SFC in Fig. 10.26. Since these figures are self-explanatory, further descriptions are omitted for the sake of brevity. It is assumed that the available raw material is a mixture of 6 wt% CH_2Cl_2, 54 wt% $CHCl_3$ and 40 wt% CCl_4 and, at steady state, the feed flowrate, temperature and pressure are kept at 10000 kg/h, 20 °C and 6.0 bar, respectively. A total of 20 equilibrium stages are chosen and the feed plate of the column is located at 10. In addition, this column is equipped with a partial condenser

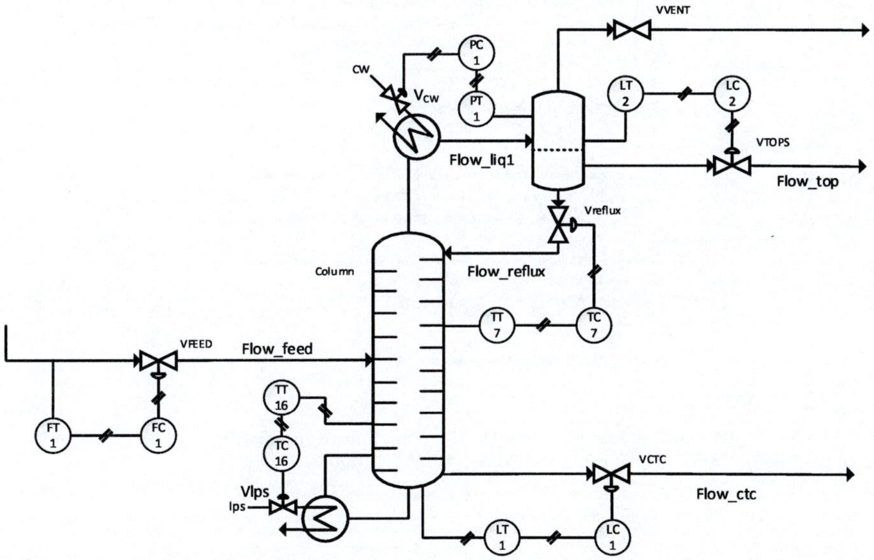

Fig. 10.25 Process flow diagram of a three-component distillation process (Reprinted with permission from Feng et al. 2019. Copyright 2019 Elsevier)

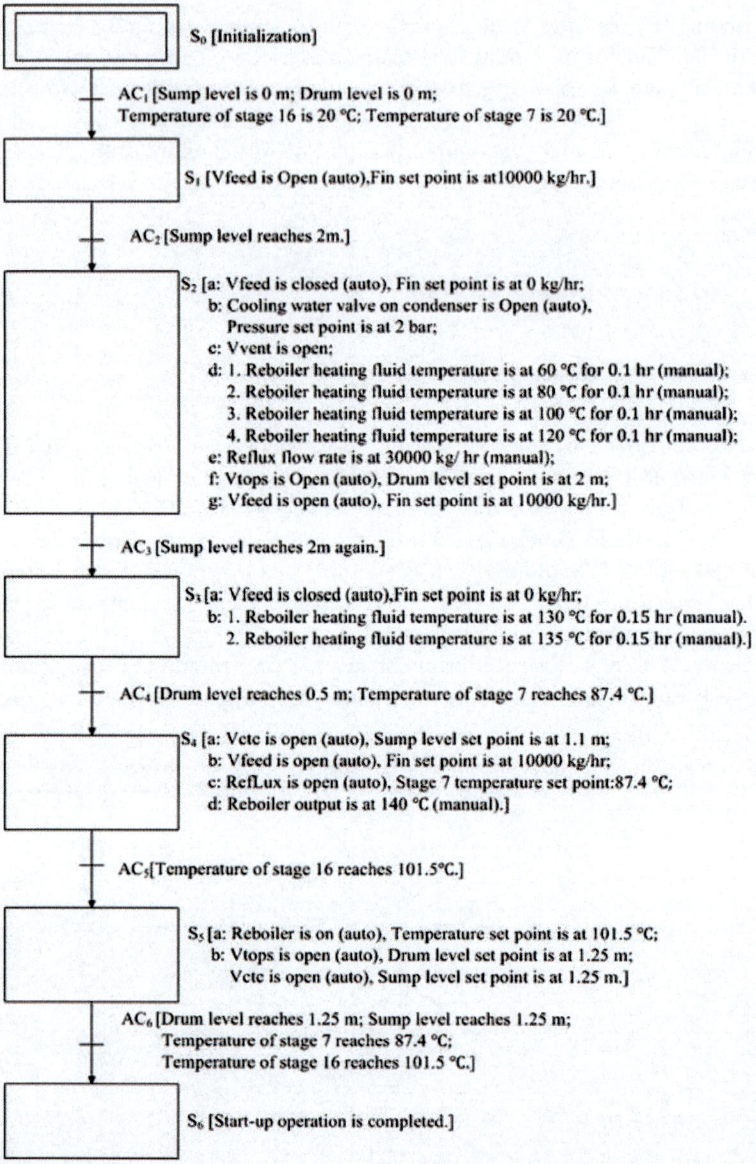

Fig. 10.26 Sequential function chart for startup operation of the distillation process (Reprinted with permission from Feng et al. 2019. Copyright 2019 Elsevier)

(stage 1) and its temperature and pressure are set at 80.3 °C and 2 bar, respectively. The pressure chosen for stage 2 is 2.02 bar and the overall pressure drop from stage 20 to stage 2 is 0.235 bar. The concentration of light key $CHCl_3$ in the overhead stream is required to be higher than 81 mol%, while that of heavy key CCl_4 in the bottom product not lower than 81 mol%. Finally, the initial conditions are chosen as follows:

- All valves are all closed, while the condenser and reboiler are both off;
- All controllers are on manual;
- Distillation column is empty and at room temperature.

For illustration simplicity, let us assume that there can be only five failure events in this example, i.e.,

- F1 (f_reboiler_failed): A partial reboiler malfunction causing a decrease in its heat transfer rate to one half of the normal level;
- F2 (f_T7controller_failed): A hardware failure in temperature controller TC7 that cuts off the reflux flow;
- F3 (f_VtopsSC): Valve Vtops sticks at close position;
- F4 (f_cond_failed): A condenser failure that disables heat removal function;
- F5 (f_VctcSC): Valve Vctc sticks at close position.

All components in this example can be classified into five levels as follows:

- Level-1 component is a PLC or operator;
- Level-2 components include six controller/actuator pairs, i.e., FC1/Vfeed, LC1/Vctc, LC2/Vtops, PC1/Condenser, TC16/Reboiler, and TC7/T7_controller;
- Level-3 components include all material and energy flows surrounding every level-4 unit;
- Level-4 components are essentially four identifiable units in the distillation system, i.e., reflux drum, bottom sump, rectifying section, and stripping section;
- Level-5 components are six online sensors for measuring the feed flowrate (FT1), the levels in bottom sump and reflux drum (LT1 and LT2), the overhead pressure (PT1) and the temperatures at plates 7 and 16 (TT7 and TT16).

Notice that, instead of the mass flowrates of cooling and heating media shown in the given PFD (see Fig. 10.25), only the heat-transfer rates (or the temperatures of cooling and heating media) of condenser and reboiler can be altered by PC1 and TC16 in Aspen environment. Consequently, to maintain consistency between automata predictions and simulation results in the validation studies, these energy flows were treated as manipulated variables and assumed to be directly adjustable with fictitious actuators in the corresponding automaton models. Notice also that, although the actuator of TC7 is a control valve on the reflux flow in the actual system (see Fig. 10.25), its failure cannot be simulated with Aspen and, therefore, the identical effects were produced by incorporating F2 in the corresponding automaton.

10.8.2 Aggregated Hybrid Model

All knowledge-based component models in the present example can be constructed with the same approach detailed in Sect. 10.2. Since an extremely large volume of corresponding component automata are bound to be generated, they are not reported here for the sake of conciseness. On the other hand, the data-based automata are delineated explicitly below to facilitate a clear understanding of the hybrid modeling strategy.

To avoid creating a set of unnecessarily complex component models, the Aspen simulated data during normal operations can be used to construct a lumped model for charactering all processing units in the fourth level, i.e., the reflux drum, the bottom sump, the rectifying and stripping sections. The discretized values of their state variables are defined in Table 10.4. The mass flowrates (in kg/h) of all inputs and outputs of each processing unit are discretized and represented with 4 qualitative values, i.e., value 0 for $\left[0, 0^+\right)$, value 1 for [10000, 20000), value 2 for [20000, 30000), and value 3 for $[30000, \infty)$, while the entire ranges of energy flowrates (in GJ/h) facilitated by reboiler and condenser are both divided into 6 intervals, i.e., $\left[0, 0^+\right)$, $\left[0^+, 2.5\right)$, [2.5, 5.0), [5.0, 7.5), [7.5, 10.0), and $[10.0, \infty)$, and these intervals are labeled sequentially from 0 to 5. Based on the above discretization schemes, the simulation data during the normal startup can be converted accordingly (see Table 10.5). An abridged version of this data set can then be produced by removing every row in which the state variables are identical to those in the previous row. All state changes during normal startup operation can be easily extracted from this abridged set and they are incorporated in the automaton model as a sequence of state-transition events. A total of twelve (12) consecutive events can be found and their guards and variable updates are listed in Table 10.6.

As described before in the flash startup example, the data-based model and its knowledge-based counterpart should both be included in a hybrid automaton to characterize the normal and abnormal modes of every processing unit, respectively. In the present example, the switch actions from the lumped model of the normal operation (which is derived from the simulation data) to the individual component models (which are built with the engineering knowledge) are triggered according to the aggregated automaton shown in Fig. 10.27. The system state denoted as database_model in this aggregated automaton refers to the lumped model of normal system behavior, while all other states denote the knowledge-based component models for predicting the same set of state variables during the abnormal scenarios. The system starts normally at database_model. If one or more failure occurs, i.e., $F1 + F2 + F3 + F4 + F5 > 0$, the switch action To_G_mode can be activated to utilize the knowledge-based models for predicting all state variables listed in Table 10.4.

Table 10.4 Discretized state values in distillation process

Value (–)	PU_Tdrum (°C)	PU_Tsump (°C)	PU_T7 (°C)	PU_T16 (°C)	PU_Ldrum (m)	*PU*_Lsump (m)
0	$[20, 20^+)$	$[20, 20^+)$	$[20, 20^+)$	$[20, 20^+)$	$[0, 0^+)$	$[0, 0^+)$
1	$[20^+, 60)$	$[20^+, 60)$	$[20^+, 60)$	$[20^+, 60)$	$[0^+, 0.5)$	$[0^+, 0.5)$
2	[60, 80)	[60, 80)	[60, 80)	[60, 80)	[0.5, 1.25)	[0.5, 1.25)
3	[80, 100)	[80, 100)	[80, 100)	[80, 100)	$[1.25, \infty)$	$[1.25, \infty)$
4	[100, 120)	[100, 120)	[100, 120)	[100, 120)	Undefined	Undefined
5	[120, 140)	[120, 140)	[120, 140)	[120, 140)	Undefined	Undefined

Table 10.5 Discretized simulation data in normal startup of distillation process

Time	PU_T16	PU_T7	PU_Lsump	PU_Ldrum	PU_Fin	PU_Ftops	PU_Fctc	PU_Fref	PU_Fliq1	PU_Fv20	PU_cond	PU_reb
0	0	0	0	0	0	0	0	0	0	0	0	0
0.156	0	0	0	0	1	0	0	0	0	0	0	0
0.318	0	0	0	0	1	0	0	0	0	0	0	0
0.512	0	0	1	0	1	0	0	0	0	0	0	0
0.807	0	0	2	0	0	0	0	0	0	1	1	1
0.852	1	0	2	0	0	0	0	0	0	2	1	1
0.897	1	1	2	1	0	0	0	3	1	2	2	2
0.915	1	1	2	1	0	0	0	3	1	2	2	2
0.958	2	2	1	1	1	0	0	3	2	3	3	3
10125	2	2	1	1	1	0	0	3	2	3	3	3
1.321	3	2	2	1	0	0	0	3	2	3	4	4
1.498	3	2	1	2	1	1	0	3	3	3	4	4
1.659	3	2	1	2	1	1	0	3	3	3	4	4
1.745	4	3	1	2	1	1	1	3	3	3	5	5
2.12	4	3	1	2	1	2	0	3	3	3	5	5
2.325	4	3	1	1	1	1	1	3	3	3	5	5
3.585	4	3	1	1	1	1	1	3	3	3	5	5

Table 10.6 State-transition events during normal startup of distillation process

	Guard												Variable				
db	PU_T16	PU_T7	PU_Lsump	PU_Ldrum	PU_Fin	PU_Ftops	PU_Fctc	PU_Fref	PU_Fliq1	PU_Fv20	PU_cond	PU_reb	d_Lsump	d_Ldrum	d_temp16	d_temp7	
db_1	0	0	0	0	0	0	0	0	0	0	0	0	0	0	0	0	S0
db_2	0	0	0	0	1	0	0	0	0	0	0	0	1	0	0	0	S1
db_3	0	0	1	0	1	0	0	0	0	0	0	0	1	0	0	0	
db_4	0	0	2	0	0	0	0	0	0	1	1	1	0	0	1	0	S2ab
db_5	1	0	2	0	0	0	0	0	0	2	1	1	0	1	0	1	
db_6	1	1	2	1	0	0	0	3	1	2	2	2	-1	0	1	1	S2c
db_7	2	2	1	1	1	0	0	3	2	3	3	3	1	0	1	0	S2d
db_8	3	2	2	1	0	0	0	3	2	3	4	4	-1	1	0	0	S3
db_9	3	2	1	2	1	1	0	3	3	3	4	4	0	0	0	0	S4a
db_10	4	3	1	2	1	1	1	3	3	3	5	5	0	0	1	1	S4b
db_11	4	3	1	2	1	2	0	3	3	3	5	5	0	-1	0	0	S5
db_12	4	3	1	1	1	1	1	3	3	3	5	5	0	0	0	0	

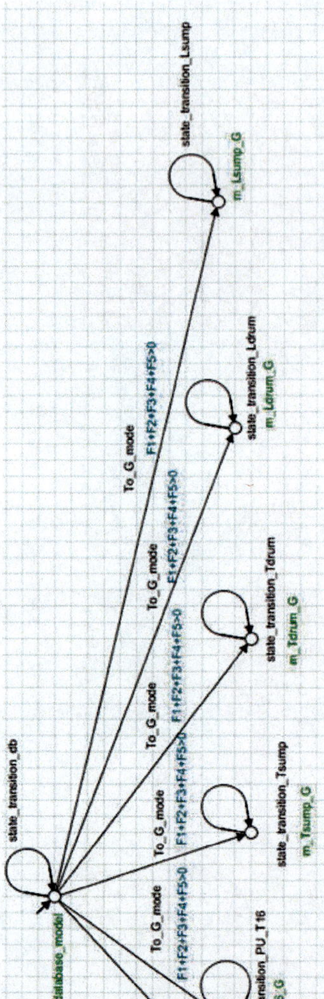

Fig. 10.27 Switching mechanism for predicting system states of the distillation process (Reprinted with permission from Feng et al. 2019. Copyright 2019 Elsevier)

10.8.3 Diagnoser

All aforementioned models were synchronized in SUPREMICA to produce the diagnoser in Fig. 10.28. Since the OETs here are labeled according to the same conventions used in Fig. 10.15, the redundant explanations are not given here for the sake of brevity. Notice that the five failures considered in the present example, i.e., Fi and

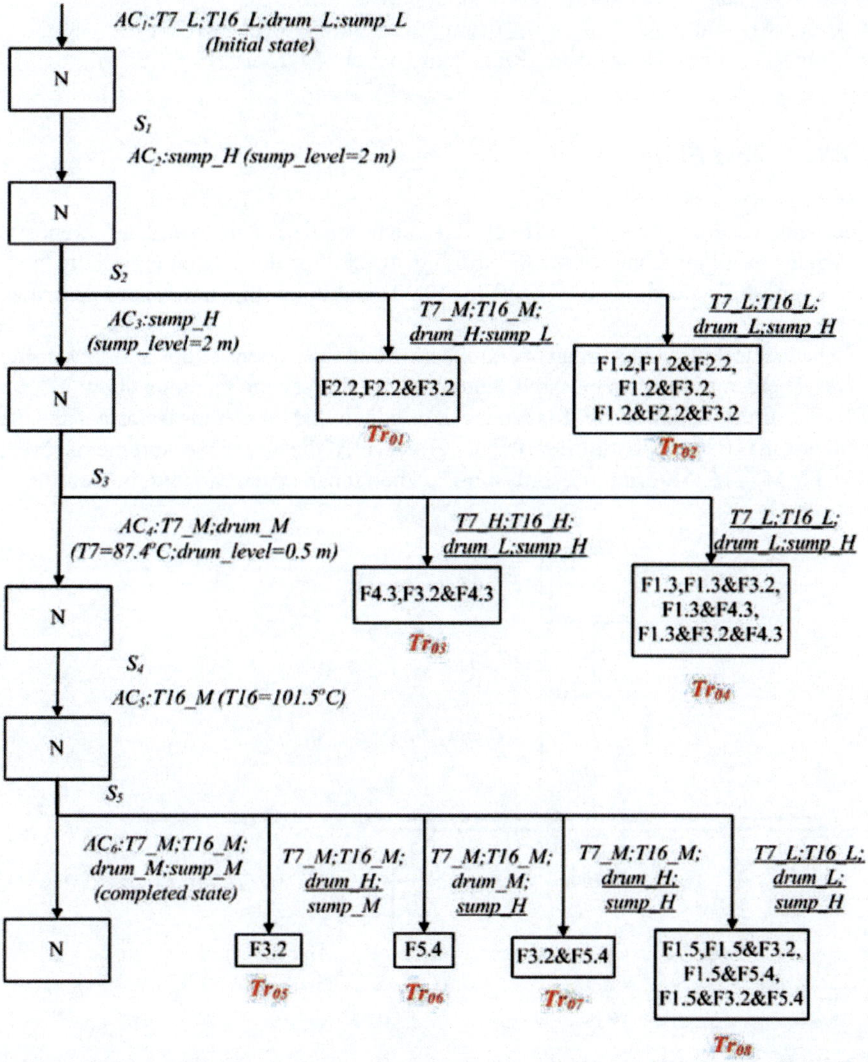

Fig. 10.28 Diagnoser in distillation startup example (Reprinted with permission from Feng et al. 2019. Copyright 2019 Elsevier)

i = 1, ..., 5, have already been defined in the beginning of Sect. 10.7. Every failure event in diagnoser is again expressed according to the format Fi.j and the attached additional index j (= 1, ..., 6) denotes the time interval between the two instances when conditions AC_{j-1} and AC_j are satisfied. Finally, the qualitative sensor readings in the OETs can be more specifically characterized as follows:

- T7_L, T7_M, T7_H ⇒ T7 $\lesseqgtr$ 87.4 °C;
- T16_L, T16_M, T7_H ⇒ T16 $\lesseqgtr$ 101.5 °C;
- drum_L, sump_L ⇒ drum_level, sump_level $<$ 0.5 m;
- drum_M, sump_M ⇒ 0.5 m $\leq$ drum_level, sump_level $\leq$ 1.25 m;
- drum_H, sump_H ⇒ drum_level, sump_level $>$ 1.25 m.

10.8.4 Test Plans

The synthesis procedure given in Sect. 10.6 can be applied again to find the diagnostic test plans of all undiagnosable OETs in Fig. 10.28. For illustration simplicity, only the diagnostic procedures of Tr_{01} (Fig. 10.29) and Tr_{08} (Fig. 10.30) are described in the sequel:

The first test plan calls for an operator action that fully opens Vtops after observing Tr_{01}. There may be two possible outcomes. If the sensor readings show T7_M, T16_M, drum_L, and sump_L, then it can be concluded that Vtops is normal and the fault origin is F2 (f_T7controller_failed). However, if the online measurements reveal that T7_M, T16_M, drum_H, and sump_L, then it can be certain that, in addition to

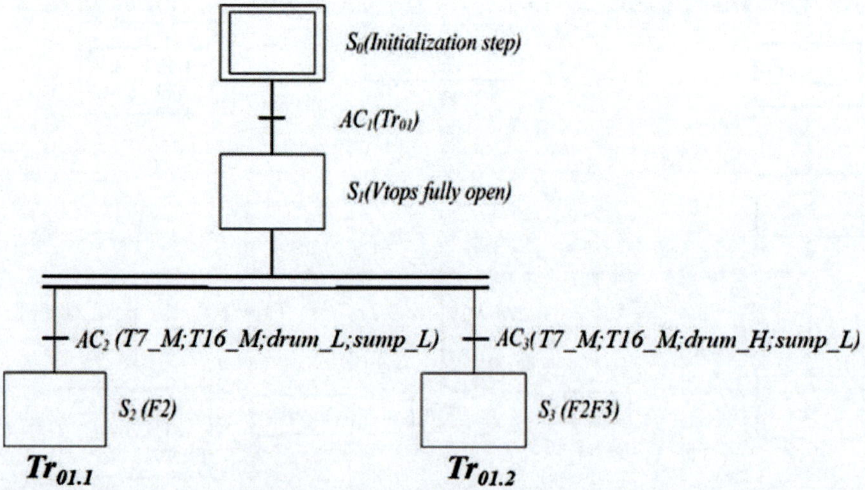

Fig. 10.29 Diagnostic test plan of Tr_{01} in distillation startup example (Reprinted with permission from Feng et al. 2019. Copyright 2019 Elsevier)

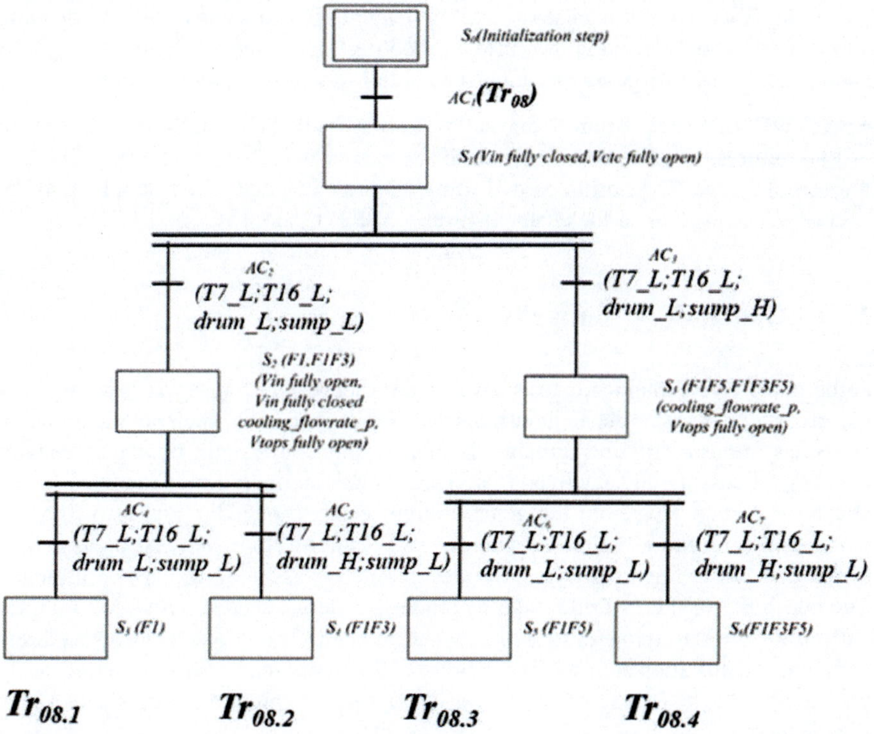

Fig. 10.30 Diagnostic test plan of Tr_{08} in distillation startup example (Reprinted with permission from Feng et al. 2019. Copyright 2019 Elsevier)

f_T7controller_failed, valve Vtops also sticks at the close position (f_VtopsSC). In other words, both F2 and F3 are present in this scenario.

The second test plan calls for two simultaneous test actions that fully closes Vin and opens Vctc after observing Tr_{08}. These tests divide the implicated fault origins into two separate groups, i.e., (1) {F1, F1 & $F3$} and (2) {F1 & $F5$, $F1$ & $F3$ & $F5$}, and they can be confirmed respectively according to the online measurements specified in AC_2 (T7_L, T16_L, drum_L and sump_L) and AC_3 (T7_L, T16_L, drum_L and sump_H). Notice that sump_L in AC_2 implies that Vctc is functional and thus F5 must be excluded from the first group of fault origins. In response to condition AC_2, steps in S_2, i.e., opening Vin and Vtops fully and increasing flowrate of cooling water, should be applied and two possible outcomes may be produced:

- AC_4(T7_L, T16_L, drum_L and sump_L) implies that the fault origin is a partial reboiler failure, i.e., F1 (f_reboiler_failed);
- AC_5 (T7_L, T16_L, drum_H and sump_L) indicates that, other than F1, Vtops is also stuck at the close position, i.e., F3 (f_VtopsSC).

On the other hand, it is necessary to implement the test steps in S_3, i.e., opening Vtops fully and increasing flowrate of cooling water, when condition AC_3 is confirmed. The corresponding diagnostic results can be summarized below:

- AC_6(T7_L, T16_L, drum_L and sump_L) implies that the fault origin consists of two failures, i.e., F1 and F5 (f_VctcSC);
- AC_7 (T7_L, T16_L, drum_H and sump_L) indicates that, other than F1 and F5, Vtops is also stuck at the close position, i.e., F3 (f_VtopsSC).

10.9 Concluding Remarks

Although effective diagnostic tests have been presented in Chap. 9 for differentiating the originally inseparable fault origins in simple processes, their applicability in realistic systems is still questionable. To address this concern, the dynamic behavior of every processing unit involved in a given sequential operation is modeled in this chapter by integrating both the generic engineering knowledge and also rigorous simulation data into a hybrid automaton. The resulting improved test plans can be synthesized according to the system model obtained by assembling all such automata. The feasibility of this hybrid modeling strategy is demonstrated with two examples concerning the startup operations of a two-component flash process and also a three-component distillation column. The validity of this approach has also been rigorously confirmed in these two examples in extensive dynamic simulation studies with Aspen Plus Dynamics.

References

Åkesson K, Fabian M, Malik R (2006) SUPREMICA—an integrated environment for verification, synthesis and simulation of discrete event systems. In: IEEE proceedings of the 8th international workshop on discrete event systems, pp 384–385

Al-Malah K (2017) Aspen Plus®—chemical engineering applications. Wiley, Hoboken, New Jersey

Benveniste A, Fabre E, Haar S, Jard C (2003) Diagnosis of asynchronous discrete-event systems: a new unfolding approach. IEEE Trans Autom Control 48(5):714–727

Cassandras CG, Lafortune S (2008) Introduction to discrete event systems, 2nd edn. Springer Science + Business Media, LLC, New York, NY, USA

Chen J, Jiang YC (2011) Development of hidden semi-Markov models for diagnosis of multiphase batch operation. Chem Eng Sci 66(15):1087–1099

Dai Y, Zhao J (2011) Fault diagnosis of batch chemical processes using a dynamic time warping (DTW)-based artificial immune system. Ind Eng Chem Res 50:4534–4544

Debouk R, Lafortune S, Teneketzis D (2000) Coordinated decentralized protocols for failure diagnosis of discrete event systems. Discrete Event Dyn Syst Theory Appl 10(1–2):33–86

Feng ST (2017) A hybrid modeling strategy to build automata for synthesizing diagnostic tests in sequential operations. MS thesis, National Cheng Kung University, Tainan, Taiwan

Feng ST, Chen YC, Chang CT (2019) An automata based hybrid modeling approach to synthesize sequential diagnostic tests. Chem Eng Res Des 145:29–47

Gascard E, Simeu-Abazi Z (2013) Modular modeling for the diagnostic of complex discrete-event systems. IEEE Trans Autom Sci Eng 10(4):1101–1123
Gomes Cabral F, Moreira MV, Diene O, Basilio JC (2015) A Petri net diagnoser for discrete event systems modeled by finite state automata. IEEE Trans Autom Control 60(1):59–71
Hsieh WC, Chang CT (2016) Timed-automata based method for synthesizing diagnostic tests in batch processes. Comput Chem Eng 84:12–27
Kang A, Chang CT (2014) Automata generated test plans for fault diagnosis in sequential material- and energy-transfer operations. Chem Eng Sci 113:101–115
Malik R, Fabian M, Akesson K (2011) Modelling large-scale discrete-event systems using modules, aliases, and extended finite-state automata. In: Proceedings of 18th IFAC world congress, vol 18, pp 7000–7005
Nomikos P, MacGregor JF (1994) Monitoring batch processes using multiway principal component analysis. AIChE J 40(8):1361–1375
Nomikos P, MacGregor JF (1995) Multivariate SPC charts for monitoring batch processes. Technometrics 37(1):41–59
Qiu WB, Kumar R (2006) Decentralized failure diagnosis of discrete event system. IEEE Trans Syst Man Cybern Part A: Syst Hum 36(3):384–395
Wang CJ, Chen YC, Feng ST, Chang CT (2017) Automata-based operating procedure for abnormal situation management in batch processes. Comput Chem Eng 97:220–241
Yeh ML, Chang CT (2011) An automaton-based approach to evaluate and improve online diagnostic schemes for multi-failure scenarios in batch processes. Chem Eng Res Des 89:2652–2666
Zad SH, Kwong RH, Wonham WM (2003) Fault diagnosis in discrete-event systems: framework and model reduction. IEEE Trans Autom Control 48(7):1199–1204

Index

A
Abnormal, 1, 26, 29, 39, 40, 48, 51, 52, 254, 257, 258, 261, 267, 275, 285, 288, 290, 296, 298, 299, 304, 320, 323, 325, 326, 338
Action, 1, 29–31, 34, 36, 39, 45, 46, 48, 49, 54, 55, 62, 70, 74, 95, 107–109, 111, 116, 117, 120, 121, 123, 130, 133, 135, 136, 142, 152, 153, 160, 171, 174, 187, 189, 190, 198, 203, 204, 207, 243, 257–259, 261–266, 270, 272, 276, 284, 287, 288, 290, 292–295, 298–300, 302, 304, 305, 308, 311, 316, 317, 319, 323, 325, 327, 328, 330, 333–335, 338, 344, 345
Activation, 1, 29, 31, 45, 70, 174, 187, 203, 225, 235, 255, 258, 273, 294, 295, 298, 305, 317, 325, 326
Actuator, 31, 32, 45, 54, 58, 62, 70, 169–172, 174, 187, 189, 190, 198, 205, 206, 236, 243, 258, 259, 261–266, 284, 288, 300, 301, 313, 314, 317, 319, 325, 327, 328, 330, 337
Analysis, 73, 113, 155, 253, 260, 283, 327
Annealing, 75, 76
Application, 10, 13, 17, 54–56, 96, 130, 137, 138, 149, 150, 155, 163, 168, 170, 175, 177, 179, 203, 217, 261, 269, 283, 323
Approach, 1, 10, 11, 22, 29, 30, 37, 39, 40, 52, 55, 60, 73, 75, 88, 95, 96, 99, 103, 110, 114, 130, 139, 150, 154, 155, 167, 168, 172, 176, 177, 183, 189, 197, 203, 217, 221, 227, 228, 234, 253, 256, 258, 262, 265, 269, 280, 283, 284, 295, 296, 302, 311, 312, 314, 315, 317, 325, 338, 346
Architecture, 32
Artificial, 96, 158, 253, 283, 290
Automata, 10, 30, 33, 34, 38, 41, 43–46, 56, 57, 62, 65, 68, 167, 168, 172, 173, 189, 191, 194, 203, 204, 210, 214, 221, 225, 228, 234, 243, 253, 255, 261, 263, 280, 283–286, 288, 295, 300, 308, 311, 314, 315, 319, 320, 323, 325, 327, 329–331, 337, 338, 346
Automaton, 30, 34–44, 46, 54–58, 60, 62, 64, 65, 168–172, 189, 194, 204–214, 234, 254, 257, 261, 262, 264–266, 285, 286, 290, 311, 314–319, 322, 323, 327–330, 337, 338, 346

B
Basic, 12, 43, 98, 113, 116, 118, 191, 228, 273, 284
Batch, 1, 95, 96, 113–115, 130, 153, 155, 162, 163, 167, 168, 177, 181, 182, 253, 283, 288, 300–302
Beer, 130, 143–145, 147, 149–154, 163, 272–278
Behavior, 13, 32, 33, 38, 43, 45, 48, 50, 54–57, 60, 76, 88, 89, 92, 98, 107, 117, 130, 132, 151, 172, 207, 227, 258, 284, 312, 315, 319, 320, 323, 334, 338, 346
Bidirectional, 135
Binary, 18, 91, 130–141, 143, 157–159, 163, 167, 286
Blockage, 97, 103, 104, 115
Blockage-based, 103, 105, 113

C.-T. Chang et al., *Process Plant Operating Procedures*, Advances in Industrial Control,
https://doi.org/10.1007/978-3-030-70978-5

Buffer, 21, 22, 24, 26, 27, 143–146, 148–150, 154, 156–158, 161, 254, 267, 269–273, 284, 285, 295, 296, 301
Bypass, 4, 256

C
Cause, 31, 38–40, 49, 60, 137, 194, 253, 283, 284, 290, 295, 298, 299, 304, 305, 308, 325
Chart, 1, 8, 9, 29, 31, 110, 114, 124, 125, 139, 162, 163, 167, 168, 203, 214, 254, 284, 312, 313, 336
Chemical, 1, 2, 11, 13, 16, 17, 29, 30, 32, 49–53, 57, 73, 78, 96–99, 101, 102, 104–108, 111–113, 115, 131, 132, 139, 144–147, 153–155, 167, 169–171, 173, 175–178, 180–183, 185, 186, 189–200, 204, 206–208, 210–216, 218, 220, 222, 223, 226–230, 232, 253, 254, 280, 283, 311
Classification, 36
Cleaning, 78, 96, 98, 99, 102–113, 138, 144, 145, 148, 150, 163, 273
Closed-loop, 54, 55
Complex, 1, 12, 96, 113, 114, 151, 155, 167, 172, 235, 283, 312, 319, 320, 338
Component, 11, 18, 30, 31, 35–40, 43, 45, 46, 49, 50, 52, 57, 58, 62, 78, 91, 95, 97, 107, 114, 115, 117, 120, 125, 130, 137, 148, 149, 156, 168, 170–172, 174, 189–193, 205–209, 214, 228, 237, 242–245, 253, 255–259, 261, 263–266, 275, 283–286, 290, 308, 312–314, 319, 328–330, 335, 337, 338, 346
Compressor, 95, 104, 107–109, 116, 117, 119, 127, 130, 132
Condition, 1, 2, 16, 22, 26, 29, 31, 36, 45, 49, 58, 70, 74, 81, 84, 86, 90–92, 95–97, 99, 109, 110, 116, 119, 127, 130, 133, 136, 149, 155, 157, 158, 167, 168, 170–174, 178, 183, 184, 187, 190, 194–198, 203–205, 207, 209–211, 213, 215, 221, 225, 234, 235, 237, 243, 254–258, 261, 263, 264, 273, 275, 276, 279, 285–287, 294, 295, 298, 299, 301, 304, 305, 312, 316, 317, 319, 320, 325–330, 333–335, 337, 344–346
Configuration, 31, 37, 38, 40, 45, 58, 60, 62, 70, 95, 98, 119, 130, 148, 184, 191, 237, 255–258, 262–264, 266, 275, 284, 292, 311, 313, 314, 319, 329
Constraint, 45, 56, 110, 131–143, 148–150, 152, 155, 159, 161, 167, 169, 203, 262, 266
Construction, 56, 99, 100, 156, 172, 205, 254, 270, 274, 284, 285, 290, 314
Continuous, 1, 13, 15, 73, 96, 103, 169, 170, 181, 187, 204, 217, 220, 221, 228, 253, 320
Continuous-time, 155
Continuous-variable, 32
Control, 9, 11–13, 17, 19, 20, 26, 29, 31, 32, 34, 36, 39, 45, 46, 48, 49, 52, 54–57, 60, 70, 74–77, 81, 83–86, 88, 109, 114, 119, 120, 123, 125, 132, 134, 136, 141, 163, 168, 169, 171–174, 183, 189, 194, 198, 203, 205, 207, 209, 210, 213, 214, 221, 234, 235, 243, 258, 261–263, 265, 290, 294, 298, 299, 302, 304, 337
Controllable, 62
Controller, 19, 20, 23, 25, 29, 30, 32, 45, 48, 51–54, 76, 78, 85, 86, 88, 89, 95, 97, 130, 169–174, 178, 179, 182, 183, 187, 189, 190, 198, 205–207, 210, 212, 213, 215, 221, 225, 227, 243, 255, 258, 284, 285, 287, 288, 290, 292, 300, 301, 305, 312–314, 319, 327, 333, 334, 337
Control-loop, 7, 11, 227, 229

D
Data, 123, 156, 208, 227, 254, 283, 311, 312, 320, 322, 323, 325, 329, 338, 340, 346
Database, 338
Deadlock, 35, 167, 258, 259, 286, 290, 320
Decision, 96, 124
Decomposition, 255, 285
Deterministic, 34
Development, 36, 115, 123, 155, 157
Device, 23, 107, 108, 113, 117, 127, 133
Diagnosability, 261
Diagnoser, 253, 258–263, 270, 271, 274, 275, 280, 284, 311, 325–328, 343, 344
Diagnosis, 253, 283, 308, 311
Diagnostic, 10, 253, 254, 256, 261–264, 266, 268, 270, 275, 280, 283, 284, 290, 299, 300, 305, 311, 312, 320, 327–334, 344–346
Diagram, 1, 3, 6, 9, 11, 29, 35, 44, 144, 167, 204, 254, 273, 312, 335
Direction, 38, 97, 98, 116, 134, 142, 207

Discrete, 15, 29, 32, 46, 54, 95, 114, 154, 155, 194, 195, 234, 284, 311
Distillation, 11, 12, 16–22, 25, 27, 73, 78, 81, 90, 92, 168, 181, 183–187, 217, 220–223, 226–230, 232, 234, 312, 335–337, 339–346
Disturbances, 13, 73, 76, 198
Dynamic, 10, 13, 14, 16–19, 32, 73–75, 77, 78, 92, 168, 172, 174, 175, 177, 179, 183, 184, 189, 191, 203, 208, 215, 221, 227, 234, 284, 311, 312, 315, 319, 320, 331, 335, 346

E

Economic, 168
Effective, 12, 13, 15, 96, 155, 172, 234, 253, 254, 266, 283, 311, 314, 319, 320, 346
Efficient, 15, 16, 65, 70, 155, 164, 167
Element, 33, 34, 48, 95, 100, 103–106, 115, 122, 123, 131, 142
Energy, 31, 73, 74, 76, 77, 86, 91, 92, 113, 168, 170, 176, 186, 187, 191, 196, 205, 209, 213, 215, 220, 236, 237, 243, 280, 312–315, 319, 321, 322, 337, 338
Engineering, 11, 17, 172, 173, 210, 234, 254, 311, 314, 319, 323, 338, 346
Equation, 13, 15, 16, 35, 41, 43, 55, 74, 85, 100, 134, 136, 137, 140–143, 148–153, 161, 169, 330
Equilibrium, 11, 76, 335
Equipment, 1, 16, 48, 51, 58, 95, 107–110, 115, 119–121, 130, 132–134, 137, 140, 142, 145, 147–149, 155, 156, 169, 273
Error, 13, 15, 29, 96, 113, 114, 154, 262, 263, 288, 311, 328
Error-free, 114
Error-prone, 1, 154, 167
Evaluation, 168, 203, 292
Evaporation, 300–302
Event, 29, 30, 32–40, 42–46, 48, 49, 51, 52, 54–56, 58, 60, 62, 70, 111, 114, 148, 154, 157–160, 168–173, 189, 190, 194–198, 203, 205–209, 213, 234, 243, 244, 253, 256–262, 265, 266, 270, 271, 274, 276, 279, 283–290, 292, 296, 298, 299, 302–304, 308, 312, 314–319, 322, 323, 325–327, 329, 330, 337, 338, 341, 344
Example, 3, 6, 8–11, 13, 17, 19, 20, 31, 33, 37–40, 43, 45–48, 52, 56–58, 60, 62–64, 68–70, 74, 97–100, 104, 105, 108–121, 123–125, 128, 130, 136, 137, 149, 154–156, 161, 163, 168–175, 177, 178, 180, 181, 183–185, 189–191, 194–200, 204, 206, 208, 217, 221, 227, 228, 234, 235, 237, 243, 254–258, 260, 261, 266, 269, 274, 280, 284–290, 292–297, 299–303, 306, 307, 312, 314, 316–319, 322, 323, 325, 326, 328–330, 332, 334, 335, 337, 338, 343–346
Extractive, 73, 78, 85, 88, 90–92

F

Fabrication, 155
Facility, 154–156, 163
Failure, 30, 38–40, 45, 46, 48, 50–53, 58, 171, 253, 255, 257–259, 261–264, 266, 269–272, 274–277, 279, 283–288, 290, 295, 299–301, 308, 311–314, 317, 319, 320, 323, 325–330, 333, 334, 337, 338, 343–346
Fault, 50, 52, 253, 261–267, 275, 277, 279, 280, 283–285, 288, 290, 292, 293, 295, 298–300, 304, 305, 308, 311, 326–331, 333, 334, 344–346
Fault-diagnosis, 10, 253, 258, 260, 283, 294, 311, 325
Fault-propagation, 253, 258, 260, 284, 288, 289, 304, 312, 320, 323
Feasibility, 109, 119, 177, 217, 280, 308, 312, 346
Feasible, 54, 56, 65, 104, 114, 121, 123, 130, 141, 157, 163, 164, 168, 173, 179, 207, 221, 254, 262, 292, 319, 320, 328, 330
Feature, 36, 43, 155, 160, 168, 173, 178, 210, 227, 257, 280, 295, 308
Feedback, 32, 54, 55
Fictitious, 43, 62, 64, 106, 124, 125, 134, 135, 140, 234, 235, 237, 254, 259, 284, 290, 317, 337
Filling, 58, 60, 144, 149, 150, 273
Filter, 143–148, 150, 151, 153, 273
Filtration, 130, 143–145, 147, 151–154, 163, 272–278
Final, 34, 56, 76, 95, 123, 125, 143, 150, 154, 156, 167, 173, 174, 183, 184, 196, 210, 213, 263, 283, 290, 299, 304, 328
Finite, 33, 36, 46, 48, 158, 168, 169, 204, 234, 256, 285, 311

Flash, 17, 168–177, 182, 183, 189–192, 194–200, 204–210, 214, 215, 221, 234, 235, 237, 243–245, 312–323, 325–329, 332–334, 338, 346
Flexible, 155, 283
Flowrate, 2, 11, 17, 19, 20, 22, 24–26, 78, 81, 83, 169, 177–179, 181, 186, 191, 205, 209, 210, 220, 312, 320–322, 334, 335, 337, 338, 345, 346
Flows, 1, 3, 16–18, 22, 29, 31, 37, 49, 73, 74, 78, 81, 83–86, 88, 89, 92, 96, 97, 103, 115, 116, 125, 132, 134, 137, 144, 167, 170, 171, 173, 174, 177, 179, 183, 184, 191, 198, 204, 205, 207, 209, 210, 221, 225, 236, 237, 243, 254, 258, 267, 273, 275, 279, 284, 301, 302, 304, 312–316, 319, 322, 327, 333–335, 337
Fragment, 96–101, 103, 107–109, 111, 114–118, 120, 121, 123, 124, 127, 131, 132, 136–144, 147–151, 153, 167
Fragment-based, 103–105, 112
Framework, 10, 29, 33, 111, 261
Function, 1, 8, 9, 29, 31, 35, 36, 46, 55, 74, 76, 78, 80, 82, 87, 88, 90, 138, 143, 150, 157, 161, 163, 167–169, 203, 204, 214, 254, 262, 284, 312, 313, 336, 337
Fundamental, 97, 114, 319

G

General, 38, 46, 50, 96, 145, 146, 148, 156, 254, 256, 262, 283–285, 323, 325
Generation, 29, 62, 108, 113, 114, 119, 154, 156, 167, 179, 253
Genetic, 155
Goal, 1, 29, 114, 136, 137, 150, 167, 173, 184, 190, 194, 203, 210, 211, 213, 221, 254, 262
Grafchart, 114
Graph, 29, 114, 154
Graphical, 322, 323

H

Hardware, 1, 29, 30, 36, 38, 156, 256, 283, 285, 301, 325, 337
Hazard, 253, 285, 311
Heater, 58, 60, 62, 70, 169, 170, 174–176, 191, 195, 196, 205, 213, 300, 301, 304, 305, 308, 312, 315, 319, 327, 333, 334
Heating, 12, 19, 20, 22, 23, 26, 58, 60, 62, 73–75, 77, 78, 80–83, 86, 113, 115, 116, 169, 173–175, 183, 184, 186, 187, 194, 196, 197, 205, 209–213, 221, 225, 235, 238–241, 315, 337
Heat-transfer, 337
Heuristic, 9, 10–13, 103, 155
Hierarchical, 30, 31, 114, 148, 194, 256, 257, 313, 318
Hierarchy, 30, 31, 130, 170, 205, 206, 237, 242–244, 263, 284, 318, 325
Housekeeping, 96

I

Identification, 65, 121, 123, 178, 183, 195, 253, 262, 263, 283, 320
Identify, 57, 70, 99, 100, 103, 106, 114, 122, 136, 155, 167, 168, 173, 203, 210, 259, 263, 296, 311, 328
Implementation, 10, 30, 103–106, 114, 120, 130, 137, 156, 168, 228, 261–263, 311
Inequality, 131, 135, 140, 149
Information, 2, 32, 35, 100, 114, 123, 173, 203, 254, 288
Initialization, 49, 70, 112
Inlet, 31, 37, 45, 48, 50, 58, 96, 97, 103, 116, 123, 143, 174, 195, 196, 198, 209, 210, 215, 221, 225, 243, 254, 258, 267, 275, 279, 284, 312, 314, 316, 319, 327, 328, 333, 334
Input, 32, 49, 98, 99, 107, 108, 117, 120, 139, 145, 158, 170, 173, 183, 184, 191, 209, 210, 213, 221, 225, 234, 236, 237, 243, 259, 304, 312–315, 322, 334, 338
Instance, 33, 60, 95, 123, 125, 140–142, 148, 157, 159, 178, 197, 205, 225, 286, 325, 326, 344
Instrumentation, 1, 6, 29, 167, 254
Integer, 130, 136, 137, 139, 142, 143, 150, 154, 156, 163, 167, 169, 171, 234, 285, 315
Integration, 13, 15, 206
Interconnection, 1, 29
Interface, 32, 95
Interlock, 167
Intermediate, 96, 173, 183, 184, 196, 210, 234, 269, 275, 295
Interval, 13, 86, 125, 207, 234–236, 243, 317, 320, 338, 344
Introduction, 1

Investment, 283, 284
Isolation, 107
Iterative, 13

J
Jobs, 156, 157, 162
Justifications, 58

K
Knowledge, 36, 60, 114, 127, 172, 173, 210, 234, 311, 314, 319, 323, 338, 346
Knowledge-based, 114, 253, 283, 312, 314, 320, 323, 325, 338

L
Label, 40, 42, 100, 123, 124, 136, 147, 158, 205, 209, 258, 259, 270, 274, 298, 325
Lagrangian, 154
Lagrangian-relaxation, 154
Languages, 10, 30, 33–35, 41, 42, 54–56, 168, 173, 210, 234, 261
Layer, 38, 206, 237, 243–245, 262, 263, 265, 266, 318, 319, 325
Leak, 45, 255, 258, 267, 269, 274, 285, 286, 288, 290, 295, 296, 300, 301, 305, 308, 316, 325, 327
Leakage, 26, 40
Level, 7, 11, 12, 17, 19, 22, 24, 26, 30–32, 34, 36–40, 45, 48, 57, 58, 60, 74, 81, 83–86, 88, 95, 130, 133, 137, 148, 168–170, 173–178, 181–183, 186, 187, 189–194, 196, 198, 205, 206, 208–211, 221, 225, 227, 229, 234, 235, 237–244, 256–258, 261, 263, 267, 269, 270, 273, 275, 276, 279, 284, 286, 290, 295, 296, 298–302, 304, 312–321, 324, 325, 328–334, 337, 338
Level-changing, 300
Level-decreasing, 333
Level-rising, 333–335
Line, 2, 31, 46, 48, 50, 87, 95, 120, 243
Linear, 29, 75, 161
Linearization, 161
Liquid, 11, 22, 26, 27, 31, 37, 38, 48, 57, 58, 60, 73, 74, 76, 169, 173–178, 182, 183, 186, 187, 191, 192, 194–196, 198, 205, 208–211, 215, 221, 225, 227, 237–241, 243, 255, 257, 258, 261, 267, 269, 270, 285–287, 290, 295, 296, 298, 299, 312, 314–316, 320, 321, 328, 329, 333–335
Liquid storage, 31, 36, 45, 48
Liquid-transfer, 8, 9, 60, 254–260, 263–269, 284, 285
Location, 78, 97, 98, 107, 117, 127, 149, 167, 204, 205, 207, 208, 225, 234, 290
Logic, 25, 26, 30, 95, 114, 130, 131, 133, 135, 142, 148, 161, 167, 169, 170, 205, 284, 288, 300, 313
Loop, 42, 54, 55, 86–88, 90, 92, 98, 99, 123, 190, 206, 207, 243, 258, 292, 293, 316–319, 325

M
Machine, 35, 129, 157, 162, 163, 204
Manipulation, 103
Manufacturing, 154–157, 163
Material-transfer, 57, 95–98, 100–109, 113–121, 123–125, 127, 130–132, 136–144, 147, 149, 163, 167, 273
Measurement, 11, 31, 34, 38, 48, 60, 95, 130, 187, 191, 193, 203, 225, 243, 254, 257–259, 262, 287, 288, 304, 311, 316, 317, 319, 325–327, 330, 333, 344, 345
Mechanical, 1, 2
Mechanism, 38, 40, 45, 50, 62, 127, 139, 151, 205, 257, 261, 284, 323, 324, 342
Medium, 19, 20, 75, 78, 81, 83, 113, 115, 116, 169, 186, 187, 205, 213, 301, 304
Method, 10, 11, 13, 15–18, 57, 74–77, 86, 88, 90, 96, 105, 110, 112–114, 119, 130, 143, 153, 155, 156, 163, 168, 204, 227, 253, 254, 275, 283, 284, 286, 308, 311, 329
Minimization, 162
Minimum-horizon, 161, 162
Mixed Integer Linear Programming (MILP), 114, 155, 167
Mixed Integer Nonlinear Program (MINLP), 155, 161
Model, 10, 13, 15, 16, 18, 29–32, 34, 36–40, 43, 45–51, 53, 54, 57, 58, 60, 62, 65, 95, 97–102, 107–121, 123, 125–127, 130–134, 136–139, 142–148, 150, 152–158, 160, 161, 163, 167, 168, 170–172, 174, 189–194, 196, 203, 205–209, 214, 234, 237, 242–245, 253–259, 261–266, 270,

274, 275, 280, 283–288, 290, 292–294, 311, 312, 314–320, 323, 325, 328–330, 337, 338, 343, 346
Modifications, 46, 108, 142, 177, 329, 330
Multi-failure, 253, 283, 299, 300
Multi-product, 96, 154, 155
Multi-purpose, 155
Multi-route, 96, 163
Multi-stage, 137, 154
Multi-step, 292, 328
Multi-task, 111, 114, 137
Multi-tool, 154
Multi-way, 253, 283

N
Nets, 98, 108, 120, 125, 139, 145, 156, 163
Network, 95–100, 102–105, 107–109, 111–115, 119, 123, 125, 137, 139, 141, 143, 148, 155, 156, 163, 167, 172, 173, 235, 253, 283, 319, 320

O
Objective, 74, 76, 78, 80, 82, 102, 114, 136, 138, 139, 143, 149, 150, 157, 161, 163, 172, 183, 221, 254
Observable, 46, 50, 54, 55, 168, 178, 253, 258–263, 270, 271, 275, 284, 287–289, 303, 308, 325
Observation, 32, 55, 56, 176, 227, 261, 304, 320, 328, 334
Operability, 58, 60
Operating, 1, 2, 9, 10, 18, 22, 29–32, 48, 49, 54, 57, 70, 73, 85–89, 92, 95, 96, 103, 105, 107–114, 117, 119–121, 123, 124, 127, 128, 130, 132, 139, 143, 149, 151–155, 162, 163, 167, 168, 170, 172–174, 178, 179, 183, 184, 187, 189, 190, 198, 203, 206, 210, 211, 215, 221, 222, 225, 226, 228, 229, 234, 235, 237, 257, 269, 271, 275, 295, 301, 304, 319, 320, 327, 333, 334
Operation, 1, 12, 18, 19, 26, 29–34, 36–38, 41–43, 45, 46, 48–52, 56–58, 60, 62, 68, 70, 73–78, 81, 83, 85–88, 95, 96, 98, 99, 102, 103, 105–114, 117, 119–121, 123–125, 127, 128, 132, 133, 135, 137–140, 142–145, 148–157, 163, 167–169, 171–178, 181–187, 189, 196, 198, 203–205, 209, 210, 213–215, 217, 220, 221, 225, 227, 234, 253–255, 258, 261, 267, 273, 275, 279, 280, 284, 285, 296, 300, 301, 304, 312–323, 325, 327, 330, 334–336, 338, 346
Operator, 13, 26, 31, 95–97, 130, 170, 172, 174, 205, 206, 209, 214, 215, 243, 261, 313, 317, 318, 326, 328, 337, 344
Optimization, 75, 159, 161, 162, 164
Origin, 52, 76, 253, 257, 261–267, 275, 277, 279, 280, 284, 285, 288, 290, 292, 293, 295, 298–300, 304, 305, 308, 326–331, 333, 334, 344–346
Outlet, 11, 23, 31, 37, 39, 40, 45, 48, 49, 58, 96, 97, 103, 116, 127, 143, 174, 195–198, 206, 209, 221, 243, 254, 273, 285, 312, 314, 319, 328, 334
Output, 107, 112, 117, 120, 145, 158, 170–173, 183, 184, 190, 191, 198, 207, 209, 210, 212, 213, 221, 234, 236, 243, 259, 262, 290, 312–314, 319, 322, 328, 334, 338
Overhead, 87, 89, 90, 175, 176, 184, 186, 187, 215, 225, 337

P
Parallel, 37, 40, 43, 46, 56, 57, 62, 255, 263, 266, 285, 290, 319, 325, 330
Path, 42, 56, 75, 80, 82, 96–99, 101–104, 106, 108–110, 116–118, 120, 121, 127, 131, 132, 136–144, 147–150, 153, 163, 168, 209, 213, 258, 259, 288, 289, 304, 312, 319, 320, 325
Pathway, 167, 172, 173, 184, 203, 210, 213, 214, 221, 235, 262
Performance, 17, 130, 155, 168, 176, 179, 182, 184, 186, 187, 215, 220, 221, 228, 234, 254, 266, 283
Petri, 10, 29, 30, 33, 46, 48–50, 52, 95–102, 107–109, 111–116, 118–123, 125, 126, 130–132, 139, 144–148, 153–157, 159, 160, 162, 163, 167, 253, 283
Petri net, 10, 29, 30, 33, 46, 48–50, 52, 95–102, 107–109, 111–116, 118–123, 125, 126, 130–132, 144–147, 153–157, 159, 160, 162, 163, 167, 253, 283
Pharmaceuticals, 113, 283
Phase, 11, 22, 112, 154, 156, 162, 168, 173, 184, 210
Pilot, 16, 17, 20, 168, 203
Pipe, 23, 97, 284

Pipeline, 31, 32, 37–40, 48–50, 95–99, 102–105, 107–109, 112–116, 125, 137, 139, 141, 143, 163, 167, 254, 273, 284, 285
Piping, 1, 6, 29, 97, 98, 114–118, 144, 167, 254
Planning, 29, 113, 114, 154
Plant, 1, 11, 15, 20, 29, 30, 54–57, 62, 96, 97, 113–121, 124, 128, 130, 143–145, 147, 151–154, 157, 167, 168, 203, 214, 253–255, 261, 272–278, 280, 283, 285, 288, 308, 311
Plant-wide, 113, 114
Plug, 17
Position, 20, 31, 36, 38, 39, 45, 48, 49, 52, 58, 60, 74, 78, 97, 98, 107–109, 112, 117, 119–121, 127, 132, 134, 135, 139, 143, 148, 156, 171, 172, 189, 195, 206, 215, 221, 225, 243, 254–256, 261, 269–274, 277, 279, 285, 286, 295, 296, 299–301, 305, 308, 312, 313, 327, 333, 334, 337, 345, 346
Prediction, 323, 325, 337
Pressure, 2, 11, 16–19, 22, 24–26, 85, 86, 88, 169, 171, 173, 177–179, 181–183, 186, 191–193, 196–198, 205, 208, 210, 220, 221, 227, 237, 243, 312, 313, 315, 335, 337
Procedure, 1, 9, 10, 12, 13, 16, 17, 20, 22, 23, 26, 27, 29–32, 48, 54, 57, 62, 65, 70, 73–75, 86, 96, 99, 102–115, 117, 119, 120, 123, 124, 127, 130, 139, 143, 144, 149, 151, 153, 154, 163, 167, 168, 170, 172, 174, 175, 179, 183, 184, 187, 189, 190, 196, 203, 206, 208, 213–215, 221, 222, 225–229, 232, 234, 235, 237, 254, 255, 258, 263, 269–271, 274, 275, 280, 283, 284, 288–296, 299, 301, 302, 304, 308, 311, 313, 319, 320, 327, 328, 333, 334, 344
Process, 1–3, 11–13, 16, 17, 19, 22, 26, 29, 30, 32, 35–38, 40, 45, 46, 48–50, 57, 58, 60, 62, 70, 73, 75, 78, 83–85, 88, 90, 91, 95, 96, 100, 104, 113–116, 121, 123–125, 127, 130, 134, 143–146, 148, 151–157, 161–163, 167–170, 173, 174, 176–179, 181–183, 186, 187, 191, 194, 204, 205, 207, 208, 210, 217, 220, 228, 236, 237, 253–255, 261–264, 266, 273, 275, 283–285, 295, 300, 311–314, 320, 322, 323, 325, 328–330, 335, 336, 339–342, 346
Production, 1, 74, 113, 154, 156, 158, 160, 161, 167, 273
Program, 130, 136, 138, 139, 142, 143, 150, 155, 156, 161, 163, 164, 167
Programming, 29, 114, 154, 156, 157, 163
Propagation, 304
Pump, 16, 17, 23, 25, 95, 97, 104, 107–117, 119, 120, 127, 130, 132, 133, 270, 295, 296, 300, 302
Purger, 143–145, 147

Q

Qualitative, 29, 171, 173, 253, 315, 322, 338, 344
Quality, 11, 103, 253
Quantitative, 253
Quantity, 150, 173, 210, 221
Queuing, 154

R

Reachability, 99–102, 104, 114, 121–123, 127–129, 154
Reaction, 13, 20, 168, 173, 177–182, 210
Reactive, 17, 19, 25, 27, 311
Reactor, 15–17, 177–179, 181, 182
Real-time, 139, 156, 213
Reboiler, 11, 17, 18, 22, 24–27, 74, 86, 181, 184, 186, 220, 221, 225, 234, 337, 338, 345
Receiver, 205–207, 209, 213, 243, 244, 286, 288, 290
Recipe, 95, 113–115, 130, 163
Redundancy, 38
Reflux, 17–19, 24, 73, 74, 77, 78, 81, 83–91, 181–187, 220, 221, 225, 227, 337, 338
Reliability, 38
Route, 96, 98–113, 115, 116, 119–121, 123–125, 127, 128, 136–138, 141, 142, 149, 155, 163, 167, 173
Rule, 9, 10–13, 15, 16, 45, 57, 95, 103, 109, 119–121, 123, 125, 135, 163, 256

S

Safe, 58, 96, 167, 168, 173, 178, 183, 210, 221
Safety, 29, 58, 60, 109, 113, 114, 119, 167, 177, 253, 311
Satisfactory, 203, 253

Scenario, 35, 50, 55, 56, 162, 163, 172, 173, 185, 196, 197, 209, 253, 257, 258, 262–264, 267, 275, 277, 280, 283, 284, 288, 290, 295, 296, 298, 299, 302, 304, 305, 308, 312, 313, 315, 320, 323, 327, 328, 331, 333–335, 338, 345
Schedule, 76, 95, 96, 111–114, 124, 125, 127–130, 139, 143, 153–157, 161–163, 167, 168
Scheduling, 113, 154–157, 162
Schemes, 11, 12, 20, 76–78, 81, 83–85, 154, 320, 322, 338
Section, 15–17, 20, 21, 23, 25–27, 57, 85, 88, 92, 116, 121, 127, 130, 177, 191, 217, 234, 237, 326, 337, 338, 344
Segment, 2, 163
Selection, 13, 87–90, 102–104, 106
Self-looping, 45, 207, 212, 213, 243, 256, 258, 259, 261, 262, 265, 266, 329, 330
Semiconductor, 78, 154–157, 163, 283
Sensor, 31, 32, 36, 38, 45, 48, 57, 58, 60, 95, 130, 170, 174, 191, 203, 205, 227, 234, 243, 244, 254, 257–259, 261, 262, 266, 267, 269, 270, 272, 275–277, 279, 280, 283–285, 287, 288, 292, 295, 300–302, 304, 305, 313, 316, 317, 325, 327, 328, 330, 337, 344
Separation, 11, 18, 20
Separator, 177
Sequence, 23, 30, 32, 33, 35, 44, 60, 62, 96, 113, 114, 148, 151, 154, 168, 173, 178, 212, 213, 253, 258, 260, 267, 270, 274, 276, 279, 283, 284, 288, 290, 298, 299, 302, 304, 322, 338
Sequential, 1, 8, 9, 29, 31, 103–105, 111, 112, 167, 203, 214, 253, 254, 280, 283, 284, 311–313, 336, 346
Setup, 16
Short-term, 155
Shutdown, 1, 12, 21, 23, 26, 27
Signal, 31, 32, 36, 48, 52, 57, 58, 125, 127, 134, 171, 172, 190, 213, 235, 243, 288
Simple, 15, 18–21, 31, 56, 57, 73, 78, 81, 90, 91, 96, 103, 161, 168, 172, 177, 204, 206, 234, 254, 284, 311, 346
Simulated, 19, 75, 155, 175, 179, 184, 203, 216, 218, 225, 227, 312, 320, 321, 331–334, 337, 338
Simulation, 10, 13, 15–19, 25, 73, 78, 92, 98, 109, 110, 112, 116, 119, 155, 163, 168, 175–177, 179, 181, 182, 184–186, 203, 206, 208, 215, 221, 223, 227, 228, 230, 232, 234, 237, 251, 311, 312, 320, 322, 323, 330, 333, 334, 337, 338, 340, 346
Single-direction, 98, 118
Sink, 96, 99, 100, 114, 121, 123, 127, 131, 132, 136, 137, 141, 147
Software, 13, 16, 18, 168, 189, 203, 259, 285, 288, 311, 319
Specification, 17, 18, 30, 56, 57, 60, 62, 73, 74, 77, 90, 125, 168, 172–174, 189, 194, 196–198, 203, 208–210, 213, 214, 234, 245–251, 261–263, 265–267, 284, 288
Stable, 13, 168, 173, 174, 177, 178, 184, 198, 210, 221, 304
Stage, 1, 16, 18, 20, 73, 78, 81, 83, 85, 86, 88, 90, 112, 130, 133, 136, 137, 139, 140, 148–152, 156, 157, 168, 172–174, 177–179, 183–187, 189, 194, 195, 198, 209–214, 221, 227, 234, 290, 293, 298, 305, 325, 335, 337
Stage-based, 130, 139, 142, 143, 148, 149, 151, 152, 163
Standard, 1, 107, 108, 114, 117, 119, 130, 132, 134, 154, 156, 167, 203, 255, 290, 326
Startup, 1, 12, 17–23, 26, 73, 74, 76–79, 81, 83–92, 168, 169, 171–177, 181–187, 189–191, 194–200, 204–206, 209, 210, 213–215, 217, 221–223, 226, 227, 229, 230, 232, 234, 235, 237, 243, 312–314, 316, 318–323, 325–328, 332, 334–336, 338, 340, 341, 343–346
State, 11, 29, 31, 32, 34–42, 45, 46, 48–52, 54, 56–58, 60, 62, 65, 95–97, 99, 100, 104, 107, 109, 114, 116, 117, 119–121, 123, 127, 129–137, 139–142, 145, 147–152, 156–158, 167–173, 184, 190, 197, 203–205, 207, 208, 210, 216, 218, 234, 236–241, 243, 244, 255–259, 261, 263, 264, 269, 285–288, 290, 295, 296, 299, 301, 302, 304, 314, 315, 317, 319, 320, 323, 325, 327, 328, 330, 338, 339, 341, 342
State-task, 113, 155
State-transition, 32, 35, 36, 38, 44, 48, 60, 148, 169–171, 173, 205–207, 209,

214, 215, 234, 235, 237, 244, 256–258, 285–287, 292, 295, 314, 322, 323, 338, 341
Steady-state, 13, 16, 18, 19, 73, 78, 84, 169, 173, 174, 181–184, 196–198, 204, 205, 207, 210, 211, 213, 215, 220, 225, 228, 234, 235, 237, 253, 312, 315, 316, 318, 335
Stirred, 13
Storage, 7, 32–34, 38, 48, 96, 115, 167
Strategy, 11, 12, 29, 30, 74, 76, 77, 79–83, 85–87, 89–91, 96, 100, 115, 154, 155, 157, 164, 167, 168, 196, 256, 258, 283, 308, 311, 320, 328, 338, 346
Structure, 19, 30, 34, 38, 46, 50, 76, 96–98, 114, 116, 118, 148, 169, 194, 204, 208, 262, 286, 313
Subinterval, 234, 235
Subsection, 37, 57, 142, 168, 184, 314, 319, 322
Subset, 33, 42, 46, 48, 54–56, 100, 104, 122, 123, 128, 129, 158, 258, 299
Substring, 33, 173, 275
Subtasks, 173, 210
Subtree, 127
Supervision, 210–214
Supervisor, 30, 54–57, 62, 63, 65, 68–70, 261–263, 266, 268
Supervisory, 29, 32, 54, 55, 114, 261, 284
Supremica, 168, 172, 174, 195, 319, 343
Switch, 22, 23, 25, 60, 70, 74, 75, 77, 85–91, 112, 113, 133, 174, 190, 207, 215, 273, 299, 300, 304, 305, 308, 323, 325, 338
Symbol, 2, 33, 49, 58, 70, 243, 256, 264, 326
Symbolic, 29, 103, 114, 155, 167
Symmetrical, 4
Synchronization, 172, 205, 234
Synthesis, 1, 10, 11, 16, 29, 57, 70, 96, 113, 114, 124, 127, 130, 167, 168, 174, 179, 184, 189, 190, 203, 213, 227, 258, 263, 270, 275, 280, 283, 284, 290, 291, 295, 299, 304, 308, 311, 327, 328, 344
System, 2, 8, 9, 12, 13, 16–18, 20–23, 25–27, 29–36, 38, 43, 45–49, 54, 62, 76, 78, 95–98, 107–109, 111, 112, 114, 115, 119–123, 125, 127, 130, 135, 136, 139, 142–144, 147, 151, 154, 155, 163, 167–169, 172, 178, 182, 184, 189, 194, 198, 203, 205, 206, 209, 210, 213, 221, 234, 237, 242–244, 253, 254, 256–273, 275, 279, 283, 284, 286, 288, 295, 296, 298, 300–302, 311, 312, 314, 318, 319, 323, 325–327, 333, 337, 338, 342, 346
Systematic, 1, 57, 95, 96, 103, 110, 154, 163, 167, 203, 284, 311

T

Table, 18, 20, 40, 58, 70, 78, 79, 91, 101–103, 109, 110, 112, 113, 121, 122, 128, 129, 135, 151–153, 156, 176, 182, 186, 187, 209, 220, 234–236, 245–251, 255, 256, 264, 322, 323, 339–341
Tank, 7, 11, 12, 15–17, 21–24, 26, 27, 31, 32, 34, 37, 38, 40, 45, 46, 48–50, 57, 58, 60, 100, 115, 122, 125, 129, 136, 138, 143–151, 167, 254, 255, 257, 258, 264–267, 269–273, 275, 284–287, 290, 295, 296, 298–301
Target, 57, 58, 62, 96, 143, 149, 168, 172, 173, 184, 194, 196, 209, 210, 215, 262, 265, 266, 288, 290, 300–302, 304, 319
Technical, 29, 203
Technology, 20
Temperature, 2, 11, 12, 18–20, 22, 23, 25–27, 57, 58, 60, 74–78, 81, 83, 85, 86, 88, 90, 91, 169, 173–179, 181–187, 191–193, 195–198, 205, 208–213, 220, 221, 225, 237, 243, 300, 301, 304, 312, 313, 315–317, 319–325, 329, 335, 337
Terminal, 99, 100, 104, 123, 127
Test, 10, 51, 52, 91, 92, 95, 107, 117, 154–156, 162, 203, 207, 208, 234, 237, 253, 256, 261, 262, 264, 270, 272, 275–279, 283, 290, 292–295, 299, 300, 305, 308, 311, 312, 320, 327–331, 333, 334, 345, 346
Test-plan, 9, 253, 254, 258, 262–264, 267, 269, 270, 272, 275–280, 284, 290, 291, 295, 299, 300, 304, 308, 311, 327–329, 331, 332, 334, 335, 344–346
Thermosiphon, 5
Time-averaged, 316
Time-based, 130, 139, 141–143, 151, 153, 154, 163
Timed-automata, 168
Token, 38, 46, 95, 97–100, 107, 109, 112, 116, 117, 119, 122–125, 127, 131, 132, 136, 141, 148, 156–160

Trace, 62, 70, 168, 203, 253, 258, 260–263, 267, 270–272, 274–278, 283, 284, 288–290, 292–296, 298–300, 302–305, 308, 320, 325, 328, 330, 331, 333, 334
Transfer, 19, 90, 96–98, 107, 113, 115, 116, 123, 145, 149, 151, 173, 184, 196, 255, 279, 280, 299, 311
Transition, 12, 35, 36, 41–43, 45, 46, 48–52, 56, 58, 95, 97–100, 107–112, 116, 117, 119–121, 123, 125, 134, 135, 139, 148, 156–160, 162, 163, 169, 171, 172, 183, 184, 190, 204, 256, 258, 260, 262, 266, 270, 274, 285–288, 290, 293, 298, 314, 315, 317, 325
Transportation, 113, 150
Tree, 99–102, 104, 114, 121–123, 127–129, 154, 163
Trigger, 45, 48, 49, 57, 111, 190, 197, 209, 257, 286

U
Uncertain, 272, 275, 276, 279
Unchecked, 291
Uncontrollable, 54, 62
Undiagnosable, 262, 270, 271, 274, 275, 280, 296, 297, 302, 308, 344
Unidentifiable, 38
Unit, 1, 2, 11, 16, 17, 31, 45, 62, 95–97, 113, 115, 130, 134, 139, 142, 145, 146, 148–151, 153, 156, 157, 162, 170, 191, 205, 206, 234, 237, 243, 244, 253, 273, 284, 285, 296, 298, 299, 302, 304, 313–316, 319, 320, 323, 325, 337, 338, 346
Unloading, 168, 173, 178, 195, 210
Unmanageable, 29, 172
Unmarked, 35
Unmeasured, 243
Unobservable, 46, 48, 55, 70, 259, 287, 325
Untimed, 10, 30, 33, 49, 167, 168, 173, 189, 203, 253, 280, 283, 314, 317, 319
Updates, 169, 171, 190, 205, 209, 245, 316, 317, 322, 338
UPPAAL, 203, 205, 213, 214, 285, 288
Upstream, 97, 108, 116, 120, 123, 131, 140, 142, 143
Utility, 2, 113, 174, 184, 221

V
Value, 13, 17, 19, 20, 73–78, 85–90, 92, 113, 133, 135, 140–142, 156, 161, 169–172, 174, 187, 190, 194–196, 198, 204, 205, 208–213, 225, 234–236, 238–241, 243, 257, 283, 304, 315, 316, 322, 338, 339
Valve-closing, 107, 117, 123
Valve-opening, 107, 117, 123
Valves, 16, 17, 19, 20, 22, 23, 25–27, 31, 32, 34, 36, 39, 40, 45, 48, 49, 52, 58–60, 78, 85, 86, 88, 95–98, 103–105, 107–123, 127, 130, 132–136, 142, 143, 145, 147, 148, 152, 167, 169, 171, 172, 174, 175, 179, 183, 184, 189, 190, 195–198, 205–207, 209, 210, 213, 215, 221, 225, 237, 243, 254–257, 269, 273, 275, 284–286, 295, 300, 302, 312, 326, 328, 329, 333, 334, 337, 345
Valve-sequencing, 96, 114
Valve-switching, 107, 113, 117, 121
Vapor, 11, 23, 169, 183, 191, 197, 198, 205, 209, 215, 221, 225, 237, 312, 314, 316, 334
Variable, 11, 13, 15, 20, 73, 76, 78, 92, 131–141, 143, 151, 157–159, 161, 169–172, 190, 191, 204–206, 208, 209, 216, 218, 234–241, 285, 286, 300, 314–317, 319, 320, 322, 337, 338
Variation, 31, 191, 227, 234, 315–317, 319, 324, 325, 329–334
Verification, 29, 114, 168, 213, 214, 285
Vessel, 26, 57, 60, 301, 319

W
Wafer, 154, 155
Wait, 26, 86, 88, 187, 215, 251, 252, 317
Water, 11, 20, 22, 23, 25–27, 78, 85, 86, 88, 90, 91, 169, 178, 184, 204, 345, 346
Work, 29, 86, 95, 97, 114, 115, 127, 130, 145, 147–152, 154, 156, 157, 159, 162, 163, 167, 168, 198
Workloads, 96, 113, 161, 162

Y
Yield, 43, 56, 182, 325

Z
Zero, 124, 141, 159, 196, 301